International acclaim for John Keegan's

INTELLIGENCE IN WAR

"Fascinating.... Beautifully written ... perceptive."
— *Los Angeles Times Book Review*

"Thought-provoking.... Keegan's book is a wise corrective, assessing just how useful intelligence has been in battle." — *The Dallas Morning News*

"Gripping.... Heart-quickening.... Lucid and very wise.... Keegan makes his points elegantly, economically and incisively."
— *New York Post*

"Crammed with the detail for which Keegan is justly famed.... Ranges through the ages to look at the value of information to generals and admirals."
— *Associated Press*

"Keegan theorizes expertly [and] summarizes brilliantly.... He brings to the literature of war a deep affection for revealing detail."
— *National Post*

"[Keegan's] case histories offer enough revelations and drama to satisfy any espionage buff." — *The New York Times Book Review*

JOHN KEEGAN

INTELLIGENCE IN WAR

John Keegan's books include *The Iraq War*, *The First World War*, *The Battle for History*, *The Face of Battle*, *War and Our World*, *The Masks of Command*, *Fields of Battle*, and *A History of Warfare*. He is the defense editor of *The Daily Telegraph* (London). He lives in Wiltshire, England.

INTELLIGENCE IN WAR

INTELLIGENCE IN WAR

THE VALUE—AND LIMITATIONS—OF WHAT
THE MILITARY CAN LEARN ABOUT THE ENEMY

---◄O►---

JOHN KEEGAN

Vintage Books
A Division of Random House, Inc.
New York

FIRST VINTAGE BOOKS EDITION, OCTOBER 2004

Copyright © 2002 by John Keegan

All rights reserved under International and Pan-American Copyright Conventions.
Published in the United States by Vintage Books, a division of Random House, Inc.,
New York. Originally published in Great Britain by Hutchinson, the Random House
Group Limited, London, in 2002, and subsequently in hardcover in the United States
by Alfred A. Knopf, a division of Random House, Inc., New York, in 2003.

Vintage and colophon are registered trademarks of Random House, Inc.

A portion of this work appeared previously in *Military Quarterly Magazine.*

The Library of Congress has cataloged the Knopf edition as follows:
Keegan, John, 1934–
Intelligence in war / John Keegan.—1st ed.
p.   cm.
1. Military intelligence—History.   I. Title.
UB250.K44 2003
355.3'432—dc21
2002044828

Vintage ISBN 0-375-70046-3

Author photograph © Jerry Bauer
Book design by Robert C. Olsson

www.vintagebooks.com

Printed in the United States of America
10   9   8   7   6   5   4   3   2   1

To Rose

CONTENTS

CONTENTS

ILLUSTRATIONS

Kent, Inflexible, Glasgow and *Invincible* leaving Port Stanley, 1914 (Robert Hunt Library)

Survivors of SMS *Gneisenau* being collected by boats from *Inflexible*, 1914 (Robert Hunt Library)

Admiral Wilhelm Canaris (Robert Hunt Library)

SECOND INSERT

An Enigma operator aboard a U-boat (The Art Archive/E.C.P.A., Ivry, France)

An Enigma machine team, Army Group Centre, Russia, 1941 (Bildarchiv Preussischer Kulturbesitz)

German paratroopers, Crete, 1941 (Bildarchiv Preussischer Kulturbesitz)

General Bernard Freyberg VC (Robert Hunt Library)

General Kurt Student (Ullstein Bild)

USS *Lexington*'s crew abandoning ship (Robert Hunt Library)

Admiral Isoroku Yamamoto (The Art Archive/U.S. Naval Historical Center)

Admiral Chester Nimitz (The Art Archive/Library of Congress, Washington, D.C.)

Captain Joseph Rochefort (The Art Archive/U.S. Naval Historical Center)

Kaga circling under attack, Midway, 1942 (Robert Hunt Library)

USS *Yorktown* sinking (Robert Hunt Library)

Admiral Karl Dönitz with Grand Admiral Raeder (The Art Archive/E.C.P.A., Ivry, France)

USS *Greer* on convoy escort in heavy Atlantic weather (The Art Archive/National Archives, Washington, D.C.)

U.S. Coast Guard Cutter *Spencer* firing a depth charge (The Art Archive/Library of Congress, Washington, D.C.)

Admiral Ernest J. King (The Art Archive/National Archives, Washington, D.C.)

A convoy conference at Derby House, Liverpool (The Art Archive/Imperial War Museum, London)

Captain F. J. "Johnny" Walker (The Art Archive/Imperial War Museum, London)

V-2 rocket at Peenemünde (Ullstein Bild)

V-1 flying bomb, France, 1944 (Ullstein Bild)

A V-1 about to impact near Drury Lane, London, 1944 (The Art Archive/Imperial War Museum, London)

General Walter Dornberger with Dr. Todt (Ullstein Bild)

Werner von Braun (© Hulton-Deutsch Collection/Corbis)

SAS troopers after the helicopter crash on South Georgia, Falklands, 1982 (The Defence Picture Library)

Argentinian Pucara ground-attack aircraft destroyed by the SAS (The Defence Picture Library)

British anti-tank missile team, second Gulf War, 2003 (The Defence Picture Library)

MAPS

ACKNOWLEDGEMENTS

I have tried to steer clear of the intelligence world all my working life. For good reason: As a young lecturer in military history at the Royal Military Academy Sandhurst I was told that any contact with intelligence organisations, those of other countries especially but also our own, would attract official disapproval (I should have answered, but did not, that I possessed no shred of information that could have been of the slightest interest to any sensible intelligence officer). Later, as defence correspondent, then defence editor of *The Daily Telegraph,* I decided that entanglement with intelligence organisations was unwise, having concluded, by that stage of my life, through reading, conversation and a little personal observation, that anyone who mingled in the intelligence world, in the belief that he could make use of contacts thus made, would more probably be made use of, to his disadvantage. I continue to believe that to be the case.

Nevertheless, and probably inevitably, given my career first in the Ministry of Defence, then as a newspaperman and continuously as a military historian, I have come to know over the years more inhabitants of the intelligence world than I would have deliberately set out to meet. Some of my Sandhurst pupils became intelligence officers; one died, heroically, at the hands of the Irish Republican Army. Some of my Sandhurst colleagues served in Special Forces, which are on intimate terms with the intelligence organisations and often act as their executive arm. Academic life, improbably as it may seem, sometimes brought me into touch with the intelligence services, though, it should be explained, with their analytic rather than operational branches. Fleet Street—where the offices of *The Daily Telegraph* were still located when I joined in 1986—had then, and still has, its own informal relations with the intelligence services; at the outset the newspaper encouraged me to get to know what were called "the contacts."

The first to whom I was sent was also the most important, the then chairman of the Joint Intelligence Committee, which oversees the work of the Secret Intelligence Service (MI6) and the Security Service (MI5), dealing respectively in foreign and domestic intelligence. It was arranged that I should meet him in one of London's grand gentlemen's clubs. I was not told how I should recognise him. I remembered, however, from the novels of John le Carré—someone with whom Fleet Street would also bring me into touch—that a good agent always sits in the corner of the room from which he can see the entrance and has access to two separate exits. On arriving, I spotted the chairman instantly.

I later met the head of MI6 and, much later still, the then head of MI5, (Dame) Stella Rimington; I have no hesitation in mentioning her name since, to the outrage of some of her colleagues, she insisted in retirement on publishing her memoirs. I met her in the company of my then editor and great friend, Max Hastings, who had invited me to join them for dinner, perhaps to assure her that the occasion was purely social and not an attempt on his part to extract information. In the event I had a strong feeling of playing gooseberry; I certainly took little part in the conversation. A few days later an outraged Max cornered me in the office. "Do you know what a friend has just told me? The morning after our dinner with Stella she posted the gist of what we discussed on the Whitehall e-mail. Can you credit it?" My memories of civil service thought processes came to the fore. "All too easily," I said, "she was getting her retaliation in first. Government servants are terrified of accusations of treating with the enemy."

I found the American intelligence services altogether more human. At an academic conference I bumped into a genial character who knew my work as a military historian and asked what aspect of it I enjoyed most. "Order of battle analysis," I replied unhesitatingly; the order of battle is the list of units involved in an operation, often surprisingly difficult to establish. "Really," he said. A little while later I got a message from him to say that he was responsible for training American government servants to whom order of battle was a matter of great importance. Could I come to Washington to give a lecture on the subject?

Washington turned out to be Langley, Virginia, and the government servants in question were trainee analysts of the Central Intelligence Agency; analysts process the information which is assembled by the Agency's other branch, its field officers. My first lecture was a success. Invited to give it a second time, I found myself delivered to CIA headquarters and put into the hands of an escorting officer. I was impressed by the Agency's attention to detail. "You'll need a pass," my escorting officer said and put before me a sheet of paper for signature. It contained more personal information about myself than I could have readily assembled. "We're going to see the Director," he said, "but not yet. Let's have a cup of coffee." He took me round the corner. The husband and wife running the coffee stall were blind.

"Now we're going to the Director's office." He set off confidently, first one floor, than another. I sensed a loss of confidence. After a bit he stopped to question a passer-by. "Wrong floor," he said, with a touch of embarrassment. Eventually arrived—all doors looked identical and had minuscule nameplates below eye level—we entered an anteroom full of muscular young men with bulges near their armpits. "The Director is expecting you," one said.

I entered the adjoining room. A very large man, who I subsequently learned was William Casey, Director of Central Intelligence, indicated a seat and began to speak. I had by this stage formed the impression, by intuition alone, that the CIA wished to communicate with *The Daily Telegraph*; the trouble was I could not discern quite what. I shuffled my chair nearer Mr. Casey's desk. He continued to speak, unintelligibly. I shuffled nearer. Eventually it dawned on me that the Director was not talking about current intelligence affairs but military history; he was a reader of my books and wanted to discuss technique, as he wrote himself. It was still difficult to understand what he was saying.

Eventually, and clearly as a sign that time was up, he rose from his desk, extracted a book from a shelf, wrote in it and said goodbye. The book turned out to be *Where and How the War Was Fought*, a surprisingly gripping account of the War of the Revolution related in terms of its geography; the warm inscription stated that my books had been of use to him in composing his. Slightly bemused, I returned to the corridor.

My escorting officer was there and so were several other senior CIA agents. "What did he say?" was their collective question. "I couldn't really understand," I answered. There was a collective burst of laughter. I had said the right thing. I later learned that the Director was known as "Mumbles" and described as "the only man in government who doesn't need a secure telephone."

The last encounter with the intelligence world I shall mention was more complex and perhaps more sinister, though I would mislead if I implied that it was really dangerous. During the nineteen-eighties I formed a connection with the American magazine *The Atlantic Monthly,* then the highest-paying publication in the world. For *The Atlantic* I went to Lebanon, during the Civil War, and later to the Northwest Frontier, during the Russian war in Afghanistan. My reason for accepting its commissions was simple. I had four children at expensive schools and the sums it paid, $10,000 an article, went far to meet the costs of tuition. My last *Atlantic* commission was to report on the security situation in South Africa, just before the collapse of apartheid.

The arrangements were made by the magazine, which had contacts in South Africa. I was grateful for its intervention because I was still on the staff of the Royal Military Academy Sandhurst and, though I would be absent during the Academy vacation, was well aware that I should not be visiting a country not then a member of the Commonwealth for a foreign magazine without official permission, which I had not secured or indeed requested. I was told I would be met on arrival. When I landed at Jan Smuts airport I found myself the only passenger whose suitcase was missing. I reported the loss and, with my guide, went on to Pretoria. During the next week, in which I bought clothes to replace those missing, I visited the Voortrekker Monument, the headquarters of the 1st Light Horse and the South African Ministry of Defence, where I was briefed by a commodore in the South African Navy. I also went to lunch in the Pretoria Club with the retired head of South African military intelligence, General du Toit. Over lunch he enquired casually, "Are you still at Sandhurst?" I felt a twinge; I thought I was travelling as a free-lance journalist. He clearly had other thoughts. The disappearance of my suitcase suddenly seemed significant.

The suitcase eventually reappeared, at Johannesburg Central Police Station, the day before my departure. In the weeks following, I forgot about the oddity of the episode. Then, a month or two later, I got a telephone call from someone at the Ministry of Defence. Could we meet for lunch? Perhaps I didn't listen very carefully. Anyhow, I jumped to the conclusion that the caller belonged to the Defence Intelligence Agency, an open and indeed very helpful body which I telephoned frequently for information when writing about wars in obscure parts of the world. We arranged to meet.

The young man from the ministry was impressive, well-dressed, well-mannered, well-spoken, what my generation would call smooth. After preliminaries, I said how pleased I was to meet a representative of the Defence Intelligence Agency, which was such a helpful source. A slight flicker passed across his brow. It was clear that there was a misunderstanding. "I'm not from the Defence Intelligence Agency," he said. Without specifying which, he indicated that he came from another agency. I realised at once that he was an officer of the Secret Intelligence Service (MI6). I wondered what on earth he wanted with me. He shortly made that clear.

With remarkable frankness, he told me that he knew me from the past, having been at Sandhurst, felt he could trust me and had read my *Atlantic Monthly* article about South Africa. South Africa was his beat. He went there frequently. Alarmed, I blurted out the question, "Do they know about you?" Imperturbably he answered, "They know some of my names. I hope they don't know them all because that would mean twenty years." While I digested that he went on, "I found your account of your interviews in South Africa very interesting. Do you think you could go back again and ask more questions? I would tell you what to ask. Everything paid, of course."

I was struck dumb and remained dumb for some time. Eventually lunch came to an end. His parting words were, "Do you think you will have to mention this conversation to your Editor?" Words returned. "You bet I do," I said. Max, when I got back to the office, erupted. "John, don't touch it with a barge pole. You haven't even been to spy school. They'd eat you for breakfast." I had already come to the same conclusion.

This terminated my one and only encounter with the inner secret world. I have not a single regret that intimacy ended there. I know with certainty that I lack any of the qualities, courage and self-confidence foremost, to serve it usefully. On the other hand, I am grateful for having met some of those who do. I do not wish to mention names. I would say, however, that among those to whom I have been introduced during my on-off encounter with the world of espionage is one of the most celebrated traitors of the twentieth century. He served the West, at great danger to himself, and is a fascinating and charming human being. About him, however, I share the feelings of my wife, in whom burns the true fire of British patriotism. "I like him," she said, after our only meeting, "but I can't suppress the knowledge that he betrayed his country. I would die rather than be a traitor."

So to her, my beloved Susanne, I make my first acknowledgement. I also wish to acknowledge the help and support of our children and children-in-law, Lucy and Brooks Newmark, Tom and Pepi, Matthew and Rose, and to thank them for the entry into our lives of their wonderful children, Benjamin, Sam, Max, Lily, Zachary and Walter. I particularly want to thank Lindsey Wood, my indispensable assistant, for her help; and my publishers, Anthony Whittome, Simon Master, Ash Green and Will Sulkin; my brilliant picture researcher, Anne-Marie Ehrlich; and the master mapmaker Reginald Piggott. Geography is the key to military history.

Without breaking confidences, I would also like to thank Alan Judd, John Scarlett, Sams Smith, George Allen, William Casey, Bill Gates, Percy Cradock, Anthony Duff, Jeremy Phipps and John Wilsey. Among my colleagues at *The Daily Telegraph* I would like to thank Charles Moore, Michael Smith and Kate Baden.

Finally, let me thank my literary agent, Anthony Sheil. He set out in life, as a rich young man, to make a fortune betting on the horses. He retired from that undertaking to backing authors, of whom I have been gratefully one. About horses he remarked, memorably, "You can never know enough." It might be the motto of this book.

INTELLIGENCE IN WAR

Introduction

This book sets out to answer a simple question: how useful is intelligence in war? The volume of literature on the subject suggests that it is very important indeed. Shelves groan under the weight of books on the German Enigma machine, on the British code and cipher school at Bletchley Park that attacked Enigma, on the American unlocking of the Japanese ciphers, on the parallel deception operations which sought to delude the enemy, on the agents who risked their lives to help make deception work or to seek to discover the enemy's secrets from within. The literature of fact is exceeded in bulk by that of fiction. The spy story became, in the twentieth century, one of the most popular of literary forms, and its masters, from John Buchan to John le Carré, grew rich and famous by their writing.

The climate created by the masters of spy fiction deeply affected popular attitudes to intelligence work. The sheer fascination of the techniques unveiled, in the use of cryptic writing, dead letter boxes, agent running, the "turning" of agents to become "doubles," surveillance, interception and a dozen other practices of the secret world, had the effect of representing technique as an end in itself. The "spy" achieved the status of hero, or sometimes anti-hero, a mysterious and glamorous figure who was made to seem significant because of what he was rather than what he did.

It is notable that very few of even the most celebrated spy stories actually establish a connection between the spy's activities and the purpose for which he presumably risks his life in the field. In *Greenmantle*, for example, John Buchan's wonderful romance of intelligence work in Turkey during the First World War, it becomes impossible for the reader to discern at the end what exactly Sandy, as Greenmantle, has done: has he frustrated a Muslim *jihad* against Britain and her allies or, contrarily, has he himself become a Muslim prophet? In *The Riddle of the Sands*, the first serious novel of intelligence to appear and still one of the best, Erskine Childers subtly suggests how the Germans may mount an invasion of Britain's east coast through the secret channels around the Friesian islands, but the dénouement of his tale does not demonstrate that his two patriotic yachtsmen actually cause the Admiralty to take appropriate precautions. In Kipling's marvellous *Kim*, ostensibly an unforgettable panorama of Indian life on the road but essentially a spy story, his hero does, unwittingly, help to frustrate a rising in one of the princely states, but the climax results in nothing more than his making fools of some Russian spies on the Himalayan border. In almost none of John le Carré's brilliantly convincing constructions of spy and counter-spy life does he show an objective outcome for what his characters do. They are fighting the Cold War; but, after all their intricate delusions and deceptions, the Cold War goes on.

The author might rightly say that he was representing reality; the Cold War thankfully did not have an outcome, certainly none in military terms, and it was the function of the intelligence services on both sides to see that it should not. They were playing a game, and the point was to keep the game going, not to win. No one would disagree with that or ought to complain, in the absence of a tangible result, that intelligence is a vacuous activity.

The intelligence services of all states originated, nonetheless, in the efforts to avert an enemy's achieving a military advantage but to achieve military advantage in return. In peacetime, intelligence services may merely tick over. In war, they are supposed to bring victory. How effective are they? How do they—or how do they fail—to do it?

The novelists of intelligence have disseminated an enormous amount of information about intelligence techniques. Some of it is accurate, some is misleading. Few of them, however, even such writers who are as personally experienced in intelligence work as John le Carré, have set out in full the essential components and sequence of effective intelligence operations. That is understandable. Much intelligence practice is mundane and bureaucratic, unamenable to treatment in readable form. Even the most mundane, however, is essential if intelligence is to be useful. There are five fundamental stages.

1. *Acquisition.* Intelligence has to be found. It may be readily available in published but overlooked form. A former director of the CIA warned his analysts against what he called the *Encyclopaedia Britannica* factor: do not waste effort in seeking information which may freely be found in newspapers, scholarly journals or academic monographs. Stalin's Russia took precautions to make information as difficult to acquire as possible, by restricting the distribution of such everyday material as telephone directories and street maps. As a general principle, however, it may be taken that information useful to an opponent is what may be called "secret" and has to be collected by clandestine means. The most usual methods are spying, in all its forms, now technically known as "human intelligence" or "humint"; by the interception of an opponent's communication, which will probably require decryption, "signal intelligence" or "signit"; by visual surveillance or imaging, through photographic or sensory reconnaissance by aircraft or satellite.

2. *Delivery.* Intelligence once collected has to be sent to its potential user. Delivery is often the most difficult stage, particularly for the transmitter of humint. The humint agent may be watched, or may rightly fear overhearing or interception, or may be vulnerable to arrest at points of meeting. Moreover, the sender is always under the pressure of urgency. Intelligence goes stale or is overtaken by events. Unless sent in timely fashion, preferably in "real time," which allows it to be acted upon, it loses its value.

3. *Acceptance.* Intelligence has to be believed. Agents who volunteer their services have to establish their credentials; they may be a plant. One's own operatives may have been turned or may have fallen under the control of an opponent's counter-espionage service. Even what they honestly offer may be wrong or only half true. Intercepts appear more dependable, but they may be bogus. Even if not, they can tell only part of the truth. Henry Stimson, American Secretary of State, rightly warned of the difference between reading a man's mail and reading his mind.

4. *Interpretation.* Most intelligence comes in scraps. For a complete

canvas to be assembled, the scraps have to be pieced together into whole cloth. That often requires the effort of many experts, who will have difficulty in explaining to each other what they understand by individual clues and who will disagree over their relative importance. Ultimately the assembly of a complete picture may require a superior to make an inspired guess, which may or may not be correct.

5. *Implementation.* Intelligence officers work at a subordinate level; just as they have to be convinced of the reliability of their raw material, so also they have to convince the decision-makers, political chiefs and commanders in the field of the reliability of their submissions. There is no such thing as the golden secret, the piece of "pure intelligence," which will resolve all doubt and guide a general or admiral to an infallible solution of his operational problem. Not only is all intelligence less than completely accurate; its value is altered by the unrolling of events. As Moltke the elder, architect of Prussia's brilliant victories over Austria and France in the nineteenth century and perhaps the supreme military intellectual of all time, memorably observed, "No plan survives the first five minutes of encounter with the enemy." He might as truthfully have said that no intelligence assessment, however solid its foundation, fully survives the test of action.

This book is a collection of case studies, beginning in the age of sail, when the supreme intelligence difficulty was to acquire information of value at any lapse of time which made it useful, and ending in the modern age, when intelligence of all sorts abounds but its volume threatens to overwhelm the power of the human mind to evaluate its worth. Its theme is that intelligence in war, however good, does not point out unerringly the path to victory. Victory is an elusive prize, bought with blood rather than brains. Intelligence is the handmaiden, not the mistress, of the warrior.

CHAPTER ONE

————◄○►————

Knowledge of the Enemy

STRATEGIC INTELLIGENCE

No war can be conducted successfully without early and good intelligence," wrote the great Duke of Marlborough. George Washington agreed: "The necessity of procuring good intelligence is apparent and need not be further argued." No sensible soldier or sailor or airman does argue. From the earliest times, military leaders have always sought information of the enemy, his strengths, his weaknesses, his intentions, his dispositions. Alexander the Great, presiding at the Macedonian court as a boy while his father, Philip, was absent on campaign, was remembered by visitors from the lands he would later conquer for his persistence in questioning them about the size of the population of their territory, the productiveness of the soil, the course of the routes and rivers that crossed it, the location of its towns, harbours and strong places, the identity of the important men. The young Alexander was assembling what today would be called economic, regional or strategic intelligence, and the knowledge he accumulated served him well when he began his invasion of the Persian empire, enormous in extent and widely diverse in composition. Alexander triumphed because he brought to his battlefields a ferocious fighting force of tribal warriors personally devoted to the Macedonian monarchy; but he also picked the Persian empire to pieces, attacking at its weak points and exploiting its internal divisions.

The strategy of divide and conquer, usually based on regional intelligence, underlay many of the greatest exploits of empire building. Not

7

all; the Mongols preferred terror, counting on the word of their approach to dissolve resistance. If duplicity enhanced their terrible reputation, so much the better. In 1258, appearing out of the desert, Hulagu promised the Caliph, spiritual leader of Islam, ruler of the Muslim empire, his life if he would surrender Baghdad. As soon as he submitted, he was strangled and the horde moved on. The Mongols, however, as a wide-ranging nomad people, also knew a great deal and, like all nomads, when not on campaign, were always ready to trade. Markets are principal centres for the exchange of information as well as goods, and it was often a demand of marauders—by the Huns of the Romans, frequently by the Vikings—that they should be allowed to set up markets on the borders of settled lands. Commerce was commonly the prelude to predation. Trade may follow the flag, as the Victorians comfortably affirmed, but it was quite as often the other way about.

Empires in the ascendant, to whom nomads were an irritation rather than a threat, adopted a different attitude. They gave and withheld permission to trade and hold markets on their borders as a deliberate means of local control.[1] They also pursued active "forward" policies. The pharaohs of the Twelfth Dynasty not only constructed a deep belt of forts on the border between settled Egypt and Nubia but also created a frontier force and issued it with standing orders. Its duty was to prevent Nubian incursions into the Nile Valley but also to patrol into the desert and report. One report, preserved on papyrus at Thebes, reads, "We have found the track of 32 men and 3 donkeys"; nearly 4,000 years old, it might have been written yesterday.

Ancient Egypt's border problem was perfectly manageable. The narrowness of the Nile Valley, amid the surrounding desert, necessitated the minimum of protective measures. The Roman empire, by contrast, was encircled on all sides by enemies, who might come by sea as well as land, and needed to be defended by elaborate fixed fortifications as well as mobile armies. At the height of their power, Rome's rulers preferred active to passive defence and maintained strong striking forces at strategic points generally behind rather than on the frontiers. It was only as their power declined and that of the outsiders grew that the border defences were thickened.

Whether on the decline or in the ascendant, Rome devoted great care to the gathering of intelligence. Caesar's conquest of Gaul was as much the result of his superior use of intelligence as the legions' superior fighting power. He took great trouble to assemble economic and regional intelligence, just as Alexander had done, and he was a coldly cynical assessor of the Gauls' ethnic defects, their boastfulness, volatility, unreliability, lack of resilience; he was equally cold in exploiting the advantage his knowledge of their weaknesses afforded. He accumulated a detailed ethnographic knowledge of their tribal characteristics and divisions, which he used ruthlessly to defeat them. Quite apart from this strategic intelligence, however, he also had a highly developed system of tactical intelligence, using short- and medium-range units of scouts to reconnoitre up to thirty kilometres in advance of his main body, to spy out the land and the enemy's dispositions when he proceeded on campaign. It was an important principle that the leaders of these units had immediate and direct access to his person.

Caesar did not invent the Roman system of intelligence. It was the product of several hundred years of military experience. Evidence for that is already given, by the time of the Gallic wars (first century B.C.), by the existence of established terms for the different categories of reconnaissance troops: *procursatores,* who performed close reconnaissance immediately ahead of the army; *exploratores,* longer-range scouts; and *speculatores,* who spied deeper within enemy territory. The Roman army also made use of local informers (*indices*), prisoners of war, deserters and kidnapped civilians.[2] If not the inventor of the system, Caesar may, nevertheless, be credited with professionalising it and institutionalising some of its most important features, notably the right of direct access by scouts to the commander in person. He also, when necessary, went to see for himself, a dangerous but sometimes essential intervention. Ultimately, the crisis of the empire in the fourth century required the almost continuous presence of one of the emperors (there were latterly two, sometimes more) with the army, a contingency that, at Adrianople in 378, led to his death on the field, progressive disaster and the empire's collapse. The emperor Valens had been in close touch with his *exploratores* on the morning of the catastrophe, and they had correctly reported the enemy's strength and dispositions. What ensued substantiates a profound and enduring truth, that "military and political survival does not depend solely on good intelligence."[3]

Systems do not, however, much change, unless circumstances change, and there was little circumstantial change throughout the five centuries of the Roman empire's greatness (first century B.C.–fourth century A.D.). Reconnaissance throughout the period was by hearing and sight, communication by word of mouth or written despatch, speed of transmission at fastest by that of a fleet-footed horse. What was true of Rome remained true of the world for another 1,500 years.

The collapse of imperial government in the West in the fifth century A.D. entailed also the collapse of organised intelligence services and such ancillary services as the publication of guidebooks and cartography (though Roman maps are strange to us, since they usually took the form of route-charts rather than two-dimensional displays of territorial features, their disappearance was a serious loss to campaigning commanders). Worse by far were the progressive degradation and eventual and complete decay of the road system. The Roman roads were built primarily for the purpose of rapid all-weather military movement and were maintained by the legions, which were as much engineering as fighting units. The dissolution of the Roman army led rapidly to the cessation of engineering work on such key elements of the Roman transport system as bridges and fords. The road network, of course, had not existed during the period of Roman conquest; Caesar had made his way through Gaul by interrogating merchants and locals and impressing guides. It was the roads, however, that had allowed Rome to defend its empire for five centuries and the break-up of their solid surfaces made long-range campaigning at speed impossible.

That was not important to the barbarian rulers who succeeded the Romans, since they sought no more than to maintain local authority. When, however, the attempt began again, under the Carolingian emperors, to reestablish wide imperial domains in the eighth and ninth centuries, the absence of roads was a serious impediment to reconquest. Things got even worse with the attempt to penetrate the Germanic regions which lay beyond the old Roman borders. In those wildernesses there were neither roads nor easily obtainable intelligence.

Some picture of the difficulties confronting medieval campaigners is conveyed by the experience of the Teutonic Knights in their effort to conquer and Christianise the Baltic shore in the fourteenth century. The Teutonic Knights, a crusading order dedicated to the conversion of the Prussians and Lithuanians, were wealthy and highly organised. They operated from a chain of strong castles built on the Baltic coast, in which they were secure from attack and could organise crusading expeditions into the hinterland. One of their principal campaigning grounds was a belt of unsettled land a hundred miles wide between East Prussia and Lithuania proper, a maze of marsh, lakes, small rivers, thickets and forest through which it was almost impossible to find a way. Local scouts were recruited by the Knights to blaze trails and report. Their intelligence was collected in a military guidebook, *Die Lithauischen Wegeberichte* (*The Lithuanian Route Guide*), compiled between 1384 and 1402. It explains, for example, that Knights wishing to get to Vandziogala from Samogitia, both near modern Kaunas in Lithuania, a distance of about thirty-five miles by today's roads, had first to cross a patch of scrub, by a track, then a large wood through which they would have to clear their way, then a heath, then another heath, then a second wood, "the length of a crossbow shot and there you have to clear your way too," then a third heath and a third wood. Beyond lay the true *Wiltnisse* (wilderness). A Prussian scout's letter describing it was copied into the *Wegeberichte*. It reads: "Take notice in your wisdom that by God's grace Gedutte and his company have got back in safety and have completed everything you sent us to carry out and have marked the way so far as 4½ miles this side of the Niemen, along a route that crosses the Niemen and leads straight into the country." The tone of the report recalls that of the Egyptian border patrol from Nubia 3,000 years earlier; the terrain described is that over which the German Army Group North advanced to Leningrad in 1941, encountering obstacles the Teutonic Knights would have found familiar.[4]

Curiously, the Holy Land Crusaders faced much less difficulty in getting to Jerusalem in the eleventh century. In 1394, the Grand Master of the Teutonic Knights had answered Duke Philip of Burgundy's enquiry as to whether there would be a Baltic crusade the following year: "It is impossible to provide a forecast of future contingencies, especially because on our expeditions we are obliged to go across great waters and vast solitudes by dangerous ways ... on account of which they frequently depend on God's will and disposition, and also on the weather." In different words, a modern intelligence officer might respond almost exactly similarly. The Holy Land Crusaders, by contrast, had found a much easier way forward, travelling either by sea or along the surviving Roman roads in Italy or inside the dominions of the Eastern Roman (Byzantine) emperor in southern Europe, where the imperial administration kept communications in repair and furnished supplies.

Once arrived at Constantinople, they were provided with guides and escorts and were able to travel on the great Roman military roads that led towards the Taurus Mountains. In what is today eastern Turkey, however, already invaded by Seljuk Turkish migrants from central Asia, they found the roads in disrepair and likewise the other conveniences of travel—cisterns destroyed, wells dry, bridges fallen, villages abandoned. It was a foretaste of how a nomadic, horse-riding people ruined a civilised countryside by rapine and neglect. The final stages of the march to Jerusalem were far harder than the departure from Europe.[5]

Campaigning inside Western Europe itself throughout the Middle Ages, the leaders of armies found conditions consistently inimical to conducting effective operations. The main problem was a chronic shortage of money in an effectively cashless society, which made the recruitment of armies difficult and their provision with food and supplies often almost impossible. Movement was laborious, because of the absence of an all-weather road system, but the lack of intelligence also impeded the efforts of rulers to deploy such forces as they could raise to the places where they were needed. That difficulty became particularly acute during the crisis of the Viking invasions in the ninth century. The Vikings, who had achieved a revolution in mobility by the development of their superbly fast and seaworthy longships, appeared without warning, overwhelmed local defenders by the ferocity of their assaults and, in the second stage of their terrorisation of the Christian lands, carried violence and pillage deep inland by learning to capture horses in large numbers at their points of debarkation. The antidote to Viking raiding would have been to create navies, but that was beyond medieval kings; another recourse would have been to maintain an intelligence system, to provide early warning, inside Scandinavia. Such sophistication lay even further outside the capabilities of ninth-century kingdoms; moreover, the Viking lands were no place for inquisitive strangers, even with money to loosen tongues. There was much more money to be made by raiding than by selling information, and the Vikings took pleasure in cutting throats.[6]

By the fourteenth century, the conditions of warfare in post-Roman Europe had altered greatly to the local rulers' advantage. The overriding need to suppress the aggression of nomadic despoilers—Vikings in the west, Saracens in the south, horse peoples in the east—had stimulated the building of fixed defences, including continuous barriers and chains of castles, which had solidified frontiers, pacified borderlands and restored the possibilities of trade, with beneficial effects on the general prosperity. Kings had money to pay soldiers; they also found the money to buy intelligence and pay agents, who moved with reasonable ease among travelling merchants and, or so at least was suspected by royal governments, under the cloak of international religious orders. It is a mark of how commonplace spying had become during the Hundred Years War between France and England that heralds, the non-partisan arbiters of propriety on the battlefield, went to great lengths to defend their reputation for impartiality; so too did ambassadors, though they were less often believed.

By the middle of the fourteenth century there were extensive networks of English agents in northern France and the Netherlands, usually foreigners working for money, with French counterparts in England, often identified by the royal government as expatriate monks or travelling friars, how accurately is now difficult to establish. What their information was worth is equally mysterious. Even more than would be the case in later ages of improved communications, messages were difficult to transmit quickly in the Middle Ages. The roads were bad, the hire of horses unreliable, the sea a barrier, particularly to the transmission of messages from France to England. The English kings tried to smooth the path. The port of Wissant in northern France, the nearest to Dover, was a usual point of departure, where crossing fees were fixed by law. On the English side of the Channel, post horses were maintained at royal expense for official messages. One piece of evidence suggests that the money was well spent. On Sunday, 15 March 1360, news was brought to the royal council, sitting at Reading, that the French had attacked Winchester, a hundred miles distant, that very day. There is no suggestion, however, that intelligence had brought advance warning.[7]

Real-time intelligence, except over very short distances, was inherently difficult to acquire in the medieval world. It simply could not be carried quickly enough ahead of the movement of enemy forces. That would remain so for centuries to come. Sometimes critical information did not travel even within the confined space of a battlefield. At Lützen, for example, on 16 November 1632, one of the most important engagements of the Thirty Years War, the Imperial (Austrian) and Swedish armies both made a tactical retreat at the end of the day. The Swedish king, Gustavus Adolphus, had been killed and if Wallenstein, the Imperial commander, had renewed the attack, the Swedes would probably have lost. Neither side, however, was aware of the other's movements. Next day the Swedes returned, captured the Imperial artillery, which had been abandoned for want of horses to drag it off, and so turned what should have been an Imperial victory into a defeat.[8]

The European armies of the eighteenth century had become much more professionalised than those of the Thirty Years War. Even so, they found real-time intelligence hard to acquire. Frederick the Great's campaign of Hohenfriedberg in 1745 was exceptional. The Imperial (Austrian) army was concentrating against him to wrest back the province of Silesia, which the Prussian king had illegally seized in 1740. He got general word of its movement but needed to put himself in a favourable position to resist its attack, by tempting it down into the Silesian plain from the surrounding hills. His first move was to use a double agent he had, an Italian clerk, in Imperial headquarters, to spread the word that the Prussians were retreating. He then concealed his army in broken ground and waited for the Austrians to appear. They made no effort to disguise their movements, and so he was able to make use of rules of observation (*indices*) which were known to provide rough-and-ready real-time intelligence when the enemy was in view. Dust was an important indicator. "A generalised cloud of dust usually signified that the enemy foragers were about. The same kind of dust, without any sighting of the foraging parties, suggested that the sutlers and baggage were being sent to the rear and that the enemy was about to move. Dense and isolated towers of dust showed that the columns were already on the march." There were other signs. The gleam of the sun, on a bright day, on swords and bayonets was open to interpretation at distances of up to a mile. Marshal de Saxe, Frederick's great French contemporary, wrote that "if the rays are perpendicular, it means that the enemy is coming at you; if they are broken and infrequent, he is retreating."[9]

Frederick, on 3 June, had positioned himself at a lookout point which commanded the level ground in front of Hohenfriedberg. Towards four o'clock in the afternoon he saw a cloud of dust, through which gradually resolved eight huge Austrian columns advancing towards the Prussian positions, illuminated by bright sunshine. As darkness fell, Frederick ordered a night march. Next morning the Battle of Hohenfriedberg began.

Despite his enjoyment of advantageous intelligence, Frederick did not win an easy victory. His army was outnumbered, and the Austrians and their allies had manoeuvred during the night to outflank him. As so often in war, it was superior fighting power that carried the day; Frederick's preliminary intelligence success was soon negated. It was his own quick thinking in the heat of action and the fierce reaction of his soldiers which turned the tide of battle.[10]

The same would most often prove to be true in wars yet to come. In their wars outside Europe, particularly in the North American forests, where Indian allies knew the ground intimately and were masters of the arts of scouting and surprise, European armies were to suffer shocking defeats in the depths of the woods. General Braddock's disaster at the Monongahela, near modern Pittsburgh, where a large British force was wiped out in a few hours in 1755, was entirely the result of walking blind into an ambush prepared by the French, led by their Native American allies, in uncharted and unscouted woodland. In what both sides came to call "American warfare," intelligence remained at a premium and usually provided the basis of victory or defeat. In the familiar campaigning grounds of Europe, during the great wars of the French Revolution and Napoleonic empire (1792–1815), intelligence rarely brought victory solely by its own account. That was true even during the British Peninsular War against the French in Spain and Portugal, 1808–14. Intelligence, however good, moved too slowly to bring a real-time advantage. Indeed, Wellington in the Peninsula depended upon exactly the same means of intelligence as Scipio in his campaign against Nova Carthago (New Carthage) in Spain in the third century B.C. Wellington, Caesar and Scipio all operated as intelligence-gatherers in exactly the same way. Their earliest concern was to discover the lie of the land (Wellington was a great collector of maps and almanacs) and the characteristics of the enemy. The collection of tactical intelligence—who was where when, what he intended and of what he was capable—was left to the month, the week, the day.[11]

Wellington had the population on his side, in both Portugal and Spain. France, the invader, was resented, and after the excesses of 1808, hated. Wellington did not have to seek intelligence. It was brought to him by the bucketload. The difficulty was to sort wheat from chaff. Much more illuminating, as an example of intelligence-gathering in the pre-electric age, was the organisation of intelligence during his campaigning days in unconquered India. Wellington (Arthur Wellesley) was in active command of armies in India from 1799 to 1804. Britain, through the East India Company, controlled large enclaves in Bengal, Bombay and Madras but huge areas of the sub-continent were under the rule of local warlords or freebooting hordes. The French, by diplomacy, bribe and direct inter-vention, sought to bring a majority of anti-British elements to their side. Wellington, operating with small armies of mixed British–Indian compo-sition, was mainly concerned with putting down such independents as Tipu Sultan and Hyder Ali, feudatories of the effete Moghul emperor, who were effectively running their own armies and states.

In order to win, Wellington needed a steady stream of up-to-date information, from both far and near, so as to anticipate the movements of his enemies and gain forewarning of shifts of alliances, the gathering of stores, the recruitment of soldiers and other signs of offensives in the making. The conventional means of securing such a supply of intelli-gence was to form a reconnaissance corps, of troops either already under command or recruited from the population. The British in India had recourse to another method. They took over a pre-existing intelligence system and made it their own.

The *harkara* system seems to have been unique to India. Because of the sub-continent's enormous size, difficult terrain and—until the build-ing of the railways and the trunk roads of the British raj—lack of long-distance routes, power tended to be local. Even when centralised under the Moghul conquerors of the sixteenth century, it remained quite dif-fuse. The Moghuls in Delhi ruled by devolution, either to mighty provincial officials or by arrangement with local princes, particularly in western and southern India. The system could be made to work only if the court was supplied with regular reports of events at the lesser courts. It came to be supplied by two groups of news-providers: writers, often scholars of high status in the Indian caste system, and runners, who car-ried verbal or written messages and reports over long distances at high speed.

Over time the system yielded a peculiarly Indian product: the newsletter, usually written in Persian, the language of the Moghul court, in a highly stylised form and on a regular, typically weekly, basis. The letters began as official documents but became, as writers and even run-ners acquired independence, a sort of private newspaper. Eventually not so private; to whom to distribute the newsletter became a decision of the *harkara*, who himself acquired a blurred identity, part intelligence-gatherer, part distributor. He also acquired odd rights, to be paid, of course, but also to be accepted as a sort of local correspondent at court, known to be working for other powers at a distant centre.

The *harkaras* survived because, through their indispensability to those at both ends of the system, they established their independent status. It was an uneasy independence; flogging or even execution could follow the provision of dubious or misleading news. The punishment, however, was personal; it was not intended to undermine the system itself. The system, by the time the British embarked on their progressive supersession of the Moghuls at the end of the eighteenth century, was deeply entrenched in the processes of Indian political and military life. Indian government could not work without it. The British, who were committed to reestablishing Moghul power on an efficient basis, ruling themselves while leaving the Moghuls nominally in charge, simply took it over. They "reconstituted under their [own] control the classic Indian intelligence system which allied the writing skills and knowledge of learned Brahmins with the hard bodies and running skills of tribal and low-caste people."[12]

Wellington could not have established himself as the leading sepoy general without the *harkaras,* whom he both cultivated and tyrannised. His successors continued to do so. Not until the arrival of the telegraph and the establishment of printed newspapers in the middle of the nineteenth century did the *harkara* system decline; and even so, training in long-distance message-running persisted into the 1920s, sustained by the Indian appetite for news, uncontrolled by official interference, which is such a distinctive feature of sub-continental life. The reason, it has been suggested, why India has become and remains the largest and only real democracy in the Third World is because of its citizens' insatiable thirst for information.

.   .   .

REAL-TIME INTELLIGENCE:
WHAT, HOW, WHERE, WHEN?

Who knows what in sufficient time to make effective use of the news—
that is as good a definition of "real-time" intelligence, the gold standard
of modern information practice, as is possible—was not often a military
consideration in the classical world or even the age of Wellington.
Alexander, Caesar and Wellington all operated within the peculiar con-
straint, to the modern way of thinking, of very slow communication
speed over any distance not to be covered by a running man or a gallop-
ing horse. The best *harkaras* were credited with a speed of a hundred
miles in twenty-four hours, but sceptics thought fifty more realistic. The
modern marathon, whose runners achieve twenty-six miles in about
three hours, gives a better indication of the nature of real-time intelligence before the coming of electricity. The armies and navies of the
pre-electricity age operated within an intelligence horizon of considerably less than a hundred miles. Hence the enormous importance attached
by the commanders of the past to strategic intelligence: the character of
the enemy, the size and capability of his force, its dispositions, the nature
of the terrain in his operational area and, more generally, the human and
natural resources on which his military organisation depended. It was
from guesses based on such factors that generals of the pre-modern
world made their plans. Real-time intelligence—where the enemy was
yesterday, in which direction his columns were headed, where he might
realistically be expected today—was arcane information, rarely to be
collected on a real battlefield. As late as 1914, ten divisions of French cavalry, beating the Franco-German-Belgian border for nearly a fortnight,
altogether failed to detect the advance of several million German troops.
French reconnaissance forces failed again in the same area in 1940. Strategic intelligence is a desirable commodity. It rarely, however, brings
advantage in actual time and space. For that, something else is necessary.
What exactly is it? How is it possible to assure that the key questions,
what, how, where and when, are answered to our advantage, not the
enemy's? That is the theme of this book.

The acquisition of real-time intelligence requires, first of all, that
the commander should have access to a means of communication that
considerably outstrips in speed that of the enemy's movement over the
ground or water. Until the nineteenth century, the margin of superiority

was very small. The marching speed of an army, reckoned at three miles per hour, was exceeded by that of a scout's horse perhaps six times; but a scout had to make an outward as well as return journey, so the margin was halved. In the interval between scouts making contact with the enemy and returning, moreover, the enemy might advance, reducing the margin still further. Little wonder that surprise was so difficult to achieve in ancient campaigns. When it was, as spectacularly by the Seljuk Turks at Manzikert in 1071, the reason was often treachery or a total failure to reconnoitre, or both. At Manzikert the Byzantine army's cavalry screen deserted, leaving the commander blind.

Manzikert was an "encounter" battle, with both armies advancing simultaneously. More typical was the situation in which an advancing army ran into the outposts of an army standing on the defence. They automatically raised the alarm and, not having to go out and back, as in encounter operations, but back only, could give early warning. Wellington, for example, during the Waterloo campaign was, though strategically surprised, not so tactically. The French ran into his outposts, allowed him to fight a delaying battle at Quatre Bras on 16 June and to retire to a previously reconnoitred main position at Waterloo two days later.

Surprise was equally as difficult to achieve at sea as it was on land until recent times. Indeed, the traditional problem in naval warfare was for opposing fleets to find each other at all. Hence the tendency for naval battles to occur in narrow waters, or "the shipping lanes," often in areas where battles had occurred before. In theory, with the invention of the telegraphic flag code at the beginning of the nineteenth century, an admiral, by disposing his ships at the maximum limit of intervisibility— intervals of twelve miles—could, if his chain was long enough, create an early-warning screen which would cover several hundred miles of ocean. In practice admirals never had enough ships; and anyhow, they preferred to keep those they had concentrated, for the danger of being brought to battle dispersed outweighed that of being surprised. An admiral surprised with his ships within range of recall could form a line of battle; an admiral with his ships scattered on reconnaissance beyond quick recall had no such hope. Not until the invention of wireless telegraphy at the beginning of the twentieth century, and its adoption by war fleets, could admirals truly begin to command the far seas. Even then, old habits died hard. Sailors like to see.

Sight is, of course, the principal and most immediate medium of real-time intelligence. It was so in the pre-telegraphic age, and it has become so again in the age of electronic visual display. In the intervening period, which embraces the invention of the electric telegraph in the middle of the nineteenth century and its supersession by radio at the beginning of the twentieth, hearing acquired a superior status. It still enjoys something like parity. Radio in all its forms is an essential tool of military communication. Strategically, it has become less important than the written message electronically communicated, by fax or e-mail. Tactically, it predominates, by immediacy and urgency. In the heat of engagement, voice-to-voice communication between commander and the front line, and in the opposite direction, is what makes battles work, a reality not altered since Caesar took personal, front-line control of the Tenth Legion against the Nervii on the river Sambre in 57 B.C. "Calling to the centurions by name, and shouting encouragement to the rest," Caesar transformed the tempo of the battle, shifting the psychological advantage to the Roman side and assuring the defeat of the Gauls.

The golden age of heard communication—the dot-dash of Morse telegraphy, the human voice of radio transmission—was comparatively short. It lasted in the military arena from about 1850 until the end of the twentieth century. It was a period punctuated by grave and frustrating blanks, particularly during the First World War, when the intensity of bombardment on the fixed fronts of west and east broke cable communication as soon as mobile operations began, and the signal services had not yet succeeded in acquiring compact radios independent of cumbersome power supplies. Intelligence in real time became an impossibility. Commanders lost voice, indeed any other contact, with their forward troops over the shortest distances, and battles degenerated into directionless turmoil.

Not so at sea. Because of the ready availability of powerful electric current in turbine-propelled warships, radio communication within fleets and between their component units had become standard by 1914. There were difficulties; lack of directionality in contemporary transmitting sets made for interference, so intense in fleet actions that admirals continued to depend upon flag hoists to control their squadrons. Nevertheless, it had become clear by 1918 that the future of naval communications lay with radio.

Not, however, with radio telegraphy (R/T), as voice broadcasts were denoted to differentiate them from wireless telegraphy (W/T) in Morse code. R/T is insecure; the enemy overhearing it is as well informed as the intended recipient. Protection is possible, through very high-frequency (VHF) directional transmission, as in the Talk Between Ships (TBS) system used by anti-submarine escorts to great effect during the Battle of the Atlantic; but it is intrinsically short-range, as a secure means of speech. The only safe way of sending messages over long distance by radio wave is through encryption, effectively a return to W/T. Paradoxically, therefore, the flexibility and immediacy allowed by voice radio was denied both strategically and, except in limited circumstances, tactically, by its insecurity. While the control of naval warfare became, as the twentieth century drew on, increasingly electronic, through the rise of such derivatives of radio as radar, sonar and high-frequency direction-finding (HF/DF), high-level, long-range communication remained stuck at the level of wireless telegraphy, because of the need to encrypt or encode, the resulting message being sent by Morse.

That imposed delay, which was why, though the same imperatives should have applied to army and air force communications, soldiers and airmen, caught up in the dynamics of close-range combat when time was too short to encrypt or encode, broadcast freely by voice radio. Tactical codes were developed—the British army's Slidex, for example—but even Slidex took time. In the cockpit of a single-seat fighter, any form of encryption was impossible. All armies and air forces therefore set up tactical overhearing services, called "Y" by the British, which listened in to the tactical voice radio transmissions of their opponents.[13] Y frequently supplied battlefield intelligence of high value. During the Battle of Britain, for example, the British intercept stations were able to anticipate the warning of air raids supplied by the Home Chain radar stations by overhearing the chatter of Luftwaffe aircrew forming up before take-off on their French airfields.

Y was nevertheless of limited and local military value. Important radio communications were, from the First World War onwards, always encrypted or encoded, and only a combatant equipped to render secret writing into plain text could hope to do battle on equal terms with the enemy. The competence of the major powers varied, between themselves and also over time. In the forty-five-year struggle between the Germans and the British during the twentieth century, for example, the Germans unknowingly lost the security of their naval codes early during the First World War and did not regain it. The British, partly by capture and partly by intellectual effort, were able to reconstruct their enemy's book codes in 1914 and thereafter to read high-level German communications at will. In the aftermath of the war, hubris—a repetitive influence on secret communication—led the British to believe their own book codes impenetrable, while the Germans during the 1930s both broke the British book codes, an intrinsically insecure means of secret writing, and adopted a machine cipher system, Enigma, which would resist the attack of its enemies' cryptanalysts—Polish, French and British—until well into the course of the Second World War. The Polish success in breaking Enigma before 1939 was negated before the outbreak of war by German variation of their machine encryption.

There are other means of acquiring intelligence in real time besides seeing and hearing, notably through the indirect sight provided by photographic intelligence and, today, satellite surveillance; human intelligence—humint, or spying—can, in certain circumstances, also convey critically urgent information. Both, however, are prone to delay and defect. Images, however acquired, need interpretation; they are often ambiguous and can cause experts to disagree. Thus, for example, photographic evidence brought back to Britain from the German pilotless weapons development station at Peenemünde during 1943 did show both the V-2 rocket and the V-1 flying bomb. The weapons went unrecognised for some time, however, in the first case because the interpretation officers did not recognise the rocket in its upright, launching position, in the second because the image of the V-1 was so small—less than two millimetres across—that it was missed. Yet the pilotless weapon photographic evidence was comparatively clear and was supported by other intelligence which told the interpreters what it was that they should be seeking to identify. They knew they were looking for "rockets" and miniature aircraft; even so they failed to recognise the evidence before their own eyes. How much more difficult is image interpretation when the interpreters do not know exactly what the evidence will resemble when they see it: the hideouts of al-Qaeda terrorists, the bunkers of illegal Iraqi weapon development centres. The intelligence of imagery is frustratingly rich—many needles but in a vast haystack.

Human intelligence may suffer from different limitations: including, first, practical difficulty in communicating with base at effective speed; and, second, inability to convince base of the importance of the information sent. The world of human intelligence is so wrapped about with myth that any clear judgement about its usefulness is difficult to establish. It does seem, that, for example, the Israeli foreign intelligence service was running an "agent in place" at a high level in Egypt before that country's attack on Israel in 1973. Because the Egyptian government dithered over the decision to attack, the agent sent a succession of self-cancelling reports, with the result that, when the attack came, the Israeli army had gone off high alert. True or not, the story leaves unresolved the question of how the agent was able to communicate in real time. The case of Richard Sorge was entirely different; he was both highly placed and well equipped to communicate by clandestine radio. His difficulty— of which he was unaware—was to secure a hearing. Sorge, a committed Communist and long-term Comintern agent, had established himself before the Second World War in Tokyo as the respected correspondent of a German newspaper. As a German native and citizen thought to be entirely patriotic, he became intimate with the staff of the German embassy, passed on information about Japanese affairs the diplomats found useful and eventually began to assist the ambassador in drafting his reports to Berlin. As a result, he was able to send the assurance to Moscow, during the terrible summer of 1941, that Japan did not intend to assist its German ally by attacking the Soviet Union through Siberia. He had earlier sent convincing warning of Germany's intention to invade and had even identified the correct date, 22 June. Stalin had received other warnings, including one from Churchill, but chose to ignore them, as he ignored that of Sorge. The idea of war was too uncomfortable; Stalin preferred to believe that he could buy Germany off by fulfilling delivery of strategic materials, including oil. As to the assurance that Japan would not invade Siberia, the evidence is that Stalin had moved a large portion of the Soviet garrison out of Siberia before the attack of 22 June and that Sorge's warning, even if heeded, was not the critical strategic intelligence it seemed to be.[14]

Other celebrated human intelligence organisations of the Second World War, the "Red Orchestra" in Germany and the "Lucy Ring" in Switzerland, also lack credibility as media of real-time intelligence, though for different reasons. Sorge is almost unique among identified operators in the world of humint by reason of his undoubted access to information of high quality and his ability to forward it speedily to base. The Red Orchestra, a coterie of left-inclined Germans of superior social standing, was a dilettante organisation led by a Luftwaffe officer with a double-barrelled name, who seems to have been animated by the excitement of misbehaviour. They transmitted little of importance to Moscow, betrayed themselves by lack of elementary security precautions and were rapidly rolled up by the Gestapo. The information supplied by the Lucy Ring—in reality an individual called Rudolf Roessler—to Moscow from Switzerland during the war seems to have been derived largely from his study of the German press. The rest was fed to him by the Swiss, who maintained contacts with the German Abwehr. The Swiss feared that a German victory would lead to the incorporation of their country into a greater Reich; the Abwehr was adept at playing a double, if impenetrable, game. Roessler may have belonged to that colourful gang of fantasists who enriched themselves, at the expense of the secret service budgets of many countries, throughout the Second World War.[15] In any case, he lacked a radio link to Moscow.

The ability to communicate, quickly and securely, is at the heart of real-time intelligence practice. It is rarely enjoyed by the agent, that man of mystery who figures so centrally in the fictional literature of espionage. Real agents are at their most vulnerable when they attempt—by dead letter box, microdot insertions in seemingly innocent correspondence, meetings with couriers, most of all by radio transmitter—to reach their spymasters. The biographies of real agents are ultimately almost always a story of betrayal by communication failure. A high proportion of the Special Operation Executive (SOE) agents in France during the Second World War were discovered by German radio counter-intelligence; the same was true of those operating in Belgium while, as has become well known, at one stage all agents dropped into Holland were collected on site by German manipulators of the SOE radio network. Even when counter-intelligence is deplorably lax, as in the notorious "Missing Diplomats" case, the certainty of Donald Maclean's guilt was finally established, even if retrospectively, by his habit of leaving Washington to meet his Soviet controller in New York twice a week.[16]

It is the intrinsic difficulty of communication, even, indeed above all, for the agent with "access," which limits his—or occasionally her—usefulness in real time. By contrast, the enemy's own encrypted communications, if they can quickly be broken, will, of their nature, provide intelligence of high quality in real time.

The history of "how, what, where, when" in military intelligence is therefore largely one of signal intelligence. Not exclusively; human intelligence has played its part, and so, latterly, have photographic and surveillance intelligence. In principle, however, it is the unsuspected overhearing of the enemy's own signals which have revealed his intentions and capabilities to his opponent and so allowed countermeasures to be taken in time.

The case studies that follow will support this argument. Not, however, in its entirety; one case study, that of the German airborne descent on Crete in May 1941, has deliberately been chosen to demonstrate that even the best intelligence will not avail if the defence is too weak to profit by it. Other studies—particularly those of the Battle of the Atlantic and the V-weapons—emphasise the parallel importance of non-intelligence factors, in the first case, and of photographic intelligence and humint, in the second, in defeating the threat. The book begins with studies of intelligence campaigns in the age before signal intelligence was available to a keen-eared listener. The examples of Nelson's Mediterranean campaign of 1798 and Stonewall Jackson's campaign in the Shenandoah Valley in 1862 may persuade those who expect too much—or too little—of modern intelligence-gatherers and -users to reconsider their opinions.

It has become part of the conventional wisdom that intelligence is the necessary key to success in military operations. A wise opinion would be that intelligence, while generally necessary, is not a sufficient means to victory. Decision in war is always the result of a fight, and in combat willpower always counts for more than foreknowledge. Let those who disagree show otherwise.

CHAPTER TWO

Chasing Napoleon

A CCURATE INTELLIGENCE in the age of sail was a scarce commod-ity. Accurate intelligence is, of course, scarce at all times, but the sea, in the years before radar and radio, let alone satellite surveillance, was an arena of the unknown. The horizon was an impenetrable barrier, the man-of-war, breasting the waves at six knots or so, a ponderous means of extending a captain's field of vision beyond it. An admiral with a fleet under command could enlarge his supervision by disposing his men-of-war and their attendant frigates at intervals of a dozen miles or so, the maximum distance at which intervisibility and so intercommuni-cation, masthead to masthead, was possible. Even thus, the transmission of news of a sighting was a haphazard business. Admiral Lord Howe, seeking to establish the whereabouts of a French convoy of 139 mer-chantmen he knew to be nearby, had to quarter the sea for eight days before he brought its escort to action on the Glorious First of June 1794. The interception was rightly regarded as remarkable. The battle took place 400 miles from land at a time when most sea engagements were fought, as they always had been, within fifty miles of the coast.

Admirals, moreover, did not like to disperse their capital ships for purposes of reconnaissance, needing to keep them concentrated for action lest the enemy were met unexpectedly. The need to gather infor-mation always had to be balanced against that of keeping firepower massed under the commander's hand. The strength of a fleet lay not in that of its individual units but in the line of battle, formed bow to stern at intervals of a few hundred feet. Ships isolated were ships exposed to "defeat in detail," or one by one, by superior numbers. Hence the importance of the capital ships' scouts, the frigates, smaller and swifter sailers which could be sent to and beyond the horizon in search of the enemy.

No admiral ever had enough frigates. The calls on their service were manifold: as despatch vessels, as commerce raiders, as convoy escorts. Such duties chronically depleted the number available as scouts and "repeater ships," which lay close to but outside the line of battle, copying the signals of the flagship, which were often masked by its near neighbours, so that they were visible from van to rear. The scarcity of frigates was odd. As they were much smaller than battleships, a third or a quarter by displacement, carried far fewer men, perhaps only 150 against 800, and cost only a fifth as much to build, it might be expected that they would have been turned out in much larger numbers. Such was not in practice the case. In 1793, at the start of the Wars of the French Revolution, the Royal Navy had 141 first- second- and third-rates, battleships with between 100 and 74 guns; of fifth- and sixth-rates, frigates of between 44 and 20 guns, it had only 145;[1] by 1798, there were still only 200. Little wonder Nelson, then the Mediterranean admiral, warned that, if he died, "want of frigates would be found written on his heart."

Essential though frigates were as scouts, their value was restricted by the limitations of the signal system then in use. It was not simply that flags—and flags were the principal means of communication—were difficult to discern at long distance, even with a telescope. No comprehensive system of arranging them to transmit information had yet been devised. Various conventions had been in use since the seventeenth century, such as then, for example, hoisting a red flag at the mizzen-top to order a particular manoeuvre. By the late eighteenth century there had been a great deal of elaboration and in 1782 Admiral Howe, then in command of the Channel Fleet, had issued a codebook, superseding many others, which allowed a commander to say 999 different things with three flags and 9,999 with four. Howe's signal book was not, however, double-entry. The recipient could work out what a signal meant by looking it up, finding the flags picture by picture on the page. The sender, however, had to know what flags he needed to hoist before he composed his message. Not until 1801, when Home Popham published his *Telegraphic Signals or Marine Vocabulary*, were sender and recipient put on equal footing. Popham, a sailor of great clarity of mind, simply had a flash of insight into the obvious which had evaded thousands of other sea officers before him. His achievement compares with that of his near contemporary Roget in compiling the first thesaurus. He analysed how language was used and saw that words might be given a numerical value, to be signalled by a set of numerical flags; 212, for example, could be made to stand for "cable," with the numerical flags 2, 1 and 2. Adding a fourth numerical flag, 3, made the signal read "Can you spare a cable?" To make the first signal, the sender looked up "cable" in his double-entry book and chose the appropriate set of flags or "hoist"; to read it, the recipient looked up 2123 and got the message.[2] By the use of special indicators it could also be signified that flags had alphabetical rather than numerical value and should be read simply as letters, spelling out an unusual word not in the vocabulary. Popham's final signal book allowed 267,720 signals to be made with twenty-four flags (of which ten doubled as numbers) and 11 special indicators.

His system remains in use to this day. It was not yet so in 1798, when the Royal Navy still sought to speak fluently between ships through the medium of Admiral Howe's single-entry book. A great deal of time was wasted, therefore, in frigates closing up to each other or to the main body in order to transmit information unambiguously or to receive precise questions or clear orders. The days when a flag lieutenant could snap his telescope shut and confidently interpret to his superior the meaning of a flash of coloured bunting glimpsed at the extreme limit of visibility on a clear Mediterranean day lay well in the future.

The fallibility of signalling was not a crucial factor in the conduct of maritime operations in the last decade of the eighteenth century. It counted for much less than want of frigates. It counted as a factor, nonetheless, particularly to an admiral like Horatio Nelson, whose mind never rested, who calculated the relative positions of ships and shore-lines as a master chess player does pieces and squares, who consumed information of every sort with the compulsion of an addict, who sought decision in battle with the relentlessness of a great financier poised to obliterate his commercial competitors. Home Popham's signalling system certainly would have assisted him in his searches had it been available; when it became so a very few years later, Nelson enthusiastically embraced it; indeed, the most famous flag signal sent, "England Expects," was made at the opening of the Battle of Trafalgar on 21 October 1805, with eight Popham hoists and "Duty" spelt out alphabetically.

On Trafalgar Day Nelson had the combined French–Spanish fleet clear in view. The encounter had come at the end of a long chase which had begun in May, taken him across the Atlantic to reach the West Indies in June, back again to the mouth of the Channel in August and finally south to the Straits of Gibraltar in September, where he blockaded Cadiz until the enemy put to sea in October. He had had a number of false starts and followed a number of false trails, but once Admiral Villeneuve had cleared the Straits of Gibraltar from the Mediterranean and set off into the deep Atlantic, Nelson had been able to make the assumption with some certainty that the French were heading for the West Indies. The campaign of Trafalgar was to prove a triumph of strategic manoeuvre. As an intelligence operation, it was not, at least in its later stages, one of complexity.

The contrast with Nelson's earlier pursuit, discovery and destruction of a French fleet was extreme. In 1798 Nelson, recently promoted to independent command, was appointed to lead a British squadron back into the Mediterranean, from which it had been absent since late 1796, and to mount watch outside Toulon, the principal enemy naval base in the south of France. It was known that General Napoleon Bonaparte was in command of an army assembling there, that transports were gathering also, under the protection of a French battle fleet, and that an amphibious expedition was planned, directed against British interests. The question was which and where? Britain itself? Ireland? Southern Italy? Malta? Turkey? Egypt? All lay within Napoleon's operational reach and some, Malta in particular, were stepping-stones to others. Beyond Egypt lay India, where Britain was rebuilding a substitute for the overseas empire lost in North America in 1783. If Napoleon could put to sea undetected, the Mediterranean would swallow his tracks and Nelson would discover where he had gone only when he had done his worst. The menace was guaranteed to perturb a watcher day and night. Nelson was perturbed. Before the French left port, he was anticipating their departure for "Sicily, Malta and Sardinia" and "to finish the King of Naples at a blow" but also perhaps for "Malaga and [a] march through Spain" to invade Portugal, Britain's longest-standing ally. After they left, in late May, he was in hot pursuit, sometimes on the right track, sometimes the wrong, sometimes behind, sometimes ahead, sometimes on the wrong continent altogether. In the end he ran his quarry to earth. The scent had died in his nostrils several times, however, and his own false calculations had led him astray. Not until one o'clock in the afternoon of 1 August, when the lookout on HMS *Zealous* reported masts in Aboukir Bay, east of the Nile delta, had Nelson reassurance that the chase begun seventy-three days earlier had been brought to conclusion. How it had makes one of the most arresting operational intelligence stories of history.

THE STRATEGIC SITUATION IN 1798

Napoleon did not yet dominate the European world as he would as Emperor of the French after May 1804. He was not yet even First Consul, which he would be appointed in December 1799. He already promised to become, however, the leading political figure of the French Republic and was unquestionably its outstanding general, at a moment when French armies dominated Europe. The First Coalition of enemies of the French Revolution, formed in 1792 by Austria and Prussia, later joined by the north Italian kingdom of Sardinia, and enlarged by the French declaration of war on Spain, the Dutch Netherlands and Britain, had progressively fallen to pieces during the 1790s. The Netherlands had been occupied in early 1795 and reorganised as the Batavian Republic, under French control. Prussia and Spain had made peace later that year; in August 1796, under French pressure, Spain actually declared war on Britain, closing its ports in the Atlantic and the Mediterranean to the Royal Navy and adding the strength of its large fleet to that of the French.

In 1796 the young General Bonaparte confirmed his growing reputation with a series of spectacular victories in northern Italy. After defeat at the Battle of Lodi in May, the King of Sardinia made peace and ceded the port of Nice and the province of Savoy to France. During the rest of the year, Bonaparte harried the armies of the Austrian empire out of its north Italian possessions by inflicting defeats at Castiglione and Arcola. Finally, after weeks of manoeuvring around the fortress of Mantua, Bonaparte won a crushing victory at Rivoli on 14 January 1797 and drove the defeated Austrians back into southern Austria. The Austrian emperor sued for peace, concluded at Campo Formio in October. The terms included the creation of a puppet French Cisalpine Republic in northern Italy and the cession of the Austrian Netherlands (Belgium) to France. In February 1798 the French occupied Rome and made the Pope prisoner; in April, France occupied Switzerland.

The outcome of this succession of conquests was to leave Britain, which refused to make peace, without any ally except tiny Portugal, and without bases, except for the Portuguese Atlantic ports, anywhere in mainland Europe. Russia, the only powerful continental state still resistant to French influence, was keeping its counsel. Turkey, ruler of the Balkans, Greece and the Greek islands and Syria, and nominal overlord of Egypt and the pirate principalities of Tunis and Algiers, remained a French ally. The old maritime Republic of Venice had been given to Austria at the Treaty of Campo Formio but would soon pass to France. The foreign policy of the Scandinavian kingdoms, Denmark and Sweden, was subservient to that of France. As a result, not a mile of northern European or southern Mediterranean coastline—except for that of the weak kingdom of Naples, Portugal and the island of Malta—lay outside French control. The Baltic was effectively closed to the British, so were the Channel and Atlantic ports, so were the Mediterranean harbours. All Britain's traditional overseas bases, except for Gibraltar, had been lost. In October 1796 the British government felt compelled to withdraw its fleet from the Mediterranean, where it had maintained an almost continuous presence since the middle of the seventeenth century, and to concentrate the navy in home waters. England was actually threatened by invasion; had it not been for Admiral Jervis' defeat of the Spanish in the Atlantic off Cape St. Vincent in February 1797 and Admiral Duncan's destruction of the Dutch fleet at the Battle of Camperdown in October, her enemies might have achieved a sufficient combination of force to fulfil the necessary conditions for a Channel crossing.

Despite the reduction in the enemy's naval strength the two victories brought about, the Royal Navy could not rest confident of its ability to contain the threat; in October 1798 Bonaparte was appointed to command an "Army of England," organised to sustain the pressure. Moreover, Britain correctly sought to pursue an offensive strategy, directed at checking invasion by forcing France to look to the protection of its own interests, rather than waiting passively to respond to French attacks. That required the maintenance of several separate concentrations of strength, a Channel fleet to defend the short sea crossing, an Atlantic fleet to blockade the great French bases at Brest and Rochefort and to keep an eye on the remains of the Spanish navy in Cadiz, detached squadrons to protect the British possessions in the West Indies, at the Cape of Good Hope and in India, and a host of smaller line-of-battleships and frigates to protect the convoys of merchantmen and Indiamen on which British trade, the lever of the war against France, depended.

Britain had the superiority. In 1797 it had 161 line-of-battle ships and 209 fourth-rates and frigates, against only 30 French line-of-battle ships and 50 Spanish.[3] The French and Spanish did not, however, have to keep the seas, but could remain comfortably in port, awaiting their chance to sally forth at an unguarded moment, while the British ships were constantly on blockade, wearing out their masts and timbers in a battle with the elements, or else in dockyard, repairing the damage. Only two-thirds of the Royal Navy was on station at any one time, while the demands of blockade and convoy even further reduced the numbers available to form a fighting fleet for a particular mission: Duncan had only 16 ships to the Dutch 15 at Camperdown, Jervis was actually heavily outnumbered, 15 to 27, at Cape St. Vincent. Moreover, both the French and Spanish navies built new ships at a prodigious rate and found less difficulty in manning than the British did. With larger resources of manpower, they conscripted soldiers and landsmen to fill the naval ranks, and while the recruits included fewer experienced seamen than the British collared by the press, they were not necessarily more unwilling. The inequity of the press, the paucity of naval pay, the harshness of life aboard, caused large-scale naval strikes in the Royal Navy in the spring of 1797— the "mutinies" at Spithead and the Nore—which, for once, frightened the admirals out of thinking flogging the cure for all indiscipline. The prospect of joining action against the revolutionary French with an untrustworthy lower deck prompted immediate improvements to the lot of the common sailor.

Just in time. By the spring of 1798, a new naval threat had arisen. Unknown to the British government, the French leadership—the Directory—had relinquished for the time being the project of an invasion of England and decided to create an alternative threat to its island enemy's strategic interests. The initiative had come from General Bonaparte. On 23 February he wrote, "To perform a descent on England without being master of the seas is a very daring operation and very difficult to put into effect . . . For such an operation we would need the long nights of winter. After the month of April, it would be increasingly impossible."[4] As an alternative, he proposed an attack on King George III's personal homeland, the Electorate of Hanover. Its occupation would not,

however, damage the commercial power of Britain. He saw another possibility: "We could well make an expedition to the Levant which would
menace the commerce of the Indies." The Levant, the region of the sun-
rise, lay across the shores of the eastern Mediterranean, in southern
Turkey, Syria but also Egypt. Egypt was not only a fabled land but also
the point at which the Mediterranean most nearly approached the Red
Sea, the European means of access to the Indian Ocean and the Moghul
dominions in India proper. France had not abandoned hopes of sup-
planting Britain as the dominant external influence in the Moghuls'
affairs, set back though its interests had been by British victories in the
sub-continent in the last thirty years. A French descent on India, principal source of British overseas wealth since its loss of the American
colonies, might deal a disabling blow to the Revolution's chief enemy.

Bonaparte, moreover, had chosen the moment shrewdly. The Mediterranean was temporarily a French lake. Even given the diminished
strength of its navy, enough warships could be found to escort a troop
convoy from France's southern ports to the Nile in safety, while its
mercantile fleet, together with those of Spain and northern Italy, would
provide transports aplenty. The withdrawal of the necessary force would
not significantly deplete that required to sustain dominance over
the defeated Austrians or to deter Russia from intervention in Western
Europe. Moreover, an expedition would not be effectively opposed.
Though Egypt was legally part of the Ottoman empire, under a Turkish
governor, there was no proper Turkish garrison in the country. Power
rested, as it had done since the thirteenth century, with the Mamelukes,
a corporation of nominal slaves, purchased on the borders of Central
Asia and trained as cavalrymen, who had usurped authority and used it
to perpetuate their privileges. Though fiercely brave, they numbered
only 10,000, and their ritualised horsemanship was tactically anomalous
on a gunpowder battlefield. The local infantry they commanded was a
half-hearted force.

Bonaparte found little difficulty, therefore, in persuading the French
Foreign Minister, Charles de Talleyrand, that an Egyptian expedition
was the next military step the Republic should take. Talleyrand enumer-
ated the advantages, which included, surprisingly in view of the long-
established Franco-Turkish entente, "just reprisal for the wrongs done
us" by the Sultan's government but also, more practically, "that it will be
easy," that it would be cheap and that "it presents innumerable advan-

tages."[5] The five Directors argued against, more or less forcefully, but were worn down one by one. On 5 March 1798, they gave their formal assent to the operation.

Preparations then proceeded apace. Toulon was nominated the port of concentration; it was base to the thirteen warships—nine of 74 guns, three of 80, one (*L'Orient*) of 120—which would form the escort and battle fleet. An order stopping the movement of merchant shipping out of Toulon and neighbouring ports quickly permitted the requisitioning of enough transports, half French, the rest Spanish and Italian, to embark the army. Fewer would have been preferable, for such a large number made a conspicuous presence, but contemporary Mediterranean merchantmen were too small to carry more than 200 men each. Some were also needed to carry horses, guns and stores. As a convoy keeping strict station a cable's length (200 yards) apart, transports would occupy a square mile of sea. In practice, the varied quality of the vessels and their masters' seamanship guaranteed straggling over a much wider area.

The Army of the Orient eventually numbered 31,000 men: 25,000 infantry, 3,200 gunners and engineers, 2,800 cavalry. Only 1,230 horses were embarked, however, Bonaparte believing that he could commandeer sufficient extra mounts in Egypt to supply the deficiency in charger and draught teams. It was a prudent decision. Horses were difficult to load, difficult to stable aboard and, however carefully tended, all too ready to die at sea. Their fodder also occupied a disproportionate amount of the cargo space, in which room had to be found for two months' food for the troops. Correctly, Bonaparte, or more probably Berthier, the future marshal who was already his trusted chief of staff, doubted the ready availability of rations in Egypt. The army was organised into five divisions, among whose officers was another future Marshal of the Empire, General Lannes. The officers of the fleet, commanded by Admiral Baraguey d'Hilliers, included Admiral Ganteaume, who would lead Nelson a dance in the months before Trafalgar, and Admiral Villeneuve, his tragic opponent in that battle. *L'Orient*, the flagship, was commanded by Captain Casabianca, father of the boy who would stand on the burning deck at the coming encounter of the Nile.

. . .

NELSON LOSES THE FRENCH FLEET

Brueys, the French admiral in Lommand, sailed from Toulon on 19 May, his 22 warships protecting a convoy of 130 merchantmen, filled with soldiers, horses, guns, stores and heavy equipment. Proceeding eastward at 37 miles a day, they headed first towards the northern point of the island of Corsica, steering to make a junction with a separate convoy of 72 ships from Genoa, which they did on 21 May. On 28 May they were joined by another convoy of 22 merchantmen from Ajaccio in Corsica, and on 30 May by the final complement of 56 ships, which had left Civita Vecchia, on the Italian mainland, on the 26th. The combined fleet, now numbering 280 transports, besides its escorting warships, set a course down the eastern side of the island of Sardinia, heading towards Sicily. It cleared the southernmost point of Sardinia on 5 June.

Nelson easily should have been up with it. He was not. The sea had sprung a surprise upon him. His flagship had been dismasted, his scouting frigates scattered, he and his crew had barely escaped from disaster. His ploy of interception had been scuppered, and he could not hope to begin reasserting control of the operational seaspace until he had completed essential repairs and found his consorts.

Nelson had left Gibraltar on 8 May, with his flag in *Vanguard,* 74 guns, commanded by Captain Edward Berry, and in company with *Orion,* 74 (Captain James Saumarez) and *Alexander,* 74 (Captain Alexander Ball). Admiral Lord St. Vincent, commander of the fleet off Spain and his superior, had given him three frigates, *Emerald,* 36, *Terpsichore,* 32, and *Bonne Citoyenne,* a sloop rather than frigate, of 20 guns. He had also assigned him another ten 74s, a 50-gun ship, *Leander,* and the brig *Mutine,* which were to join later.

Nelson's departure did not go unnoticed, and *Alexander* was actually struck by a shot from a Spanish shore battery. He arrived nevertheless, apparently undetected, 70 miles south of Toulon on 20 May, "not discovered by the enemy, though close to their ports . . . and exactly in the position for intercepting the Enemy's ships,"[6] as Captain Berry wrote to his father. Moreover, *Terpsichore* had captured a prize, from which it was learnt that Bonaparte had arrived at Toulon and that fifteen warships were ready for sea, and, though it was not yet known when or whither

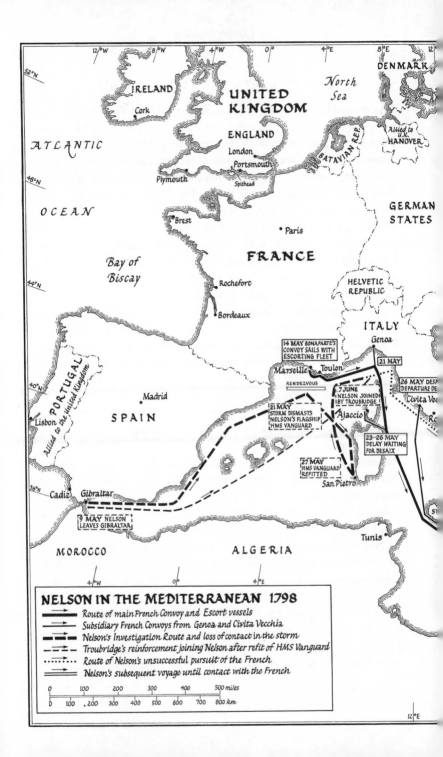

NELSON IN THE MEDITERRANEAN 1798

→ Route of main French Convoy and Escort vessels
→ Subsidiary French Convoys from Genoa and Civita Vecchia
━▶ Nelson's Investigation Route and loss of contact in the storm
–▶ Troubridge's reinforcement joining Nelson after refit of HMS Vanguard
····▶ Route of Nelson's unsuccessful pursuit of the French
⟹ Nelson's subsequent voyage until contact with the French

0   100   200   300   400   500 miles
0 100 200 300 400 500 600 700 800 km

14 MAY BONAPARTE'S CONVOY SAILS WITH ESCORTING FLEET

21 MAY

26 MAY DESP DEPARTURE DE

7 JUNE NELSON JOINED BY TROUBRIDGE

21 MAY STORM DISMASTS NELSON'S FLAGSHIP HMS VANGUARD

23–26 MAY DELAY WAITING FOR DESAIX

27 MAY HMS VANGUARD REFITTED

9 MAY NELSON LEAVES GIBRALTAR

RENDEZVOUS

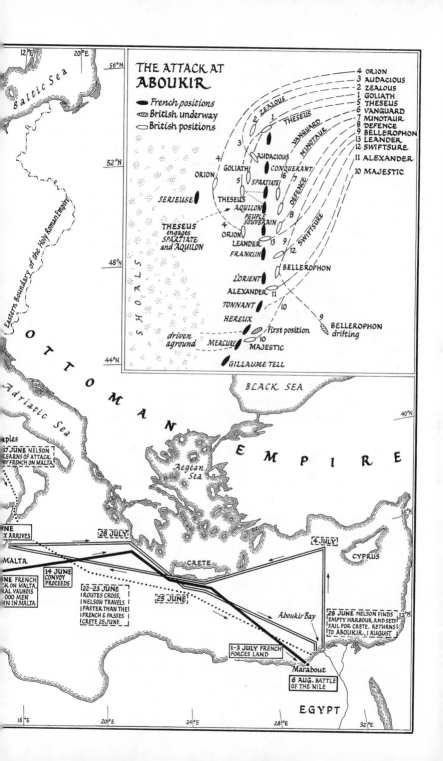

THE ATTACK AT ABOUKIR

● French positions
⟿ British underway
○ British positions

4 ORION
3 AUDACIOUS
2 ZEALOUS
1 GOLIATH
5 THESEUS
6 VANGUARD
7 MINOTAUR
8 DEFENCE
9 BELLEROPHON
13 LEANDER
12 SWIFTSURE
11 ALEXANDER
10 MAJESTIC

2 ZEALOUS
1 THESEUS
VANGUARD
MINOTAUR
ORION
GOLIATH
AUDACIOUS
CONQUERANT
SPARTIATE
SERIEUSE
THESEUS
AQUILON
DEFENCE
THESEUS engages SPARTIATE and AQUILON
PEUPLE SOUVERAIN
ORION
LEANDER
FRANKLIN
SWIFTSURE
BELLEROPHON
L'ORIENT
ALEXANDER
TONNANT
HEREUX
first position
driven aground
MERCURE
MAJESTIC
GILLAUME TELL
BELLEROPHON drifting

S H O A L S

Eastern Boundary of the Holy Roman Empire

Baltic Sea

O T T O M A N   E M P I R E

BLACK SEA

Adriatic Sea

Aegean Sea

ples

17 JUNE NELSON LEARNS OF ATTACK BY FRENCH ON MALTA

CRETE

CYPRUS

4 JULY

UNE X ARRIVES

28 JULY

MALTA

14 JUNE CONVOY PROCEEDS

NE FRENCH K ON MALTA. AL VAUBOIS 000 MEN N IN MALTA

22-23 JUNE ROUTES CROSS, NELSON TRAVELS FASTER THAN THE FRENCH & PASSES CRETE 25 JUNE

25 JUNE

Aboukir Bay

28 JUNE NELSON FINDS EMPTY HARBOUR, AND SETS SAIL FOR CRETE. RETURNS TO ABOUKIR, 1 AUGUST

1-3 JULY FRENCH FORCES LAND

Marabout

6 AUG. BATTLE OF THE NILE

EGYPT

they would sail, the intelligence gave Nelson and his captains assurance that they were in the right spot, ahead of time.

Then the wind began to freshen. *Vanguard* had sent up its topgallant masts, usually sent down when bad weather threatened. In the early morning of 21 May, *Vanguard*, still under topgallant masts, lost its main topmast, and with it two men, one swept overboard, one killed by falling to the deck. By daybreak, the mizzen topmast had gone as well, the foremast altogether, and the bowsprit was sprung in three places. The ship was almost unmanageable, could be sailed only on a broad reach—at right angles to the wind, which was approaching Force 12 on the Beaufort scale—and was driving towards the rocky west coast of Corsica, on which, unless brought about by some means, she would shortly dash to pieces.

The situation demanded any remedy, however unpromising. The rigging of a spritsail under the creaking bowsprit, an antique device not in naval use for many decades, succeeded in bringing up her head. Very slowly, she was worn round with the wind until she was pointing away from Corsica and so, during the course of the morning, as spars and standing rigging were hacked into the sea, clawed off the lee shore. On 22 May, as the hurricane abated, *Alexander* was able to pass a tow and began to drag *Vanguard* southward towards the west coast of Sardinia. By late afternoon, with the wind moderating, a safe haven between Sardinia and the island of San Piétro was in sight; but the danger of driving ashore still threatened. Nelson signalled an order to *Alexander* to cast off the tow. It was refused and, very gradually, *Vanguard* was brought to anchor on the morning of 23 May. The captain of *Alexander*, Alexander Ball, of whom Nelson had hitherto had a very guarded opinion, became henceforth one of his most valued advisers.

Vanguard at once undertook repairs, using some of its own spare spars and others sent from *Alexander* and *Orion* to replace its lost lower, top and topgallant masts. After four days it was ready to sail. Next day, 24 May, a Marseilles vessel was encountered. It told that Napoleon's fleet—which had been outside the track of the storm—had left Toulon on 19 May but gave no indication as to its destination.

Nelson therefore decided to retrace his course rather than press on into the uncertainties of the wider Mediterranean. He had lost touch with his three accompanying frigates during the great gale. He had not yet made contact with the squadron St. Vincent had allotted him. His

judgement was that prudence demanded a return to his
where he could concentrate his forces, gather in his fr
fresh intelligence of the enemy's movements. By 3 June
Toulon, where on 5 June the brig *Mutine* appeared, bearing
Troubridge's squadron of ten men-of-war would soon join. The *Mutine*
was commanded by Thomas Hardy, of "Kiss me, Hardy" at Trafalgar,
already a favourite of Nelson. His information brought reassurance. On
7 June, Troubridge appeared. Nelson's command now numbered thirteen 74s and a 50, quite enough to defeat the French if they could be
found. To find the French, however, Nelson needed frigates. Where had
the frigates gone?

Terpsichore, Emerald and *Bonne Citoyenne* had been scattered by the
storm that dismasted *Vanguard. Bonne Citoyenne* had sent down her topgallant masts and ridden out the storm; she was a weatherly little ship,
much admired for her sailing qualities. *Terpsichore* had also struck her
topgallants and eventually her topmasts also, after three of her foremast
shrouds had broken. She was alone for two days, 20–21 May, during the
height of the storm but found *Bonne Citoyenne* again in the afternoon of
the 22nd. Both were then well south of Toulon. *Emerald* had been driven
even farther south, but also east, so far away from her two sister frigates
that early in the morning of 21 May she caught a glimpse of *Vanguard* off
Corsica in her dismasted state. She was not in a position to render assistance, and the two ships lost each other in the tumult.

Emerald's captain, Thomas Waller, then decided, as the weather
abated, to head towards the coast of Spain, in the hope of picking up
prizes, desirable in themselves, but also to gather information from
them. Without luck; although he intercepted two merchantmen, he got
no news of either Nelson's or Bonaparte's whereabouts. On 31 May, however, he fell in with another British frigate, *Alcmene,* commanded by Captain George Hope, which St. Vincent had sent after Nelson on 12 May. It
was in company with *Terpsichore* and *Bonne Citoyenne,* which it had met
two days earlier. They had told Captain Hope of the great storm but
had, of course, no news of Nelson. Captain Waller went aboard *Alcmene,*
told Hope of his sighting of the dismasted *Vanguard* and thus set in train
a sequence of events which was to deprive Nelson of his scouting group
for the next two and a half months.

Nelson had left instructions for his frigates to obey in the event of
their separation from the flagship. That was a common and sensible

ghteenth-century precaution designed, in the absence of anything but spoken or visual communication, to allow contact to be re-established by designating a rendezvous. His instructions laid down that, if lost, they were to cruise on a line west to east and back again, due south of Toulon to within 60 to 90 miles of Cap St. Sebastian near Barcelona. When, "not having heard from me for ten days, to return to Gibraltar." The scheme should have worked. Hope in *Alcmene* began to work the patrol line on 23 May, sailing back and forth on latitude 42 degrees 20 minutes north as instructed. He continued to do so after *Terpsichore* and *Bonne Citoyenne* joined. Had he kept on until 3 June, only one day more than the stipulated span, he would have been found by Nelson, who himself arrived on station that day.

On 31 May, however, Hope had detached *Terpsichore* and *Bonne Citoyenne* to search for Nelson between Sardinia and North Africa. On 2 June he met *Mutine* and was told by Hardy that Troubridge, with ten men-of-war, was close behind him, also looking for Nelson. There were now four separate British forces in the western Mediterranean, all looking for Bonaparte but also for each other: Nelson approaching his designated patrol line, *Alcmene* and *Mutine* on it, *Terpsichore* and *Bonne Citoyenne* heading for Sardinia, Troubridge south of all of them but heading north and anxious to make touch. If Hope had kept *Alcmene* and *Bonne Citoyenne* in company and stayed on station with *Mutine,* he would inevitably have met Nelson, and Troubridge later, thus forming a junction of heavy ships and scouts which, with the merest addition of luck, would have intercepted the slow-sailing French in the central Mediterranean within the month at most. The destruction of the French fleet, and with it a major portion of the best of the French army, would have followed, Bonaparte would have been a beaten man and none of his most famous victories, Marengo and Austerlitz foremost, would have been won. The First Coalition might have been revived, the Revolution contained, the French Empire never founded, the future of Europe changed altogether.

As it was, Hope decided on another course. *Emerald's* report of the extent of damage suffered by *Vanguard* was decisive in forming his mind. He concluded that its severity would require the flagship to enter dockyard for repairs. The only available were at Naples and Gibraltar. To look for Nelson at Gibraltar required a retrogression, which would add in both space and time to Bonaparte's head start; in any case, Hope had been told by Hardy when he had left in the *Mutine* that Nelson had not returned to Gibraltar. He also decided against seeking out Troubridge, a bad mistake, since Troubridge shortly found Nelson himself, and had he been able to bring Hope's frigates with him, would thereby have added enormously to the fleet's powers of reconnaissance. Hope instead made the calamitous decision to mount a search for the French by himself. Having already detached *Bonne Citoyenne* and *Emerald* to Sardinia, he sent *Terpsichore* to search the north Italian ports while sailing *Alcmene* round Majorca and Minorca, then to Sardinia and eventually towards Naples, picking up his detached consorts on the way. The pattern of search would have been justifiable had either Nelson or the French armada been standing still. Nelson, however, was cruising on the patrol line while the French were heading steadily east and south, opening up irrecoverable searoom with every day that passed. Had Nelson known of Hope's movements and orders, his anguish at "want of frigates" would have been even more acute than it was.

Nelson, back on his rendezvous line off Toulon, now at least had the consolation of picking up the ships that were to constitute his fighting force, first *Mutine*, then Troubridge's ten 74s, on the afternoon of 7 June. Then the weather again intervened. A calm fell, so that it was not until 10 June that *Orion* and *Alexander*, of his original three, which had been detached to chase merchantmen in hope of news, rejoined and the fleet was fully assembled. Nelson, with thirteen 74s, the 50-gun *Leander* and the nimble *Mutine* could now turn in pursuit of the enemy. Where to head?

Troubridge had brought orders from St. Vincent which recapitulated the strategic situation. Nelson was requested and required to proceed "in quest of the Armament preparing by the enemy at Toulon and Genoa, the object whereof appears to be, either an attack on Naples and Sicily, the conveyance of an army to some part of the coast of Spain, for the purpose of marching towards Portugal or to pass through the Straits [of Gibraltar] with the view to proceeding to Ireland." However, in additional instructions, he was also authorised to pursue the French fleet "to any part of the Mediterranean, Adriatic, Morea [southern Greece], Archipelago [Greek islands] or even into the Black Sea, should its destination be to any of those parts." He was to supply himself from the ports of "the Grand Duke of Tuscany, the King of the Two Sicilies [Naples], the Ottoman Territory, Malta and ci-devant [former] Venetian Dominions now belonging to the Emperor of Germany [Austria]." He might also expect help from the Bey of Tunis, the Bashaw of Tripoli (modern Libya) and the Dey of Algiers, three nominal but effectively independent possessions of the Ottoman empire.[7]

Portugal? Ireland? Naples? Sicily? No mention of Egypt. The only inference Nelson could make, as he assembled his fleet, was that Bonaparte must be assembling his also, which meant bringing together the Toulon and the Genoa elements. He concluded that Toulon would go to Genoa, rather than vice versa, and decided accordingly to search the north Italian coast; implicitly, he thereby discarded the notion of Portugal and Ireland as destinations and thought more of Naples and Sicily. He began, having cleared the northern point of Corsica, by looking in Telamon Bay (Golfo di Talamone), south of Elba, thought by him a suitable mustering place for the Toulon and Genoa convoys. *Mutine*, having explored the bay and run between the offshore islands of Montecristo and Giglio, reported no sight of the enemy; at this stage Nelson still believed that "not all the French troops had left Genoa on the 6th."[8] Next day, 13 June, he went to look for himself, sailing the whole fleet between Elba and the islands of Pianosa and Montecristo, a laborious detour. Had his frigates been with him one could have been sent to do the work, while Nelson pressed forward. *Mutine* was not fast enough to perform detached duty and keep up with the fleet. He might have used one of the 74s as a scout, but that would have diminished his fighting power; he had told St. Vincent, before leaving Gibraltar, that he intended to keep "the large Ships complete, to fight, I hope, larger ones."

On 14 June the clouds lifted a little. Near Civitavecchia he spoke a Tunisian warship, which told him it had spoken a Greek on 10 June which had "on the 4th, passed through the French Fleet, of about 200 Sail, as he thought, off the N.W. end of Sicily, steering to the Eastward."[9] It is not clear if that meant it was moving along the north coast of Sicily or had passed Trapani and was off the south coast. If the former, it was just possible it might be making for Naples; if the latter, it had some other objective; but in either case it might land troops on Sicily, eminently worth occupying in itself. In any case, Bonaparte's Armament had been nearly three hundred miles ahead of him ten days earlier and, even allowing for its sluggardly rate of advance, might have made another three hundred miles since. The cloud of unknowing, even if it had lifted a little, still concealed most of the future.

In the circumstances, Nelson decided to go to Naples. There were good reasons for doing so. The long-serving British ambassador, Sir William Hamilton, in post for thirty-four years, had important sources of information, drawn from diplomatic, political and commercial contacts all over the central Mediterranean. The kingdom of Naples, or the Two Sicilies, as it was known, was well disposed towards Britain, and in fear of France, whose armies were just over the border in the Papal States. It might lend Nelson's fleet supplies and assistance. Its Prime Minister, General Sir John Acton, a cosmopolitan man of affairs, held the title of a British baronet and had some loyalty to his ancestors' country. Nelson hoped for both intelligence and material support.

Arrived at the Ponza islands off Naples on 15 June Nelson sent Troubridge ashore in the *Mutine*. He landed on the morning of the 17th. Thomas Troubridge was a trusted subordinate, a colleague of twenty-five years and a no-nonsense fighting-ship captain. St. Vincent thought him "the greatest man in that walk the English Navy has ever produced." A veteran of the Glorious First of June and the Cape St. Vincent battles, his attitude to command was straightforward. "Whenever I see a fellow look as if he was thinking," he gave as his opinion after the widespread outbreaks of indiscipline in 1797, "I say that's mutiny." Taken to see Hamilton and Acton, he came straight to the point. Hamilton recorded, "We did more business in half an hour than should have been done in a week in the official way here . . . Now being informed of the position and strength of the enemy" and having extracted an order from Acton authorising the governors of every Neapolitan port to supply "the King's ships with all sorts of provisions," Troubridge "brightened up and seemed perfectly happy." Putting Acton's order in his pocket, he departed for the fleet offshore, which he reached on 18 June.

Fighting is one thing. Intelligence is another. Each requires different qualities, not often found in the same person. The Royal Navy was to rediscover that, on a similar occasion, on 31 May 1916, when a seaman officer asked the wrong question of the intelligence staff on the morning of the Battle of Jutland. The fault then was superciliousness; he disdained to explain why he asked the question he did, not deigning to take the cryptologists into his confidence. Troubridge was not supercilious. He, Hamilton and Acton seem to have got on like a house on fire. His fault was bluntness. He wanted supplies for the ships, almost a naval officer's first thought. He wanted the freshest news available of the enemy's whereabouts. Acton's order ensured the first. Hamilton's hard information—that the French were going to Malta—supplied the second. No wonder Troubridge departed wreathed in smiles.

What he should have extracted from Hamilton, and might have done had he not stuck so directly to the point as he saw it to be, was softer news. It might have emerged in speculative or even general conversation, clearly not Troubridge's strong point. The news was the indication that the French Armament was bound farther afield than Sicily or Malta. On 28 May Acton, whose first language was French (he had been born at Besançon), had told Hamilton that the French ambassador at Naples had told him "that the grand expedition from Toulon . . . was really destined for Egypt." Hamilton appears to have suspected that he might be dealing with disinformation. As a result, although he minuted Acton's report to the Foreign Office in London, he did not pass its content on to Troubridge nor put it in writing to Nelson.[10]

WHAT LONDON KNEW

London may indeed have been better informed than Nelson was. The Foreign Office, the Admiralty and the War Office all collected intelligence, from professional agents, consular officials, well-disposed or garrulous travellers and foreign newspapers, among other sources. As early as 24 April Lord Spencer, the Foreign Secretary, had noted the destination of "the Toulon ships" as "Portugal—Naples—Egypt." Two days later "61'78'71" (the designation of an agent) "believes," he wrote, "the object to be Egypt incredible as it seems." Henry Dundas, Secretary of War and a member of the board of the East India Company, was meanwhile telling the Admiralty of news passed by an American recently in France of French plans to invade the Channel Islands, to send an expeditionary force to Ireland (which came about in August), to raise revolution in Naples and Poland (both blows against Austria), but also of a "strange scheme respecting Egypt," by which 400 French officers were to be sent overland through that country to assist Tipu Sultan against the British in India.

The Admiralty had its own man in the Toulon Armament's operational zone, Lieutenant William Day, sent to Genoa to sell three Navy Board transports marooned there since the withdrawal from the Mediterranean in 1796. Day's reports, sent overland via the normal route through Germany to Hamburg and then by sea to London, the transmission time being anything from three to five weeks, first suggested that Spain was the destination. By 1 May, however, when he himself arrived in London, he brought news that indicated the eastern Mediterranean as a possibility. It was that the Armament was embarking 4,000 ten-hooped barrels, without bungholes, the purpose of which was judged to be to buoy warships over shallows. The First Lord deduced that they were needed for the passage through the Dardanelles to the Black Sea. It is indicative of how defective was the Admiralty's contemporary information that it believed a waterway navigable by modern container ships was not by men-of-war a fraction of their draught.

Other information available in London was, however, better. French newspapers, often acquired within a week or less of publication, were remarkably indiscreet. During late March, April and early May, *L'Echo, Le Surveillant* and *Le Moniteur* all printed material which amplified the picture the government was forming of the Toulon Armament's strength, provisioning and even destination. *Le Moniteur,* under government control, tried to muddy the water by printing deliberate misinformation but the trend of the news remained unmistakable: a big fleet was preparing for a long-range military operation. Gossip helped to refine the picture. Some of the academics who were to accompany the expedition began to boast, a notorious failing of clever men leading unimportant lives. De Dolemieu, a mineralogist, wrote to de Luc, Professor of Natural History at Göttingen, that books about Egypt, Persia, India and the Black and Caspian seas were being shipped and that rumour had it the objective was Egypt and the purpose to intercept Britain's commerce with India. De Luc, unfortunately, was both a member of the household of Queen Charlotte, George III's wife, and a Foreign Office agent. He passed the word on 7 May.[11]

The best intelligence received in London came, however, through official channels. It had been assembled by what would become classic spy-novel methods. The consul in Leghorn (Livorno) in northern Italy, Udney, had a well-informed contact in a local British merchant, Jones, who maintained commercial correspondence with other trading houses throughout the Mediterranean. His sources led him to overestimate somewhat the size of the Toulon Armament, but he got its departure date roughly right and its destination and purpose uncannily so. Its intermediate stop was to be Malta, which would be surrendered, and then Alexandria (though perhaps alternatively the Black Sea), with the object of landing troops to march overland to the Persian Gulf or sail down the Red Sea to attack the British East India Company's possessions in India. Udney's report, dated 16 April, was passed by the Foreign Office to the Admiralty on 24 May.

For a while London chose to discount the information. There were other dangers nearer home that a great French amphibious expedition threatened, a descent on Portugal, in concert with the Spanish, an offensive against Britain itself, perhaps via Ireland, where rebellion broke out in May. What may have been deliberate French disinformation suggested that the rumours about Egypt were a cover story to conceal the real strategic purpose of the Toulon Armament. On 1 June, the Foreign Secretary wrote to Lord Mornington, Governor General in India, that "Bonaparte has at last embarked at Toulon with the project of attacking Ireland . . . taking or not taking Portugal in his way."

New information soon dispelled these misapprehensions. Some came from the French press, more, and more compelling, from the gossipy academic world. A French scholar, Faujas de St. Fond, was reported from Frankfurt, in the occupied German territories, as affirming that the Armament was sailing for Egypt; had Bonaparte known of their stream of leaks he certainly must have regretted the decision to encumber the expedition with so many professional talkers. St. Fond's indiscretion was received in London by 13 June. On the 11th a despatch from the diplomatic mission in Florence had brought an even more credible report: the French general Carvoni had revealed that the expedition, which he was to accompany, was going to Egypt and then India. Two days later the Foreign Secretary wrote to his brother, "It really looks as if Bonaparte was after all in sober truth going to Egypt; and Dundas [Secretary of War] seems to think the scheme of attacking India from thence not so impractical as it may appear. I am still incredulous as to the latter point, though as to the former I am shaken. But as Bonaparte on the 23rd was still off Toulon [wrong] and as Lord St. Vincent must have detached [Troubridge's ships] on the 21st at latest, there is real reason to hope that Nelson may destroy all these visions."[12]

NELSON RECOVERS THE SCENT

That was certainly London's hope, but it was strictly circumscribed by its inability to communicate to the central Mediterranean either what it wanted or what it knew. On 13 June, when Lord Spencer wrote his intelligence summary to his brother, Nelson was still in the Tyrrhenian Sea, between Corsica, Sardinia and Sicily. Orders had been sent from London to India, and points in between, to sail ships towards Suez, in particular to Commodore John Blanchett, in the *Leopard, 50* guns, on his way to India, to organise a small squadron in the Red Sea. It was anyone's guess when word might reach him. It was equally difficult to estimate when either fresh orders or information might be got to Nelson. St. Vincent, off Cadiz, had instructions and good reason to stay there, blockading the Spanish and guarding the Straits of Gibraltar. He had sent all the fast sailers at his disposal to Nelson already and could spare no more. He could forward messages by neutral ships, but they were few, and his own rear link to London was tenuous and slow. He did not even know, from week to week, where Nelson was; after mid-June, when Nelson sent back the brig *Transfer* from Naples with despatches, he did not know at all.

Nelson, by contrast, may have known something of Udney's intelligence from Leghorn, since his papers contain a copy of an Udney letter which he may have picked up while on his way back to the Toulon rendezvous line after the dismasting; but it tells only of the Toulon Armament's strength, not its destination. Soon after he left Naples on 18 June, however, he got firm news that it was sailing for Malta. On 20 June, when he was in the Strait of Messina, between Sicily and the toe of Italy, the British Consul at Messina came aboard "to tell me that Malta had surrendered," but not before he had written to the Grand Master of the Knights of Malta, urging him to put the island into a state of defence, while he hurried to help.

His message left too late. Malta had already been surrendered, as Consul Udney had warned it would be on 26 April. The Knights had caved in. That should not have come as a surprise. The Sovereign Military and Hospitaller Order of St. John was no longer what it once had been. Founded originally to care for sick Christian pilgrims to the Holy Land, it had become during the Crusades an order of military monks, who built and defended castles all over the crusader states. Driven under Muslim pressure step by step from Jerusalem, Acre and the island of Rhodes, the Knights eventually ended up in Malta, where they found a new vigour. In 1565, under the leadership of Grand Master de la Valette, they defeated a major Turkish effort to capture the island and push into the western Mediterranean. For the next 200 years the Knights harried the Ottoman fleet, liberating Christian galley slaves and taking Turks to be their own. There was no nonsense about loving thine enemy in the Knights' version of the Christian creed. The catafalques of the Grand Masters, in their headquarters church in Valetta, are supported on the bronze shoulders of turbaned Turks, chained to and bowed under their burden.

Grand Master Hompesch lacked Valette's resolution. When Bonaparte's armada appeared on 9 June, he quickly came to terms—a pension for himself, resettlement for the remaining Knights. Such resistance as was shown came from the ordinary Maltese, though they had little love for the decayed Order. By 18 June, Bonaparte was off, having installed a French administration and garrison, proclaimed various civil and ecclesiastical reforms and thoroughly looted the churches of treasure. It was a characteristically Napoleonic irruption, not least by its alienation of the Maltese, one of the most Catholic people of Europe. Had the Knights only shown more backbone, and encouraged the islanders to prolong resistance, the outcome would have been very different. Nelson, only a hundred miles behind and pressing onward, would have caught the Armament at a total disadvantage, with its commander and amphibious force ashore and its warships dispersed about the island's periphery. Disaster would have been unavoidable.

Nelson, however, was misreading the signs. On Wednesday, 20 June, when he had written to Grand Master Hompesch from off Messina, he promised to be at Malta by the 22nd. So he was, or nearly. He was still convinced, however, that Sicily was the French objective and that Malta was to be used only as a base for its capture. His thoughts, therefore, misled him. He was shortly misled by objective misinformation.

Early in the morning of 22 June, when he had promised to be at Malta but was actually just south of Cape Passaro, the south-east point of Sicily nearest the island, he was brought fresh news of the French from two different sources in quick succession. The first came from Hardy, who came aboard *Vanguard* from *Mutine* at 6:25 a.m. to report stopping a brig from Ragusa (modern Dubrovnik, on the Adriatic coast), with the news that Malta had fallen. The second was a sighting report from *Leander* of four strange ships to the east-south-east.

Nelson now decided, uncharacteristically, to consult his captains as to what to do. His conferences before Trafalgar would launch the legend of the "Band of Brothers"; but then he was expressing what he intended to do, telling not asking. Yet by 1798 he had already acquired a reputation for decisiveness. It was odd that at this moment he felt the need for moral support. Still, it was a highly complex situation. The Ragusan brig had told that Malta had fallen the previous Friday, and the French fleet sailed the following day, 15 and 16 June respectively. It was now the 22nd. Nelson must have calculated that, if the French had gone to Sicily, they would have arrived, and news could not have failed to reach him of their arrival in the intermediate six days. As there was no news, they had gone somewhere else. Given the current direction of the wind, which was westerly, the Armament was most likely to be heading east, which might mean towards the Dardanelles and the Black Sea but almost certainly meant Egypt. It was a compelling conclusion; but he needed reassurance.

The four captains for whom he sent were senior and trusted—Saumarez of the *Orion*, Troubridge of the *Culloden*, Darby of the *Bellerophon* and Ball of the *Alexander*. In *Vanguard*'s cabin, he put to them the following assessment: "with this information [of the "strange ships" and from the Ragusan brig] what is your opinion? Do you believe under all circumstances which we know that Sicily is [Bonaparte's] destination? Do you think we had better stand for Malta, or steer for Sicily? Should the Armament be gone for Alexandria [Egypt] and got safe there our possessions in India are probably lost. Do you think that we had better push for that place?"

He got a variety of answers. Berry, of the *Vanguard* itself, was for going to Alexandria, Ball agreed that the French were heading for Alexandria, Darby thought that probable, Saumarez and Troubridge emphasised the importance of protecting Alexandria, without stating an opinion about the French destination. Still, collectively, they made Nelson's mind up, with regrettable consequences.

Resolved now to press on at best speed to Egypt, Nelson dealt peremptorily with the sighting reports of the "strange ships." His own sent to follow them kept up a stream of signals. At 5:30 a.m. *Culloden* reported that they were running, with the wind behind them. At 6:46 a.m. *Leander* signalled "strange ships are frigates," and *Orion* repeated it to the flagship so that there could be no mistake. Four frigates made a sizeable force, likely to be part of a larger one. It was not an unreasonable guess that they might belong to the Armament. Soon after 7 a.m., however, Nelson ordered the "chasing ships" to be called back. His thoughts, which he was outlining at the time to his five captains in

Vanguard's cabin, admitted only two lines of decision: to go back to Sicily or make for Malta; alternatively, to race to Egypt, with the favourable wind. He did not raise, perhaps even to himself, the option of disposing the fleet in scouting formation and running down the course taken by the "strange ships" to see if they were in company with others. Captain Thomas of the *Leander* clearly could not understand and scarcely bear his superior's refusal to follow up such an obvious pointer to the enemy's whereabouts. At 8:29 a.m. he signalled again, "ships seen are frigates." Nelson was unmoved. *Leander*, *Orion* and *Culloden* were obliged to rejoin the fleet which crowded on sail for Alexandria.

The episode brings to mind the exchanges between Admiral Nagumo and the aircrew of the cruiser *Tone*'s reconnaissance aircraft in the early morning of the Battle of Midway, 4 June 1942—with this difference. Then it was the admiral who was desperate to know what sort of ships the airmen had spotted, they who were slow to respond. Their first signal reported that they had sighted the enemy, their second that the enemy ships were cruisers and destroyers, no threat to Nagumo at all, only their third, sent nearly an hour after the first, that "enemy is accompanied by what appears to be a carrier," a very serious threat indeed. Despite the differences, there is this similarity: had the commander and his reconnaissance force in each case been in tune, the enemy would have been destroyed.

Nelson might nevertheless have heeded his scouting ships had he possessed one vital piece of information: the actual date of departure of Bonaparte from Malta. The "Ragusan brig" had said Saturday, 16 June. In fact he had not left until Tuesday the 19th, and on the 22nd, when the "strange ships" were sighted, had been at sea only three days. Nelson was harder on Bonaparte's heels than Nelson guessed; may, indeed have been only thirty miles or so behind him. That night, in the mist, the French heard bells striking and signal guns firing, which surely must have been aboard Nelson's ships. The French Armament, however, warned by the frigates seen earlier that day, was sailing in silence, closed up tight for mutual protection. By the time day broke, Nelson had passed ahead and was over the horizon. The chance of a decisive encounter had been lost.

. . .

BACK AND FORTH

The captain of the Ragusan brig may have been mistaken; he may equally have been misunderstood. We do not know what language he spoke, perhaps Italian, perhaps Serbo-Croat, perhaps another Mediterranean tongue. As Alfred Thayer Mahan suggests in his life of Nelson, had Nelson done the interrogation himself, he might have found out more, for he was a shrewd questioner, and his intellect was sharpened by anxiety, and by constant dwelling upon the elements of the intricate problem before him; but by the time Hardy came aboard *Vanguard,* it was two hours since he had stopped the Ragusan, which was then beyond reach. Nelson, in any case, was in a fever to get forward. The wind was in his favour and over the next six days he made exceptional progress, sometimes covering 150 miles in twenty-four hours. On 28 June he had Alexandria in sight and he spent the night taking soundings off shore; the Royal Navy had few charts of the eastern Mediterranean. It was disquieting, however, that there was no sign of the Armament, and when Hardy returned in *Mutine* next morning after a passage inshore, his fears were confirmed. Hardy had failed to find the British Consul, to whom Nelson had written, and could not have done, for he was absent on leave; but the Ottoman fortress commander, who eventually appeared, told him that the French had not arrived, that the Turks were not at war with France and that the British, though they might water and store their ships, according to custom, should go away. Nelson did not linger. On the morning of Saturday, 30 June, he set sail. He had decided he had made a mistake and that the Armament had gone elsewhere, perhaps to Turkey proper. Four days later, having left Cyprus to starboard, he was in the Gulf of Antalya.

Had Nelson only contained his impatience, the French would have sailed into his hands. Twenty-five hours after he departed Alexandria, the Armament anchored to the east of the city and began to send the army ashore. This was Nelson's second, perhaps third, even fourth near-miss. But for the gale, he might have caught Bonaparte coming out of Toulon. But for his anxiety to protect Naples, he might have devastated the Armament at Malta. But for his refusal to follow the "strange ships," he might have slaughtered the Armament at sea on 22 June. Had he but waited a day at Alexandria, he certainly would have destroyed it, or forced its surrender, in the delta of the Nile. As it was, he was now hastening away from his quarry, while Bonaparte and a clutch of his future battle-winning marshals—Berthier, Lannes, Murat, Davout, Marmont—were being rowed ashore to take possession of Egypt, more or less at their leisure.

Nelson, by contrast, was in a frenzy. "His anxious and active mind," wrote Captain Ball, "would not permit him to rest for a moment in the same place." Where to go? He decided first to "stretch over to the coast of Caramania" (southern Turkey), as he later wrote to Sir William Hamilton. His conclusion, made ten days earlier, that the French were going east, seems to have left him with the conviction that, if they were not in Egypt, then they must be somewhere else in the Turkish Sultan's dominions. He had noticed the preparations the military commander at Alexandria had been making—"the Line-of-Battle Ship . . . landing her guns," "the Turks preparing to resist," as he later wrote to St. Vincent and Sir William Hamilton respectively—but in the absence of the French, he must have interpreted those signs as elements of a general Ottoman alert. That, or else his premature decision to depart implies an uncharacteristic moment of mental confusion, poor analysis, general jumpiness, not traits which he normally displayed.

He arrived in the Gulf of Antalya on 4 July and, seeing nothing, turned west again, heading first to cross the track of the Armament if it were still on its way to Egypt, then steering south of Crete, briefly north towards mainland Greece, eventually direct once more for Sicily, which he reached on the 20th. Off Syracuse, where he proposed to water and take on stores, he wrote three letters on 20 July, to his wife, to Sir William Hamilton, to St. Vincent. His few short words to Lady Nelson were a *cri de coeur*: "I have not been able to find the French Fleet . . . however, no person will say that it has been for want of activity." To Hamilton he regretted again his "want of frigates," from which "all my misfortune has proceeded," and made arrangements for his letters to be forwarded to the Foreign Secretary and to St. Vincent. They, of course, had no more idea of his whereabouts than he of the French. To St. Vincent, supplementing a recapitulation and justification of his wandering since the *Vanguard*'s dismasting (written on 29 June, which Captain Ball had urged him not to send), he raised again the issue of lack of frigates, "to which must be attributed my ignorance of the movements of the Enemy," and then outlined his next plan: "to get into the mouth of the Archipelago [the Aegean], where, if the Enemy are gone to Constantinople, we shall hear of them directly; if I get no information there, to go to Cyprus, when, if they are in Syria or Egypt, I must hear of them."

He ended, however, by retailing "a report that on the 1st of July, the French were seen off Candia [Crete], but near what part of the Island I cannot learn." Leaving Syracuse on 24 July, his last word to Hamilton was "No Frigates!—to which has been, and may again, be attributed the loss of the French Fleet." Frigates or not, Nelson's luck was about to change. On 28 July, when south of the Greek mainland, he sent the *Culloden* into the Gulf of Coron (modern Messenia, the large western inlet into the Peloponnese), from which he was brought news that "the Enemy's Fleet had been seen steering to the S.E. from Candia about four weeks before." The news came from the Turkish governor, who had heard, from Constantinople, that the French were in Egypt. *Culloden* also brought in a French brig, which hailed from Limassol in Cyprus and endorsed the Turkish governor's report. It was further confirmed by the master of a merchantman stopped by the *Alexander*. Nelson's fleet had by now stopped 41 merchant vessels during its toing and froing and would have stopped more had not the French admiral captured any stray ship he found in the Armament's path, no doubt a fruitful counter-intelligence measure.

The visit to the Gulf of Coron effectively ended the intelligence famine. Nelson now had good reason for believing that Bonaparte was not at Corfu, the most likely destination had he headed for Greece, was not going to Constantinople and was not on the south coast of Turkey, nor in Cyprus. The Armament might possibly have landed in Syria, a term that contemporaneously embraced modern Israel and Lebanon as well, but if so, its ships would be within easy sailing distance of Alexandria and would certainly be heard of there. For Alexandria, on 29 July, he accordingly made all sail and during the next few days achieved very rapid passages; in the 24 hours of 31 July the fleet covered 161 miles, at an average speed of nearly eight knots, very fast going for line-of-battle ships.

Landfall on 1 August brought a brief repetition of the disappointment of 30 June. The harbour was empty. A short eastward cast along the coast set fears at rest. At 2:30 in the afternoon, *Goliath*'s signal midshipman, aloft in the foremast, spotted a crowd of masts in Aboukir Bay. Desperate to be first with the news, he slid to the deck to tell his captain but then broke a halyard as he made his flag hoist to *Vanguard*. So it was *Zealous* that got the signal first to Nelson: "Sixteen sail of the line at anchor bearing East by South."

The report was not quite accurate. Admiral Brueys commanded 13 line-of-battle ships, but also four frigates, two brigs, two bomb vessels and a collection of smaller gunboats. It was the thirteen heavy ships that mattered, the enormous 120-gun *L'Orient*, three 80s and nine 74s. They were variously armed, one with 18-pounders instead of 32-pounders, and some were old, as much as fifty years old, and less strongly built than the British. Still, *Victory*, which was to be Nelson's flagship at Trafalgar, was then forty years old. Neither age nor even weight of metal counted really among the decisive features. Seamanship, ship-handling and bloody-mindedness did. The British were masters of their craft, to a degree that the relatively inexperienced French, officers and men alike, were not; the code of revolutionary correctness had robbed the French navy of many good officers, conscription to the army of much of its manpower. The diet of victory on land in particular had sapped the French navy's will to win. Victory at sea was not essential to France. It was crucial to the British as a people and to the Royal Navy as a service.

Bonaparte, as Sir Arthur Bryant, the great popular historian of Britain's role in the wars of the French Revolution and Empire, was to remark, never saw and therefore could not imagine "the staggering destructive power of a British ship of the line in action." The Royal Navy had been a ferocious instrument of war ever since the seventeenth century. Its defeat in the American War of Independence, however, had infused it with a ruthless killer instinct. It had been outraged by the French and Spanish seizure in 1780–81 of command of the sea, its birthright, as it saw it, and had not relented since the resumption of hostilities in 1793 in the determination to humble its enemies. Bonaparte, the mastermind of the Egyptian expedition, was now far from the fleet, winning new victories over feeble enemies in the interior of Egypt.* Had he been nearer, he might have sent his fleet away, to be out of danger, perhaps at

---

* Bonaparte's army stormed and took Alexandria on 2 July. He then led it up the Nile to Cairo, where on 21 July he fought 60,000 Mamelukes and followers in the Battle of the Pyramids outside the city. His victory was total, Cairo was taken and he moved north again to occupy Syria, since Nelson's destruction of his fleet ended any plan to invade India. A Turkish expedition, supported by the British, defended the Syrian coast and its ports. At Acre the defence was conducted by Captain Sir Sidney Smith of the Royal Navy. When plague broke out in his besieging force Bonaparte retired on Cairo but returned to the coast when a British

Corfu, from which it could have been recalled quickly at need, and where it would have constituted a threat to Nelson's lines of communication. The concept, however, of a "fleet in being," affecting events by doing nothing, may have been alien to Bonaparte's active and aggressive mind. He therefore ordered Brueys to remain in Egyptian waters but to put the fleet under the guns of Alexandria. It was then anchored in Marabout Bay, where the landings had been staged, a clearly unsatisfactory roadstead. Alexandria, however, was a difficult harbour, shallow and easily blocked. It was therefore eventually decided to transfer the ships to Aboukir Bay, nine miles to the east.

Brueys had anchored his ships in a position he thought made a successful attack by the British—which he expected—impossible. They lay in a shallow crescent formation, bows on to Aboukir Castle with Aboukir (Bequières) Island to starboard and shoal water between them and the land to port. They could be approached from only two directions: from below Aboukir Island, though the northerly wind denied the British that course; or through the gap between the island and the castle. Brueys had apparently judged the gap impracticable, believing that, even if negotiated, the water beyond was too shallow for the British to pass on either side of his ships; that is, between his line and Aboukir Island or between his line and the shoreward shoals. He had strengthened his defences by having cables run between most of his ships, which were about 175 yards apart, and by ordering springs to be attached to their anchor cables; springs, ropes taken to the capstan, could be tightened to swing the ship by the bow or stern, so that they were manoeuvrable even though at anchor. Not all the French captains, however, had attached springs by the time the battle began.

Nevertheless, the French position was formidable enough to deter a cautious enemy; but the British were not cautions, nor were they unobservant. Foley, captain of *Goliath*, had one of the only two charts of the coast in the fleet, and a good one; it showed the depths of water right up to the shoreline.[13] More important, Foley made a snap judgement about the way the French were anchored. Nelson himself would shortly come

naval escort brought another Turkish expeditionary force to Aboukir. Bonaparte defeated it on 25 July, but assured of French control of Egypt and concerned about his personal position at home, sailed on 23 August for France, where he seized power. The French army in Egypt was subsequently defeated by combined British–Turkish forces and it was repatriated by treaty to France in 1801.

to the same conclusion, saying to Berry, his flag captain in *Vanguard*, "where there was room for an enemy's ship to swing, there was room for one of ours to anchor."[14] Foley saw that instantly as he passed the gap between the castle and the shoals and so pointed *Goliath* inshore, to pass round the *Guerrier* at the head of Brueys' line, and so down the inside of the anchored enemy.

Foley had intended to anchor alongside *Guerrier*, into which he fired as he rounded her bow, but his crew ran out too much cable. *Goliath* ended up farther down the French line, opposite *Conquérant* and *Spartiate*. The mistake did not really matter, for the British ships next astern were following fast, *Zealous, Audacious, Orion* and *Theseus*. They also joined in the cannonade against *Guerrier*—which collected fire from all of them as they passed by and was quickly dismasted—while *Theseus* positioned herself to fire into both *Spartiate* and *Aquilon*; Miller of *Theseus* was a New Yorker, one of two North American Loyalists among Nelson's captains.

The head of the French line was now solidly engaged by anchored opponents. *Vanguard*, which was following *Theseus*, took a different course, steering to pass on the seaward rather than inshore side of the French and to anchor opposite *Spartiate*, which was thus taken between two fires. *Minotaur* engaged *Aquilon*, also caught between two fires, while *Defence* stopped opposite *Peuple Souverain*, which was being fired into by *Orion* on the other side.

The centre of the French line was composed of the heaviest ships, *Franklin*, 80 guns, *L'Orient*, 120, and *Tonnant*, 80; the other 80, *Guillaume Tell*, was some distance away, third from rear. Darkness had fallen as the centre's British opponents began to appear, first *Majestic*, which was mishandled and ended up opposite another 74 farther down, then *Bellerophon*, then *Alexander*, then *Swiftsure*. The last two, positioning themselves skilfully in the gaps astern of *Franklin* and *L'Orient* respectively, were able to do serious damage without suffering heavily themselves. *Bellerophon*, coming alongside *L'Orient*, suffered terrible damage and loss by choosing to engage the heaviest ship present. In an hour of fighting she lost her main and mizzen masts, while her foremast also was damaged. By ten o'clock her ordeal began to abate as fire from *Swiftsure* and *Alexander* raked the French flagship from bow and stern. They did terrible slaughter; Admiral Brueys, badly wounded, insisted on remaining on deck until struck by a shot that killed him. Below decks the spaces were full of wounded, including Captain Casabianca's young son. They were also cluttered by flammable stores, Lieutenant Webley, of *Zealous*, noted when *L'Orient* took fire. *Swiftsure*'s captain ordered his crew to fire into the seat of the blaze to stop the French crew from fighting the flames. Soon it became obvious that *L'Orient*'s magazine would be set off, and both her British and French neighbours cut their anchor cables to reach what was hoped to be a safe distance. *Alexander* drifted off, so did *Tonnant, Heureux* and *Mercure*, either to anchor again or to ground in shallow water. *Swiftsure*, close ahead of *L'Orient*, was judged by its captain to be safer where it was; he calculated that the coming explosion would pass over his ship.

So it did; the enormous detonation sent the debris of broken timbers, masts, cordage and bodies hundreds of feet into the air, to rain down detritus into the waters of the bay for a mile around, while the noise, heard in Alexandria nine miles away, temporarily brought the battle to a stop. When it resumed, after a quarter of an hour, the scene of battle had been decisively altered. The disappearance of *L'Orient* and the shift of *Tonnant*, which had drifted dismasted towards the rear, left a large gap in the middle of the French line, widened by the falling away of *Heureux* and *Mercure*, which had cut their cables also and gone aground, though their crews continued to serve the guns. The French were thus in almost total disarray, with their admiral dead, flagship destroyed and surviving ships separated into two groups. In the forward group, *Guerrier*, whose crew had fought heroically while her captain had refused to surrender twenty times, at last struck after three hours, dismasted and devastated. *Conquérant*, after another valiant passage of resistance, had also at last struck. *Spartiate*, third in line, had surrendered after two hours, the first French ship to give up, but with 200 dead and wounded aboard and the survivors pumping to keep the ship afloat. *Aquilon* surrendered a little later, with 87 dead aboard and 213 wounded. *Peuple Souverain*, fifth in the order of battle, had drifted out of the line, perhaps because her cables had been severed by gunfire. *Franklin*, still in line, had ceased to fight after being set on fire four times, the last by burning debris from the explosion of *L'Orient*. By early in the morning of 2 August, therefore, the French fleet consisted of a shattered and defeated van, a central void and a rear in disarray. *Franklin*, anchored ahead of *L'Orient*'s original position, did recommence fire after the great explosion but was swiftly brought to surrender. Aft of the gap, some of the French ships continued resistance for several hours, *Hereux* and *Mercure*, which had gone aground after cutting their cables, from inshore. Admiral Villeneuve, in *Guillaume Tell*, eventually decided, however, that it was his duty to escape, cut his cable and sailed out of the bay, followed by *Généreux* and the frigates *Justice* and *Diane*. He left behind the dismasted *Tonnant* and *Timoléon*, which, with heroic but pointless obstinacy, continued to work their guns into the afternoon of 2 August. *Tonnant* eventually hauled down her colours but *Timoléon's* crew left theirs flying when they set fire to the ship and rowed ashore to escape capture.

Nelson had won a crushing victory, in its completeness never exceeded during the days of sailing-ship warfare and equalled in naval history only by Japan's annihilation of the Russian fleet at the Battle of Tsushima in 1905. Of the enemy's thirteen line-of-battle ships, two had escaped, but two had blown up and the other nine had been captured in action or driven ashore. Nelson had lost none of his ships. *Culloden*, which had grounded during the approach, to the fire-eating Troubridge's fury, had been floated off; *Bellerophon* and *Majestic*, the hardest hit, survived. Nelson's casualties—he himself had suffered a nasty scalp wound early on—numbered 208 killed and 677 wounded. The French, by contrast, had surrendered more than a thousand wounded while their dead came to several thousand, a thousand in *L'Orient* alone.[15]

It was the nature of the battle that determined the scale of the slaughter; ships anchored broadside to broadside, firing into each other at point-blank range, caused ghastly carnage among their crews. Engagements in the open sea, when ships had the freedom to manoeuvre, were much less costly in human life. Yet at Copenhagen, a battle Nelson was to fight in almost identical circumstances in 1801, Danish casualties were only 476 killed, 559 wounded. A killer instinct was at work at the Nile, a determination among the British to prevail, among the French not to be overcome.

What animated the French is the harder to estimate; revolutionary fervour no doubt, certainly Bonapartist inspiration, perhaps also the determination not to return to the traditional state of inferiority prevailing before their naval renaissance in the American War of Independence. Analysis of the British mood is more straightforward. Victory was a way of life for the Nelsonian sailor. He believed all races inferior to his own, and expected to beat them, and would fight unremittingly to ensure that he did. Moreover, the fleet had been led a merry dance by

Brueys for nearly three months. Cornered at last, he and his sailors became the object of their enemy's pent-up frustration.

No one in Nelson's fleet had been more frustrated than Nelson himself, sleeping badly, eating little, railing in every letter he wrote against the bad luck which had him in its grip. Want of frigates, want of help from those he believed owed it him, were his constant themes. He also came to believe that the fates were against him, that he had consistently made the right choices, but that some malign spirit had intervened to disappoint his best intentions. In his letter to St. Vincent, composed at the nadir of the campaign, the letter Captain Ball had urged him not to send, he had itemised his setbacks. It was written off Alexandria during his first visit, when he found the harbour empty.

> The only objection I can fancy to be stated is, "you should not have gone such a long voyage without more certain information of the Enemy's destination": my answer is ready—who was I to get it from? The Government of Naples and Sicily either knew not or kept me in ignorance. Was I to wait patiently till I heard certain accounts? If Egypt was their object, before I could hear of them they would have been in India. To do nothing, I felt, was disgraceful: therefore I made use of my understanding, and by it I ought to stand or fall. I am before your Lordship's judgment (which in the present case I feel is the Tribunal of my Country), and if, under all circumstances, it is decided I am wrong, I ought, for the sake of our Country, to be superseded; for, at this moment, when I know the French are not in Alexandria, I hold the same opinion as off Cape Passaro (south-east point of Sicily, 21–22 June)—viz; that under all circumstances I was right in steering for Alexandria, and by that opinion I must stand or fall.[16]

It is almost impossible not to sympathise with Nelson's analysis of his own decisions and actions. He made mistakes during his seventy-three days of chase, between the great storm of 18 May and his bringing of Brueys to battle on 1 August, notably in deciding not to chase the French frigates sighted off Sicily on 22 June and in not waiting off Alexandria on the 29th when the signs were that the Turks expected trouble; had he then reined in his impatience for twenty-four hours, he would have won what might have been the most decisive naval battle in history. On the other hand, as an essay in pure intelligence operations by a commander on the spot, Nelson's Nile campaign is difficult to fault. The restraints under which he worked are clear to enumerate: no reconnaissance force ("want of frigates"), no means of communication with land-based sources of information except by going to get it himself, no reassurance that any such information gleaned was reliable, even from friendly sources (Hamilton's and Acton's economy with the truth should be remembered), no access to the central intelligence resources of his own home base (three to five weeks' delay in communication between the Mediterranean and London in the inward direction, therefore twice that two-way), no certain home intelligence even if sent. Other restraints were an active disinformation campaign conducted by the enemy (manipulation of the official press) and energetic denial of local sources of intelligence (Brueys' commandeering of all merchant shipping encountered during the voyage to Alexandria).

Nelson had to work, therefore, by optimising local intelligence acquisitions (particularly the interrogation of Turkish officials in the Peloponnese and merchant captains off Crete after his first passage to Alexandria), which were offset by misinformation (the report that the French had left Malta three days earlier than was the case) and by his own "understanding."

Can we reconstruct the picture of the strategic situation Nelson must have formed in his mind once he knew that Bonaparte had sailed from Toulon after the great storm of 18 May? He early and correctly discarded the idea that Bonaparte was making for Spain, to attack Portugal, or sailing out of the Mediterranean to invade Ireland (the presence of St. Vincent's fleet at Gibraltar nullified that threat in any case). He therefore had to picture where Bonaparte might land his army once he was certain that he was heading eastward. There were really only three destinations. The Mediterranean is not one but two seas, separated from each other by the Sicilian–Tunisian narrows, where it is only 200 miles wide. In the political circumstances of 1798, the only objectives worthwhile to the French west of the narrows were Sicily itself and its parent kingdom of Naples; the capture of Malta was an alternative aim, but only as a preliminary to a descent on Sicily/Naples. East of the narrows the objectives widened, but not irreducibly. The dead end of the Adriatic could be discounted. Its waters were already controlled either by the French, or by Austria, with which France was not at war, or by Turkey, which was not an enemy.

The rest of the eastern Mediterranean was also Turkish and it might be, as Nelson calculated, possible that the French Republic, despite a historic alliance with the Ottoman emperor, had decided to invade his territory, not to overthrow his rule but to strike through his lands against British interests farther to the east. One route, if the French were to pass by the Dardanelles to his capital at Constantinople, lay across Anatolia to the Persian Gulf. The other, via Alexandria, gave on to the Red Sea and so to the Indian Ocean from another direction. In either case, Britain's rich possessions in the Indian sub-continent were the objective.

Sicily/Naples (with Malta as a subordinate target); Constantinople; Egypt: those were the three destinations which Nelson had to juggle in his mind. By 22 June, when he knew that the French had taken Malta but passed on, he had convinced himself that Bonaparte's point of disembarkation must be Egypt. Sicily/Naples required a retrogression, which the winds and the intelligence had argued against. Constantinople was too roundabout a route to the Indian sub-continent. From Alexandria, however, the path stretched forward. After the conference with his captains aboard *Vanguard* on 22 June, he knew that he would find the French fleet in Egypt, and he was right. Only contingencies, and two misjudgements, denied him the decisive fruit of his intelligence assessment.

Nelson's chase to the Nile compares well with another chase in Mediterranean waters 116 years later, when the French and British Mediterranean fleets tried to bring the German battlecruiser *Goeben* and its escorting cruiser *Breslau* to action and allowed both to escape to the Turks in Constantinople. Technology had, over a century, altered the conditions of chase greatly to the pursuers' advantage, allowing intelligence to be passed almost instantaneously, always supposing it was accurate, which, in 1914, proved to be scarcely more so than in 1798, and discounting, too, unhelpful intervention by admiralties. The speed of pursuit had greatly increased, tripling from about 8 to over 24 knots. On the other hand, the need to refuel, often at a few days' interval, tied ships to ports or to rendezvous with coalers, limiting their freedom of action to a degree which Nelson, taking his means of motion from the wind, would have found frustrating. Even allowing for the tendency of winds to fail or blow in the wrong direction, sailing-ship fleets had an operational autonomy not to be regained by automotive navies until the development of nuclear power.

Admiral Souchon, commanding *Goeben* and *Breslau*, had been in the

Mediterranean since 1912, using Austrian Adriatic ports as his friendly bases and otherwise resupplying at Italian or Spanish ports. Early on 4 August 1914, Souchon bombarded the French North African ports of Philippeville and Bône, doing little damage but reminding Lapeyrène, the French Mediterranean commander, that he had the ability to interrupt the transport of the XIX Army Corps from Algeria to France. Souchon then made off for the Straits of Messina (where Nelson had missed Brueys' fleet in June 1798), intending to coal. En route he encountered the main elements of the British Mediterranean fleet, the battlecruisers *Indefatigable* and *Indomitable.* Their orders were to close the Straits of Gibraltar to hostile ships, as pressing a concern to the British government in 1914 as it had been when Ireland was under threat of invasion by the Toulon Armament in 1798.

Captain Kennedy, the battlecruiser squadron commander, at once reversed course but, since Britain was not yet at war with Germany (or would not be until midnight), kept his distance. Souchon cracked on all speed and shook Kennedy off. As the British ships had made 28 knots on trials, and *Goeben* was limited by boiler defects to 24 or even 22 knots, that was not a creditable outcome.

It was even less creditable that, with Souchon coaling at Messina in the twenty-four hours' grace custom allowed (what Nelson had got at Alexandria on his first visit in June 1798), the commander of the British Mediterranean Fleet, Admiral Sir Berkeley Milne, so disposed his forces as to allow Souchon a bolt hole. With a correct but over-scrupulous regard for Italian neutrality, he kept his ships away from the Strait of Messina, deploying them west of Sicily against a resumption of their interference with the French troop convoys. He recognised that Souchon might turn in the other direction, to join the Austrian fleet in the Adriatic, but counted on a subordinate squadron, Admiral Ernest Troubridge's four armoured cruisers, currently off the west coast of Greece, to block a move in that direction; unfortunately, he passed on to Troubridge (a descendant of Nelson's Troubridge) an Admiralty warning not to attack a "superior force."

The Admiralty, which in this case was synonymous with its political chief, Winston Churchill, meant that Milne's fleet should not engage the dreadnoughts of Austria or Italy, the latter a signatory of the Austro-German-Italian Triple Alliance (from which it was not yet disengaged). Milne, and so Troubridge, unfortunately took the signal to mean that they should fight shy of the *Goeben,* a very powerful unit in its own right. As a result, Troubridge delegated the duty of following *Goeben* first to the light cruiser *Gloucester*—which was hopelessly outgunned, though it bravely attacked all the same—then to another light cruiser, *Dublin,* which failed to find the Germans. Troubridge had a clear picture of Souchon's options: he knew, from *Gloucester,* that Souchon was steaming towards Greece and the Aegean; he calculated, correctly, that the Germans would either continue on that course or turn tack towards the Adriatic and junction with the Austrians. Worried that his obsolete armoured cruisers might meet and be outgunned by *Goeben,* Troubridge called off the chase and steered away.

It was a disastrous misjudgement; the consequences were heightened by further misleading signals from the Admiralty and more bad decisions by Milne. Although on the night of 6 August he learnt that the French were now organised to prevent any attack by Souchon on the troop convoys, so freeing him to crack on all speed after the enemy, he decided to coal *Indomitable* in Malta. While there, he heard from the British naval attaché in Greece on 8 August that Souchon was already in the Aegean (where he, too, was coaling). That gave him the chance to make up lost ground. Almost simultaneously, however, a signal arrived from the Admiralty stating that Austria had declared war on Britain; the signal was wrong: Austria would not do so until the 12th. Milne accordingly decided that the priority was to guard the exit from the Adriatic, by which Austrian dreadnoughts could enter the Mediterranean, and turned back. A further passage of order, counter-order and misinformation from the Admiralty ensued. It was not until 9 August that Milne got clear instructions "to chase the *Goeben* which passed Cape Matapan [southern Greece] on the 7th steering north-east."

Shades of 22 June 1798, when Nelson was misinformed by three days of the date of Bonaparte's departure from Malta. Nelson had then, all the same, pressed forward at best possible speed. Milne, toying with the possibilities that Souchon might have doubled back to renew his attacks on the French North African ports, to enter the Adriatic, to make for Gibraltar or even to raid Alexandria and the Suez Canal, did not accelerate. As a result, Souchon enjoyed sixty trouble-free hours in the Aegean and eventually anchored in the mouth of the Dardanelles on 10 August. Milne, believing that the Dardanelles was mined against the passage of any warship, was subsequently astounded to learn that the

Goeben and *Breslau* had been conducted by the Turks to Constantinople where, by a diplomatic device, they would become units of the Turkish navy and the agency by which the Ottoman empire was brought into the war on Germany and Austria's side.[17]

By strict comparison, the management of the chase to the Nile was a superior, even if more protracted and intermittent, exercise in the use of intelligence than the pursuit of Souchon. The absence of intervention by the Admiralty, which in 1914 twice seriously misled commanders on the spot, was a positive advantage; Nelson was not bothered by London or by intermediate authorities and, though he made his own mistakes, was spared the misjudgements of others. The Admiralty of 1914, although having available to it information in quantities and of an accuracy denied to eighteenth-century governments, together with the means to communicate with subordinates almost instantaneously, by wireless, instead of at a delay of weeks by courier or sailing despatch vessel, first sent Milne an ambiguous order—to avoid contact with a superior force which deflected him from engaging *Goeben* when he could have done so—then wrongly informed him that Austria had entered the war, when it was not to do so for another five days, so turning his back towards the Adriatic when he should have been making best speed into the Aegean.

Milne also seems to have lacked Nelson's ruthless ability to stick to the main issue. By 22 June 1798, when he knew of Bonaparte's capture of Malta and departure for another destination, Nelson discounted every other consideration that had distracted him thus far—Spain? Portugal? Ireland?—and decided, correctly, that the enemy was heading for Alexandria. His reasoning was that Egypt, and beyond it India, was the object of the highest strategic value and an intention to land the army there the only explanation of the voyage of the Toulon Armament to the central Mediterranean. Milne, by contrast, after having lost the *Goeben*, continued to confuse his thinking with the possibilities of the Adriatic, the French troop convoys and Egypt as well as the Aegean. He was a man in a muddle which, after 22 June, Nelson never was. Nelson's failure to find Bonaparte in Egypt the first time caused him acute anxiety but not doubt of his reasoning. Milne appears not to have thought rigorously at all.

The Nile campaign demonstrates that, to Nelson's many other qualities, which included inspirational powers of leadership, lightning tactical verve, ruthless determination in battle, incisive strategic grasp and a revolutionary capacity for operational innovation, all combined with complete disregard for his own personal safety in any circumstances, must be added the abilities of a first-class intelligence analyst. Few dispute that Nelson was the greatest admiral who ever lived. The range and depth of his powers suggests that he would have dominated in any age.

CHAPTER THREE

——◄O►——

Local Knowledge:
Stonewall Jackson in the Shenandoah Valley

IN THE MEDITERRANEAN in 1798 Nelson had been much mystified and often misled, at least twice to his crucial disadvantage, even though he enjoyed a superiority of force, intimately understood the geography of his theatre and was pursuing an enemy with a severely restricted choice of manoeuvre. All had come right in the end, but his victory would have been even more complete had he made identifiably different decisions on several occasions. The Mediterranean was a closed and circumscribed strategic arena in which, in optimal circumstances, a fleet commander might have achieved total domination.

In 1862, in a sort of mirror image of Nelson's Mediterranean campaign, though on land rather than at sea, Thomas "Stonewall" Jackson, also operating within a closed strategic arena, the Shenandoah Valley of Virginia, had consistently mystified and misled his enemy—his catchphrase was "always mystify and mislead"—even though usually inferior in force to the Union armies chasing him and despite severe geographical limitations on his room to manoeuvre. In 1798 Nelson had never, until the last stages of his pursuit, had quite enough intelligence. In 1862 Jackson enjoyed ample intelligence and exploited it to win a succession of victories his objective weakness should have denied him. Better information, keener anticipation and cleverer judgement combined to make Jackson's Valley campaign a model of an active intelligence victory.

The situation of the Southern Confederacy, at the outbreak of the American Civil War, was intrinsically weak. By every material measure—population, industrial capacity, miles of rail track, to mention but a few essential indices—its capacity to wage successful war was greatly exceeded by that of the North. Of the United States' thirty-two million people, only five million lived in the eleven seceding states (and four million blacks whom, as slaves, the Confederacy would not arm); of the country's 30,000 miles of rail track, 22,000 ran in the Northern states; the North produced 94 per cent of the country's manufactured goods and the vast proportion of its raw materials, including iron, steel and coal. The South was a rich region nonetheless, but it was rich in cotton, tobacco, rice and sugar, crops which brought their planters income in overseas sales and the export of which the North could and did interrupt, as soon as secession was declared, by blockading the Confederacy's coasts.[1]

Had the war been a contest between economic systems alone—as Winfield Scott, Union General-in-Chief, hoped to keep it—the South would have quickly collapsed.[2] The Southern people, however, were resolute in their determination to preserve "States Rights," the legal issue over which they had declared separation, and soon showed themselves equally resolved to sustain the deprivations that economic isolation brought them. Hardy and frugal in their rural way of life, they rapidly made it clear that they would have to be beaten in battle if they were to be brought to surrender. President Abraham Lincoln was quick to grasp that, quicker than Winfield Scott. The question was, where to fight? The South might be materially weak, but it was strategically and geographically very strong. Protected on two sides by sea and ocean, it was also shut off from the rest of the United States by unsettled semi-desert to the west and by mountains to the north. Its paucity of internal communications, which in any case connected poorly with those of the North, was a positive strategic advantage. Moreover, in the great valley of the Mississippi, it enjoyed the protection of a sort of secondary internal water frontier, denying Northern armies any easy way forward into its heartland. Above all, the South's enormous size—its eleven states covered an area as large as Europe west of Russia—was in itself a strength. Even if its outer crust could be penetrated by the Union, there still remained the difficulty of covering the vast distances inside the South between the point of entry and an objective of any value. To get from anywhere to anywhere within the Deep South was a problem in peacetime—there were few railroads, appalling or nonexistent roads, while the inland rivers were too short and usually ran the wrong way. In wartime the problem seemed designed to defy the efforts of a general of genius.

The South, on the other hand, should it choose to attack, faced no such problem. From its Virginian frontier with the Union, Washington, the Federal capital, lay only forty miles distant; not much farther lay the great city of Baltimore. Also within striking distance lay smaller but still desirable urban targets and the rich farmlands of Maryland and Pennsylvania. A successful thrust into the North would also bring menace to industrial New Jersey and perhaps even New York. The South's strength was the widespread dispersion of its centres of population and production; the North's weakness was the concentration of similar objectives in the Middle Atlantic coastal corridor and that corridor's vulnerability to a Confederate offensive.

Critically, the South had a way in, the Shenandoah Valley. The dominant geographical feature of Atlantic America is the Appalachian Mountain chain, which runs roughly parallel to the coast, at diminishing distance, from Alabama to Maine. The Appalachians shut off the enormous interior of the continent from the coastal strip for hundreds of miles and had been used by the French, when they ruled Canada and what they called Louisiana, to deny the Ohio country and the Mississippi Valley to the English colonists in Virginia, the Carolinas and Georgia.

The defeat of the French in 1763 had opened up the trans-Appalachian wilderness to the English and thereby set in train the events that led to the division of 1861. Virginians, Carolinians and Georgians, migrating westward, had taken slavery with them into Mississippi and Tennessee. New Yorkers, Pennsylvanians and New Englanders had established the Midwest as non-slave territory. Disputes over the states of the westerly borderlands had generated the constitutional conflict that resulted in the crisis of secession. Slave or free? That was the issue over the new lands opening up to settlement in the old French region of "Louisiana." When it could not be settled by debate, the Southerners chose separation.

What would then have happened had the North chosen to adopt Winfield Scott's passive Anaconda Plan, and had the South chosen to sit inside its formidable natural frontiers, challenges easy speculation. There might not have been a civil war at all. Neither eventuality occurred. The South, as James McPherson convincingly argues, was spoiling for a fight.[3] The outraged North, outraged both by the challenge to the Constitution and by the South's defiant defence of the sin of slavery, was adamant for an offensive. Inspired by the cry "On to Richmond," the Virginian capital of the Confederacy, the North launched the manoeuvre that led to encounter and defeat at the First Battle of Bull Run (Manassas) in July 1861.

In the aftermath, the Northern leadership pondered a better way. In the west, beyond the Appalachians, local generals tried to open up a new front on the river approaches to the Mississippi. Along the coast, Union admirals began to close off the Confederates' outlets to the wider world. In Washington, however, Lincoln and his government sought a more direct means to strike at their Confederate enemies. The way, as they recognised, was barred by a succession of water obstacles, the short rivers running off the Appalachian chain between the mountains and the Atlantic that furnished one of the South's best strategic defences. The course of the Rappahannock, the Mattapony, the York, the James, might have been designed by a friend of slavery to frustrate the advance of Northern armies to the seat of rebellion. At twenty-mile intervals or less, river after river, each easily defensible, stood between the Union forces and the enemy capital.

A solution to the infuriating strategic difficulty was proposed in the spring of 1862 by the man who had recently become Abraham Lincoln's favoured general, George McClellan. Convinced that a repetition of the "On to Richmond" effort by the overland route would stumble again, at one or other of the water obstacles, McClellan persuaded Lincoln to let him put the Army of the Potomac, the Union's main force, into troopships, sail it down Chesapeake Bay from Washington and land it at the point of the Virginian Peninsula, between the York and James rivers. There he would enjoy the security of a firm base, Fortress Monroe, one of the great stone citadels of the coast-defence programme known as the Third System, and still in Union hands; from it by easy marches Richmond lay only seventy miles distant. McClellan was confident that he could make the amphibious operation work. As a junior officer marked for promotion, he had been sent in 1855 as an observer to the Anglo-French expedition to the Crimea, so had seen an amphibious operation at work with his own eyes, and had also witnessed the military use of the newly invented telegraph.[4] As a railroad company executive, which he

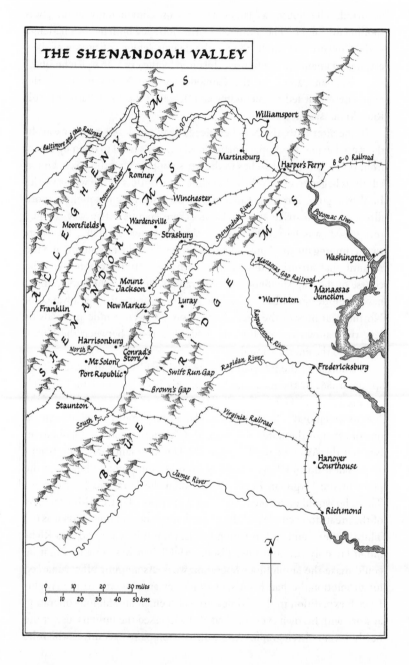

THE SHENANDOAH VALLEY

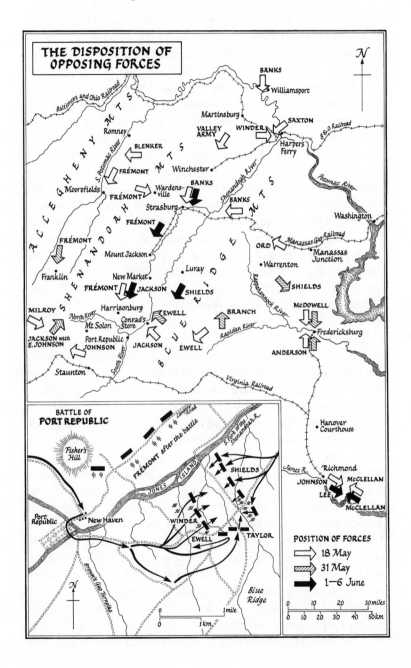

THE DISPOSITION OF OPPOSING FORCES

BANKS
Williamsport

Baltimore and Ohio Railroad

Martinsburg
SAXTON
B & O Railroad

ALLEGHENY MTS
Romney
BLENKER
VALLEY ARMY
WINDER
Harpers Ferry

FRÉMONT
S. Potomac River
Winchester
Potomac River

Moorefields
FRÉMONT
Wardensville
BANKS
Strasburg
BANKS
Washington

FRÉMONT
Manassas Gap Railroad
ORD
Manassas Junction

FRÉMONT
Mount Jackson
Warrenton

Franklin
New Market
Luray
SHIELDS
FRÉMONT
JACKSON
SHIELDS

Rappahannock River

MILROY
Harrisonburg
Conrad's Store
EWELL
BRANCH
McDOWELL

North River
Mt Solon
Rapidan River
Fredericksburg

JACKSON with E. JOHNSON
Port Republic
JOHNSON
JACKSON
EWELL
ANDERSON

Staunton
South River

Virginia Railroad

BATTLE OF PORT REPUBLIC

Lewis Road
Fisher's Hill
FRÉMONT after the battle
S. Fork of the Shenandoah R.

Hanover Courthouse

JONES ISLAND
SHIELDS

James R.
Richmond
JOHNSON
McCLELLAN
LEE

Port Republic
New Haven
WINDER
EWELL
TAYLOR
McCLELLAN

Browns Gap Turnpike

N

Blue Ridge

POSITION OF FORCES
⬜ 18 May
▨ 31 May
⬛ 1–6 June

1 mile
1 km

0 10 20 30 miles
0 10 20 30 40 50km

71

left the army to become in 1856, he had learnt more about the telegraph as a means of control and also about bulk supply over long distances; both telegraphic control and efficient logistics were to be of central importance in the running of the Peninsula Campaign.

The arrival of the Army of the Potomac at Fortress Monroe greatly alarmed the Confederate high command. Entrenchments were hastily dug across the nose of the Peninsula, in places following the line of earthworks constructed during the British defence of Yorktown in 1781. General Joseph E. Johnston's Army of Northern Virginia was withdrawn from the proximity of Washington, and Richmond was put into a state of defence. The Confederates were right to be alarmed, despite the temporary security these measures provided. They were greatly outnumbered on the spot, by 105,000 to 60,000 in the vicinity of Richmond, and potentially by an even larger number. Three other Northern armies hovered nearby, that of Frémont in West Virginia, that of McDowell near Washington and that of Banks in the Shenandoah Valley. If they could be brought into combination with McClellan's, Joseph E. Johnston's Army of Northern Virginia would be overwhelmed and the fate of the Confederacy sealed.

There were only two points of light for the Confederacy amid the encircling gloom. The first was McClellan's capacity for procrastination. Though objectively superior in strength to the enemy, he was constitutionally incapable of accepting the evidence, constantly petitioned Lincoln for more troops and issued frequent warnings of his inability to proceed unless reinforced. Instead of pressing forward, he hung back, professing to see dangers visible only to himself, thus conferring on his enemies opportunities to strengthen their position which they should not have been allowed. He had landed at Fortress Monroe on 22 March 1862. He then spent a month, 4 April to 4 May, besieging the weak Confederate position at Yorktown. Not until 5 May, after the Confederate garrison had withdrawn, did he advance to fight his first proper battle, at Williamsburg, and not until the 25th did he draw near to Richmond, his proper objective. He had taken over eight weeks to cover seventy miles and had inflicted no damage on the enemy at all. Joseph E. Johnston's army stood intact and remained to be brought to battle.

The other point of light was the existence of a Confederate diversionary force, poised to unsettle both McClellan and Lincoln, though in different ways. McClellan could be thrown into anxiety by any move threatening to deprive him of the reinforcements he craved. Lincoln, more tellingly, was prone to alarm at any prospect of a Confederate advance against Washington. The Confederates around Richmond lacked the capacity to level either threat. General Thomas Jackson, "Stonewall," far away though his small army lay in the Shenandoah Valley, was equipped, equally by location and capability, to organise both. Any thrust northward he might make would menace Washington, which Lincoln increasingly believed had been selfishly stripped of troops by McClellan to bolster his Peninsula adventure. Such a thrust would simultaneously lessen the likelihood of Lincoln's agreement to the redeployment of the covering armies of Banks and McDowell from the Shenandoah Valley vicinity towards McClellan at Richmond. Stonewall Jackson, in the spring of 1862, suddenly found himself in a "swing" position, capable of altering the course of the war, if he handled his force correctly, with decisive effect.

The Shenandoah Valley was a strategic corridor, which worked as a critical anomaly in the military geography of the Civil War. The heartland of the Confederacy, as originally constituted, between the sea, the Mississippi and the mountains, was virtually impenetrable. McClellan had cracked the carapace by finding a maritime point of entry on the Virginian Peninsula, but to enlarge the breach he would need to show a determination and single-mindedness his contemporaries of the West Point class of 1846 had good reason to doubt he possessed. Otherwise, as long as the lower course of the Mississippi was held, there was only one other way in: down the Shenandoah. The Valley is the easternmost feature of the central Appalachian chain. Its southern exits lead into the plains of Virginia, the Carolinas and Georgia, its northern exits into Maryland, Pennsylvania and towards the outskirts of Washington. In the circumstances of the Civil War, it could be used offensively or defensively. Theoretically, the North could use it as a way into the Confederate heartland; in practice, the lack of a north–south railroad within the Valley made that manoeuvre logistically too difficult to undertake, though it was one against which the South always had to be on guard. On the other hand, the South could much more easily use the upper mouth of the Valley as a sally port from which to surprise Northern armies near their major cities. During the course of the war, it was the South which better exploited the strategic potentialities of the Valley and never more so than in the spring of 1862.

There is a large and small strategic geography of the Valley. The large is that of a corridor leading either in or out of what between 1861 and 1865 was the Confederacy; the small is that of its internal features which, if correctly understood, can be put to decisive military use. About 120 miles long and 30 wide, from the headwaters of the South River to the confluence of the Shenandoah and Potomac at Harper's Ferry, and from the crest of the Alleghenies to that of the Blue Ridge Mountains, the Valley is an enclosed environment of what, in 1862, was rich, cleared farming land. Down the centre, however, runs a dividing ridge, the Massanutten Mountain, which itself divides the Shenandoah River into a North and South Fork. Joined near Front Royal, the forks become the Shenandoah proper, to run northward for forty miles to enter the Potomac at Harper's Ferry. The tail of the South Fork separates at Port Republic into three minor streams, the North, Middle and South rivers.

Many rivers mean many bridges, and there were, in 1862, at least twelve of military significance, providing crossings at the Valley's principal townships and villages; like others of that period in the American settlement of the interior, most were wooden and easily burnt. Besides bridges, there were other passages of strategic significance, fords here and there, but also gaps in the surrounding mountain chains. Those leading out into West Virginia, dominated by the North, were few and of lesser military importance. Those giving on to the lowlands of Virginia proper were more numerous—there were eleven in all—and of altogether greater significance, since they provided a Confederate army inside the Valley with the means to dodge back and forth. Equally important was the gap through the central Massanutten ridge between New Market and Luray and the east–west links around the Massanutten's headlands at Front Royal and Port Republic.

The Valley's internal geography determined its road network. It was better west of the Massanutten, where the Valley Pike—an all-weather macadamised road of impacted gravel—led from Williamsport on the Potomac, via Winchester, Strasburg, New Market and Harrisonburg, to Staunton, between the South and Middle Rivers. East of the Massanutten, an inferior road ran from Front Royal through Luray to Port Republic and eventually to a junction with the Valley Pike at Staunton.[5]

Few, if any, in the Northern armies understood the Valley's geography. There were two reasons for that. The first was that in peacetime, the

Valley's communications with the outside world had been almost exclusively by river, up and down the Shenandoah and its branches to Harper's Ferry; so important were the waterways that Valley people described the northward passage to the confluence as going "down," the southward as "up." The North therefore knew the Valley only as a river system, and then at its external points of connection. The second was that there were virtually no Valley maps. That was a prevailing condition of warmaking between the Union and the Confederacy. The Federal government had, before 1861, invested considerable sums in mapping the United States' coasts; one of the branches of the United States Army, the principal instrument of the government's internal administration, was a Corps of Topographical Engineers. It had also sponsored a major exploration of the west, as a support to its sponsorship of settlement beyond the Mississippi. It had done nothing similar in the old Thirteen Colonies or the eastern states founded since 1782. The result was that the generals of the Civil War embarked on their operations with wholly inadequate cartographic resources.

> No accurate military maps existed. [The Union] General Henry W. Halleck was running a campaign in the western theatre in 1862 with maps he got from a book store. With frenetic haste, the general set topographical officers and civilian experts to work, making maps, but the resulting charts were generally incorrect. Benjamin H. Latrobe, the civil engineer, drew a map for a general going into Western Virginia, but the best he could promise was that it would not *mislead* the expedition. General George B. McClellan had elaborate maps prepared for his Virginian expedition of 1862 and found to his dismay when he arrived on the scene that they were unreliable; "the roads are wrong," he wailed. Not until 1863 did the [Northern] Army of the Potomac have an accurate map of northern Virginia, its theatre of operations.[6]

At the root of the trouble lay the cartographic backwardness of the United States. That might be thought understandable: the United States was an enormous country, still largely unsettled, by no means completely explored and without a central mapping agency; the army had its Corps of Topographical Engineers, the navy a Hydrographic Office and the federal government a Coast Survey, but they were all tiny.[7] The basis for accurate survey, a comprehensive triangulation of the land mass, was absent. Yet it had been done elsewhere. The British Isles had been triangulated and a comprehensive series of high-quality maps published, at one inch to the mile, by the Ordnance Survey, beginning in 1791; a small undertaking, certainly, but magnificently accomplished. Impressive by any standards was the work of the Survey of India; India, though smaller than the United States, is topographically even more complex, because of the height and extent of the Himalayan chain. Beginning in 1800, the Survey, under the direction of a succession of military engineers, had embarked on a complete triangulation. Triangulation, which supplies measured distances between a series of intervisible points, allowing for the curvature of the earth, provides the grid from which accurate maps can subsequently be drawn. It was largely complete by 1830 but was subsequently extended and corrected, notably under the leadership of Sir George Everest, after whom the world's highest mountain is named. His team of surveyors and trigonometricians was never more than a few hundred strong but, largely inspired by the challenge of the enterprise itself, they succeeded within seventy years in producing a complete series of accurate maps of a sub-continent equivalent in size to that of the United States west of the Mississippi.[8]

By 1861, no triangulation of the United States had been undertaken. It was a strange blind spot in the American attitude to their magnificent country. George Washington was by training a surveyor; so was Abraham Lincoln. Thomas Jefferson, most intellectual of presidents, had a passionate interest in exploration and sponsored the Lewis and Clark transcontinental expedition to the north-west in 1804. He made it clear, however, that its purpose was to discover "the most direct and practicable water communication across this continent . . . for commerce." Route-finding, first for commerce, then for settlement, then railroads, defined American official interest in continental geography. In 1836 President Andrew Jackson sent a U.S. Exploring Expedition, under Lieutenant Charles Wilkes, to investigate the territory of the United States, but it was seaborne and largely committed to investigating the coasts. The earliest major exploration of the interior was authorised in furtherance of the Pacific Railroad Act of 1853, "to Ascertain the most Practical and Economical Route for a Railroad from the Mississippi River to the Pacific Ocean," which designated five possible lines, all to be surveyed by the army's Corps of Topographical Engineers. The routes were mapped, but no comprehensive and accurate survey of the United States resulted. That lay in the future.[9]

There were, of course, already many local maps of the United States, made necessary by the need to define sub-division of the land for farming under the pressure of immigration and westward expansion. Flatness, so characteristic of the American landscape in the Midwest east of the Mississippi and in the Great Plains beyond, allowed accurate delineation of property boundaries by reference to astronomical observations of latitude early on and longitude, by telegraphic time calculation, by the 1860s. Such mapping, however, was piecemeal. Without comprehensive triangulation, local maps did not accurately connect with one another, nor did they, in the hilly areas of the Appalachian chain and in the coastal regions to the east, usefully depict height, or contour. No wonder that, as late as 1864, Colonel Orlando Poe, General William Sherman's chief engineer, should complain that the maps of North Carolina he was able to find "vie with each other in inaccuracy."[10] Traditionally maps had been military secrets, those of one's own country to be kept from the enemy, those of his to be made with stealth; with reason, for mapmaking was rightly regarded as espionage. Frederick the Great in 1742 established a secret map room (*Plankammer*) in his palace at Potsdam, which contained maps both of Prussia and of surrounding territories, such as that of Silesia, which he had made before his invasion that caused the Seven Years War.[11] The Survey of India ran what was effectively a widespread espionage network in the countries bordering the Indian empire to the north, including Tibet, Nepal, Afghanistan, China and Russian Central Asia, staffed by Indians who were trained to measure distances by counting their steps on strings of prayer beads. Hurree Chunder Mookerjee, one of Kipling's most delightful creations in the cast of characters in *Kim,* was such an agent, but he had his models in real life. The most famous, as he became when allowed to emerge from obscurity, was Nain Singh, known as "the Pundit" or sage, who between 1864 and 1875 twice visited Lhasa, then a closed city, covered 1,200 miles of previously unsurveyed country and followed the course of the great Tsangpo River for 600 miles from its source. On retirement from the Survey of India, he was rewarded with a grant of land, the rank of Commander of the Order of the Indian Empire and the gold medal of the Royal Geographical Society, at which he lectured to rapt attention when he visited England.[12] He was perhaps lucky to survive to retiring age. Nepal, when threatened with invasion by the East India Company in 1814, carefully disguised the mouth of the main road leading into the country and the government threatened death for its betrayal.[13]

The South, threatened with invasion in 1861, could not disguise the mouth of its internal roads, since they connected with those of the North. Thereafter, however, their course was often poorly reproduced on such maps as were available to Northern generals, or inaccurately represented, or not marked at all. Local knowledge often counted far more than the plates in a shoddy bookshop atlas. It was much more readily available, inside the South, to Confederate defenders than Union invaders. Without it, confusions accumulated. Even quite good maps could be out of date, while there was no guarantee that the mapmaker's choice of place-name was that used by locals. "Cold Harbor, Virginia" (the site of one of General Ulysses Grant's battles in 1864) "was sometimes called Coal Harbor, and there was also a New Cold Harbor and a 'burned' Cold Harbor. Burned Coal Harbor was known by the locals as Old Cold Harbor. Many of the roads were known by one of two names: the Market or River Road; the Williamsburg or Seven Mile Road; the Quaker or Willis Church Road. To add to the confusion, there were sometimes other nearby roads with the same or similar names that ran in completely different directions."[14]

Locals knew; invaders did not. That was to confer an almost consistent advantage on the South, which, for most of the war, was campaigning within its own territory and defending it very often with locally raised troops. That was particularly the case in the Shenandoah Valley in 1862. Stonewall Jackson, the commander of the Valley Army, was a Valley man. After retiring from the regular army, he had become a professor at the Virginia Military Institute, the private military academy at Lexington at the Valley's southern end. Many in the Valley Army were Valley men, particularly those of the Stonewall Brigade, which had won its name at the First Battle of Bull Run, and of the Rockbridge Artillery, largely recruited from students at Washington College, also in Lexington. Perhaps the most important Valley man in the Valley Army, however, was a civilian, Jedediah Hotchkiss. A schoolmaster, he had set up his own school at Staunton in 1847. It flourished and, though a New Yorker, he stayed. He also began to pursue the hobby of mapmaking. General Robert E. Lee employed him as a mapmaker in his campaign in the Alleghenies, west of the Valley, in 1861. In 1862, home after illness, he attached himself to the Valley Army and was introduced to Jackson. The latter was impressed by his local knowledge and on 26 March added him, though he was a civilian and would remain so, to his staff. His first order to Hotchkiss was "I want you to make me a map of the Valley from Harper's Ferry to Lexington, showing all the points of defense and offense between those points."[15]

Hotchkiss set to work. He was untrained in cartography but methodical. He first surveyed the terrain from horseback, making sketches and notes as he moved around the terrain, then worked his observations up into a finished product. His 1862 map of the Valley still exists.[16] The course of the rivers is shown in pale blue, the road network in red, hills (uncontoured and without spot heights) in black, by hatching. There is no scale though, as the bottom of the sheet has been torn off, it may simply be missing. As a map, it reflects all the defects of those of the Civil War period: the appearance is messy, there is both too much and too little detail, and it has an unfinished, amateur look. Compared to the clear and elegant map of Yorktown, on the Virginia Peninsula, drawn by Thomas Jefferson Cram from an original by one of the French *ingénieurs géographes* of Rochambeau's army in 1781, it is a very inferior thing.[17] It is perhaps unsurprising that Jackson disliked the drawing classes at West Point more than any other subject. It seems probable that mapmaking was badly taught at the Academy, and if Federal military mapmaking, which in Europe and particularly Britain set the standard, was defective, it would follow that American mapmaking in general was unsatisfactory.

Still, Hotchkiss provided Jackson with a map based on local knowledge and derived from contemporary observation, and that put the general on a superior footing to his Union opponents. As late as 1864, during Jubal Early's resumption of the Confederate offensive in the Valley, the Northern general Philip Sheridan was found to be conducting operations against him from an inaccurate civilian map thirty years old. Hotchkiss' map told Jackson at least plain essentials: where the gaps in the mountains were, distances between inhabited places, compass orientations, crossing points over the waterways, the course of paved routes. It was better than nothing and would serve him well. Positively bad maps of the Valley would lead his Northern enemies into serious error.

The Valley campaign of 1862 opened at a moment of strategic equilibrium between Union and Confederacy, after one Union offensive had been checked in the west, but before McClellan's began on the coast, in the east. During 1861 the Confederacy had lost much territory west of the Appalachian chain, which marked the physical boundary between the two theatres of war. Most of the state of Missouri, largely Southern in sentiment, had been lost in August, despite a technical Confederate victory at Wilson's Creek. Kentucky, also pro-South, was held for the North by a well-timed advance organised by the junior but aggressive General U. S. Grant. He would be encouraged by his success to embark on an advance into Tennessee, bringing the capture of the strategic river forts of Henry and Donelson, but then leading to the costly pitched battle of Shiloh in April 1862. The Confederates' western front, consolidated by a new overall commander for the theatre, Albert Sidney Johnston (who died at Shiloh), would be held largely intact for the rest of the year.

In the east, where the Union had begun a campaign, progressively crippling to the South, to secure control of the Confederacy's coastline, little dry land had changed hands in 1861. Following the Confederate defensive victory at Bull Run, Joseph E. Johnston's Army of Northern Virginia had remained close to Washington, threatening the Federal capital. Its presence caused constant anxiety to President Lincoln, particularly because its size was consistently exaggerated by his new General-in-Chief, George McClellan. In March 1862 Joseph E. Johnston withdrew it south of the Rappahannock, one of the west–east waterlines that defended Richmond. That move somewhat relieved Lincoln's concern for the security of his capital; but it objectively complicated McClellan's plan to take the Confederacy's by his seaborne invasion, since it put the South's largest army closer to his ultimate objective.

On a large-scale map—paradoxically, in mapmaking, the larger the scale, the less the detail shown; one mile to one inch, small scale, is much more informative than ten miles to one inch, large-scale, though the latter is the more useful for strategic planning—the situation in March 1862 would have looked thus: Joseph E. Johnston, with 40,000 men in the Army of Northern Virginia, stood on the Rappahannock, forty miles north of Richmond; McClellan, with 155,000 men in the Army of the Potomac, was sailing it down that river to land at Fortress Monroe at the tip of the Virginian Peninsula, sixty miles from Richmond; various Northern detachments, under the command of Nathaniel Banks, amounting to some 20,000, protected Washington. In the Appalachian Mountains to the west, other Union generals deployed detachments of various strength. Implanted in the middle of the theatre, confronting but also threatened by the Union forces in the mountains and around Washington, Stonewall Jackson deployed fewer than 5,000 men to protect Joseph E. Johnston's flank, to hold the Federals in the mountains at bay and to deter Banks from bringing the Northern defenders of Washington down to assist McClellan in his seaborne advance on Richmond.[18]

In unbroken country—the flat, unforested, unwatered terrain of the Great Plains, say—Jackson's position would have been untenable. He would have been swept up during a few days of fighting in a concentric advance by Banks and the Northerners to the west. Jackson, however, was not in that vulnerable position. He had the mountains and rivers of the Shenandoah Valley on his side and, by employing the accidents of geography, natural and man-made, to his advantage, might overcome the odds confronting him. In the months of March, April, May and June 1862, he defied every probability in the most brilliant exercise in manoeuvre warfare, depending wholly upon superior use of intelligence, in the broadest sense, perhaps ever achieved.

The Valley Army (formally the Army of the Shenandoah Valley District) began its virtuoso campaign of diversion at the head of the Shenandoah Valley, where it had spent a hard winter near Romney, Jackson's boyhood home. His orders were to avoid pitched battle but to operate in such a manner as to prevent Banks, outside Washington, from reinforcing McClellan as he advanced on Richmond. As events unfolded, he was to fight several pitched battles but nevertheless achieve the spirit of his instructions.

Though tied to Washington, Banks was also under orders to clear the northern end of the Valley and in late February he crossed the Potomac River where it joins the Shenandoah at Harper's Ferry, then he advanced south. His purpose was to protect the two strategic lines of communication, the Chesapeake and Ohio Canal (connecting the sea to the Ohio River system beyond the Appalachians) and the Baltimore and Ohio Railroad (a principal rail route westward through the Appalachian Mountains), from Confederate interference. Jackson at first proposed attacking his advance guards at Winchester, where a railroad spur terminated at the Valley Turnpike, believing that he could inflict a defeat on the Union forces while they remained dispersed. The plan, however, defied Joseph E. Johnston's order to decline action; while Johnston was withdrawing his army from Manassas to protect Richmond, he was particularly anxious not to risk a defeat anywhere that would allow Banks to bring his army to reinforce McClellan's. Jackson's plan also frightened his subordinates, who were sure they would be beaten. After a heated debate in a council of war, his first, on the evening of 11 March, Jackson gave up the argument. As he rode away into the darkness, he burst out to Dr. McGuire, his chief medical officer, "That is the last council of war I will ever hold."

He was to be as good as his word; indeed, better. It is a military catch-phrase that "Councils of War never fight"—the phrase was to be President Theodore Roosevelt's but the idea is as old as antiquity—and, after the timidity shown by his brigadiers at Winchester, Jackson withdrew into himself.[19] Famously taciturn even in his cadet days at West Point and much more given to private prayer than conversation, he henceforth kept his thoughts to himself, revealing his intentions only at the last moment and then in peremptory, often cryptic orders. That was not a deliberate security measure, more a reflection of his introverted nature; but it had the highly desirable effect, in what was to be a campaign of repeated surprises, of shrouding the unexpected in silence.

Between 11 and 20 March, the Valley Army retreated southward down the Valley Turnpike, covered by the cavalry force under Turner Ashby. Ashby was a born cavalier, untrained in formal cavalry tactics but a horseman to his fingertips and a dasher and doer. At times during the campaign, his and his troopers' lack of discipline would infuriate the professional Jackson, but his relentless aggressiveness always restored him to his general's favour. Meanwhile, as the retreat lengthened, Jackson was pondering his strategy. "Mobility was the essential factor in the Valley Army's future."[20] The army could manoeuvre successfully in the face of a superior enemy, however, only if it made correct use of the Valley's geography, forced the enemy to make mistakes and denied Banks the use of essential links in the communication chain. It was a crucial factor in Jackson's calculations to know that his opponent was not a professional soldier, indeed not a soldier at all; a leading type of the Civil War "political" general, appointed for party reasons, Banks had been a Congressman, Speaker of the House of Representatives and most recently Governor of Massachusetts. Jackson's calculations essentially turned, nevertheless, on objective, not subjective factors: roads, bridges, rivers, hills. Now that Banks was inside the Valley, he had to keep him there, but without fighting battles he might lose. He also had to keep at a safe distance from the Union forces to the west, in the Allegheny Mountains. Finally, he had to keep open his line of withdrawal eastward towards Richmond, should Joseph E. Johnston send for him to assist in the defence of the city against McClellan's army in the Peninsula.

His first thought was of bridges: those to be denied to the enemy, those essential to his army's ability to manoeuvre. There were many in the Valley, most wooden and easily combustible, but some of critical importance. Two were railroad bridges, one over the South River at the southern end of the Valley, which Jackson needed if he were to escape by rail to Richmond, and one at Front Royal on the Manassas Gap Rail Road, a main line in the Northern supply chain. It had already been burnt by Jackson's headquarters guard, and he had sent the rolling stock beyond it south to prevent Banks from using the wagons in a subsequent advance.

Of the road bridges, the headquarters guard had also burnt the one at Front Royal, to impede Banks' advance down the Luray Valley, east of the Massanutten Mountain into which the North Fork flowed. The three bridges at Luray were essential to Jackson, however, were he to decide to slip across the central mountains through the Massanutten gap, and he also needed to preserve the spans at Port Republic and Conrad's Store, both crossing the South Fork or its tributary, which carried roads leading through the Blue Ridge gaps and so to Richmond. Finally, there was a wooden bridge at Rude's Hill, where the Valley Turnpike crossed the North Fork, which was perhaps the most important of all. If destroyed, with Banks to the north and Jackson to the south, its loss would stop a Northern advance dead at that point. Equally, its destruction behind Jackson's back would terminate his chance of opening a counter-offensive up the Valley west of the Massanutten.

A dispassionate observer, taking his stance in mid-March 1862 at Staunton, Jackson's main base at the extreme south of the Valley, would have assessed the situation thus: Banks, having failed to follow up Jackson's retreat from Winchester with energy, was stuck between that place and Strasburg but retained the option of moving down either the North or South Forks; the latter manoeuvre would require bridging at Front Royal but that was within his army's capability. Jackson, at Mount Jackson on the North Fork, had two choices: he could reverse his retreat and move up the Turnpike to find and fight Banks near Winchester; or he could cross through the Massanutten Gap to enter the Luray Valley and open a new offensive front.

The second choice, however, would take the Valley Army off the macadamised Turnpike onto dirt roads, limit its mobility and expose the main base at Staunton to Federal attack. Jackson therefore decided, even though he thereby kept himself further from contact with Johnston at Richmond and nearer to the remaining Federal forces in the Alleghenies, to retrace his steps and bring Banks to battle at Winchester. Moreover, he was encouraged to reverse his course by Johnston, who, retreating towards the Richmond river lines from Manassas, now expressed the anxiety that Jackson had got too far away from Banks. "Would not your presence with your troops nearer Winchester prevent the enemy from diminishing his force there? . . . I think it important to keep that army in the Valley, and that it should not reinforce McClellan. Do try and prevent it by getting and keeping as near as prudence will permit."[21]

He had implicitly not encouraged Jackson to seek battle, but Jackson was not prudent when he scented the chance of a successful fight. On receipt of Johnston's despatch, he immediately turned north again, marched through unseasonal snow on 22 March and, on the 23rd, found contact with Banks' advance guard at the village of Kernstown, five miles short of Winchester.

Ashby's cavalry opened the engagement, skirmishing forward during the morning with infantry in support. As the Union troops opposite began to form a line of battle, he fell back, to meet Jackson bringing up the main body. Ashby may have sent word to Jackson that he was opposed by only four regiments; alternatively, the intelligence may have come from local spies. In either case, Jackson was misinformed. The Federals were in much greater number, about 10,000 to Jackson's 4,000, and with plentiful artillery, which, quickly brought into action from well-chosen positions, began to cause casualties.

Despite his inferiority in strength—and despite the day being a Sunday, on which the pious Jackson always sought to avoid fighting—he decided in the early afternoon to attack. The Northerners were deployed on both sides of the Turnpike but in greater strength to the west, where ridges and hillocks gave commanding views. It was there that Jackson made his effort. To assist him in directing the battle he summoned an officer of the 2nd Virginia, Major Frank Jones, "who knew the country-side: he could look across the Pike and see his front porch."[22] Local knowledge would not on this occasion, however, get Jackson out of a spot. He was about to bite off more than he could chew. Worse, his temperamental taciturnity added to the difficulty of the situation. He issued an unclear command and then lost control of events by leaving his central position to gallop about, trying to restore order. His leading brigade lost direction, came under heavy artillery fire, took cover and then fell back. Jackson brought up guns of his own—of which he had nearly as many as the enemy—and infantry reinforcements, but after a final and bitter exchange of volleys at short range, his men were beaten; many had run out of ammunition. Jackson himself wrote a few days later, "I do not recollect of ever having heard such a roar of musketry"; but the Federal fire was the heavier and at about six o'clock in the evening the Valley Army began to slip away and retreat down the Turnpike.[23]

The Battle of Kernstown was a Confederate defeat. Southern losses were 455 killed and wounded, 263 taken prisoner; Union losses were 568 killed and wounded. Proportionately, the Valley Army had come off much the worse. On the other hand, the strategic effect was to its advantage. Even though the enemy had advanced when attacked, they formed only part of Banks' army; another division had already left to join McClellan at Richmond, and Banks had gone to Washington. McClellan himself ordered Banks, who returned from Washington posthaste on the Kernstown news, "Push Jackson hard and drive him well beyond Strasburg." He amplified his instructions on 1 April, emphasising that the Kernstown battle had forced a change of plan, requiring Banks to stay in the Valley instead of leaving it and, once the railroad was repaired, to advance to Staunton, Jackson's main base, at the bottom of the Valley, so as to force "the rebels to concentrate on you and then [you to] return to me."[24]

What he did not do was to offer Banks more troops. Lincoln's anxiety to protect Washington, the pull on his resources exerted by operations in and west of the Alleghenies, all combined to reduce his striking power against Richmond. Given McClellan's specific orders to Banks to advance down the Valley Turnpike, west of the dividing barrier of the Massanutten Mountain, he thereby spared Jackson the anxiety that he might have to defend the Luray Valley to the east of the Massanutten also. Indeed, once he became aware of the pattern of Northern deployment, Jackson recognised that the opportunity was opened to use the Luray as an avenue for a counteroffensive of his own. He was to take full advantage. Although he was to spend the rest of March and much of April falling back west of the Massanutten, he was already contemplating countermeasures which would take him up the corridor to the east, where he could reopen attacks towards Harper's Ferry and Manassas— and so heighten Lincoln and McClellan's anxieties.

Before he would be free to act in that way, however, there was to be much action at the south of the Valley. Jackson, following his retreat from Kernstown, had brought the Valley Army into defensive positions near Mount Jackson, on the North Fork of the Shenandoah, where he reorganised. Banks, following slowly, occupied Woodstock. The actual outpost line between the two armies, from 3 to 17 April, was along a minor stream called Stony Creek. The two sides skirmished across it during two weeks of inactivity, Jackson content to keep Banks in play, Banks hesitating to advance lest Jackson slip through the Massanutten Gap to Luray and strike at his line of communications higher up the Shenandoah. Eventually, however, Banks perceived—with a rare flash of inspiration—that if Jackson could make geography work his way, it could be made to work for him also. He saw that, given the very small distance involved, he might, by a brisk advance down the Turnpike, drive Jackson past New Market, the entrance to the Massanutten Gap, and harry him on south to Harrisonburg or even Staunton. At dawn on 17 April, Union infantry launched a surprise attack, cavalry following. The Confederate defences were driven in, and when Ashby's troopers tried to stop the Northern advance by burning the bridge at Rude's Hill, where the North Fork runs in an impassable trench, the Union cavalry were upon them quickly enough to put the blaze out. The Valley Army, outnumbered nearly two to one, had no option but to leg it south as quick as it could go. Two days of forced marching took it out of reach of the pursuit; but, following Kernstown, Jackson knew that he had suffered a local reverse.

Strategically, however, he was still in the ascendant. Joseph E. Johnston, increasingly hard pressed by McClellan near Richmond, had actually sent orders for him to be ready to leave the Valley; his new quarters, in Swift Run Gap, one of the key passes through the Blue Ridge, positioned him to do so. Jackson, however, became increasingly persuaded as April drew out that he could protect Richmond better by staying where he was and using Swift Run as a secure base—the high ground on two sides protected him against surprise attack—from which to strike at Union forces in the vicinity. He calculated that they numbered 160,000 altogether, spread out across eastern, northern and western Virginia, and that most were successfully pinning down their Confederate opponents: McClellan had Joseph E. Johnston fixed at Richmond, McDowell was facing Anderson on the Rappahannock at Fredericksburg, Frémont, in the Alleghenies, menaced the small force of Edward Johnson. Jackson alone had freedom to manoeuvre, for, though the Federals now appeared to dominate the southern Valley, he was confident that he could outwit them in a mobile campaign. The question was whether Banks presented the most profitable target.

What tipped the decision eventually was growing evidence that Frémont was emerging from the Alleghenies to strike at Edward Johnson's small and isolated force near Staunton, Jackson's main base, crammed with war supplies and with produce from the Valley farms. To go to Johnson's aid would require a march of fifty miles along bad roads and across the front of Banks' army, still stationed near Harrisonburg, on the Valley Turnpike, after its advance from victory at Kernstown the previous month. The risk was sustainable, however, for Banks lay behind the North River, the bridges over which had been burnt on Jackson's orders to cover his retreat to Swift Run Gap. Hotchkiss was therefore sent to locate Edward Johnson's exact position and to reconnoitre a route towards him. On 30 April the Valley Army set out.

It would have reassured Jackson had he known that Banks believed the Valley Army was already leaving Swift Run Gap to go to Richmond. His mind, however, was set on his course, so much so that when torrential rain—"great sluices of water running along the road for hundreds of yards"—blocked the route Hotchkiss had chosen, Jackson turned his column about, marched it back into Brown's Gap, gave his men a night's rest and then started them west again along a more southerly route. It had the advantage of running parallel to the Virginian Central Railroad, onto which Jackson loaded his sick and stores. Piecemeal by rail and road the Valley Army concentrated at Staunton on 6 May, left the next day to join forces with Edward Johnson, who was marching to meet it, and then pressed westward towards a tiny place called McDowell (also, confusingly, the name of the Union general commanding on the Rappahannock north of Richmond).

During the afternoon of 8 May skirmishers from the two sides found each other and a battle began to develop. Jackson had reconnoitred the heavily broken ground and formed a plan to fall unawares on the Northern force, a detachment of Frémont's army commanded by General R. H. Milroy. Milroy, however, had got wind of his approach and, though outnumbered, moved to the attack. In the confused fighting that followed, his men inflicted the heavier toll of casualties. Jackson reported to Lee "God blessed our arms with victory," and in the sense that Milroy broke off the action, and retreated, the Confederates were the winners.[25] It was a costly victory, nonetheless, and Jackson later reproached himself for bad management of the battle. It was the last mistake he would make in the Valley campaign.

Its pace was about to quicken. Lee, in Richmond, was increasingly concerned to keep the Union forces surrounding the Southern capital separated; so was Joseph E. Johnston, and both counted on Jackson to operate in a way that would pin Banks west of the Blue Ridge and keep Frémont in the Alleghenies. After the battle at McDowell, therefore, Jackson decided that he must pursue Milroy, meanwhile taking steps to impede Frémont's ability to manoeuvre. He sent Hotchkiss, with a scratch force of cavalry, to block the routes from the Alleghenies into the southern Shenandoah, while himself following up Milroy's retreat. By 12 May he had got as far as the small town of Franklin, deep in the mountains, but had not caught up. He decided accordingly to break off the pursuit and return to the Valley. His purpose as before was to keep Banks from leaving, but he also intended to rejoin his subordinate, Ewell, and combine forces so as to confront the enemy in superior strength.

The Valley Army was now adapting to the extraordinary exertions Jackson expected of it. On 8 May, the day of the Battle of McDowell, the Stonewall Brigade had marched, from breaking camp to contact with the enemy, and then from leaving the battlefield to regaining camp, thirty-five miles. Such marches would, in the month that was to follow, become normal practice. Despite dreadful roads, shortage of food and deficient footwear—marching barefoot, often for dozens of miles, became a common experience—the Valley Army would rise to the challenge. Though Jackson concealed his intentions from even his closest subordinates, the Army came to understand during the month of May 1862 that his strategy was to mystify and mislead the enemy by achieving speeds over distance quite outside the capacities of normal infantry. They came to call themselves "Jackson's foot cavalry" and, on many days, justified the title by marching for as long as horsemen could ride.

On 17 May, after a hard trek out of the Alleghenies, Jackson's men re-entered the Valley near Harrisonburg, west of the Massanutten. Banks had been there the previous month, his army facing southwards along the North River, but had since departed to Strasburg at the northern end of the Valley, in preparation to move to Fredericksburg. He had already sent ahead Shields' division. It remained, as before, Jackson's duty to hold him where he was. In his favour was a shift in the balance of forces; the departure of Shields had left Banks with only 12,000 men; Jackson now had, either directly under command or readily to hand, about 16,000 if the division of Ewell, in the Luray Valley, was included. Also in his favour was the deteriorating quality of Northern intelligence—Banks was unsure of the Valley Army's dispositions, and his information would get worse. By 21 May he was placing Jackson eight miles west of Harrisonburg, Ewell in the Swift Run Gap, forty miles apart, with the gap widening. In fact, by then, Jackson had transferred to the Luray Valley, via the Massanutten Gap, Ewell had joined him and the combined army was pressing northwards against a weak detachment of Union troops at Front Royal, guarding the Manassas Gap railroad bridges east of Strasburg.

The realignment had not been achieved without difficulty, even creative disobedience. Mid-May was an awful time for the Confederacy. During March and April, defeat had followed defeat all around its frontiers, in the far west, on the Atlantic coast. By early May the defensive line across the Peninsula had been abandoned, the Battle of Williamsburg outside Richmond had been lost and McClellan was laying siege to the defences of the city itself. Between 15 and 18 May, Robert E. Lee and Joseph E. Johnston, both a hundred miles away from the Valley in Richmond, and in touch at a delay of only two to three days, despite having the telegraph and a relay of fast despatch riders at their disposal, had sent a variety of conflicting orders, the impact of which was, nevertheless, to separate Ewell from Jackson and send him to watch McDowell at Fredericksburg. Neither Jackson nor Ewell wished to conform, since to do so would be to rob the Valley Army of its temporarily decisive superiority over Banks, without any guarantee that success could be won elsewhere by the separation. Covertly, they agreed to play on the ambiguity of the orders they were receiving and to use the delay in their transmission to stay together and march on Banks.

Jackson moved on 19 May. His bridge-burning at Harrisonburg, which had protected his sortie into the Alleghenies, now ought to have blocked his own recrossing of the North River into the Shenandoah Valley proper but Hotchkiss, effectively operating as his intelligence officer, discovered a number of large wagons that, positioned to straddle a ford, allowed passage even though the river was in flood. By 20 May, Jackson had reached New Market at the western end of the Massanutten Gap, by the 21st he had passed through the mountain to join Ewell at Luray and by the 23rd his vanguards were on the outskirts of Front Royal. By a forced march of seventy miles in three days, he had arrived in Banks' rear and was ready to strike a decisive blow.

He then had a stroke of pure luck, though brought by the circumstance of fighting in friendly territory. Advancing to contact, but unaware of the strength of the Union defence at the Front Royal bridges, one of his officers was met by a breathless girl, Belle Boyd, a pretty eighteen-year-old who had just walked through the enemy camp, charmed an officer and discovered that only one Northern regiment was present. "Tell him [Stonewall]," she urged, "to charge right down and he will get them all."[26] In the confused fighting that followed, most of the Northern infantry got away but the Confederate cavalry saved the bridges, which Jackson needed for the next stage of the operation against Banks at Strasburg, and destroyed the telegraph lines which would have warned Banks of the defeat.

On the evening of 23 May, Jackson pondered the situation that his rather ragged victory at Front Royal had won. He correctly concluded that Banks would feel exposed to a further Confederate attack in his position at Strasburg, where his numbers could be calculated to have fallen to about 10,000. He might fall back on Frémont, in the Alleghenies, but that was unlikely, since one of his duties was undoubtedly to protect Washington, which lay in the opposite direction. He might, improbably, go over to the offensive and attempt to recapture Front Royal, perhaps calculating that Jackson would set out northwards towards Harper's Ferry, assuming that the Northerners were beating a retreat in that direction also; or he might simply do the obvious thing and retreat anyhow.

Eventually Jackson decided, correctly as it turned out, that Banks would go back towards Harper's Ferry. He therefore ordered his army to follow the presumed line of Banks' retreat, up the Valley Turnpike towards Winchester, by a converging route along the less good road leading through Cedarville and Ninevah. The distance each had to cover was about twenty miles, but while Banks was encumbered by a large wagon train, crammed with stores the Confederates coveted, Jackson was able to cover the countryside with a cloud of reconnoitring cavalry. On the morning of 24 May, the Confederate cavalry found Banks' wagons, almost unprotected by fighting troops, jammed nose to tail in the Valley Turnpike at a point where it ran between stone walls. The Southerners brought guns up to fire into the mass, and Ashby's cavalry charged in. Though the Federals set fire to as many of their wagons as they could, the Confederates captured a rich prize. Meanwhile, Jackson pressed the pursuit. On the evening of the same day, his army was arrayed outside Winchester, tired and footsore but prepared to give battle.

The appearance of the Valley Army outside Winchester, only twenty-five miles from Harper's Ferry, only seventy from Washington, caused acute alarm in the Federal capital. It seemed to threaten a direct attack at worst, at least the need to dilute the campaign against Richmond. Lincoln, like Jackson, was studying a map—a less good map than Hotchkiss'—of the theatre.[27] Between four and five on the afternoon of 24 May, he ordered Frémont to abandon his plan to move west out of the Alleghenies against the rail centre of Knoxville, in Tennessee, and to march east to the relief of Banks. He also ordered McDowell, who was preparing to join McClellan in the Peninsula, to send half his army to the assistance of Banks as well. "At that moment, 5 p.m., May 24, the Valley Army won its Valley campaign."[28]

Jackson still, however, had much fighting to do, both up and down the Valley. On the morning of 25 May, in thick early mist, his advanced guard found Banks' men positioned outside Winchester on hills that protected the town from the south. Jackson's local intelligence had for once failed him. He thought the Union forces were behind, not ahead, of him, and he was expecting to cut them off from Harper's Ferry. In the confusion that followed the initial encounter, brought on not by Jackson's own brigades but by those of Ewell, marching to meet them, the Union troops at first inflicted heavy losses. Their batteries were well positioned on high ground. As the Confederate concentration grew, however, the Northerners found themselves outflanked to both left and right, their batteries brought under direct rifle fire and their infantry forced to fall back. Soon Banks' men were in full retreat. They tried to make a stand in the streets of Winchester itself, but the townspeople, producing hidden weapons and shouting information to the advancing Southerners—many in the 5th Virginia came from the town in any case—undermined their resistance. By noon Banks' army was streaming up the Valley Turnpike towards Harper's Ferry with Jackson's infantry—his "foot cavalry"—hot on their heels.

Had Jackson had his full force of horsemen under his hand at that moment, the destruction of his enemy might have been complete. Ashby, his cavalry leader, was elsewhere at the critical moment, on some cavalier venture of his own, the besetting fault of Southern riders. Banks, as a result, got clean away, managing to keep just ahead of Jackson's vanguards until he reached Harper's Ferry, where he crossed the Potomac on the night of the 25th, leaving the Valley in Confederate hands.

For how long? Lincoln, acutely alert to the dangers of the changed situation, and accurately reading it, was determined to prevent Jackson from disrupting the Union convergence on Richmond. He accepted General McDowell's analysis: "Jackson will paralyse a large force with a very small one." By correct disposition of his own forces, however, he hoped to crack the paralysis and re-establish the dominance that the North's superiority in numbers ought to confer. Jackson's advance to Harper's Ferry appeared, on the map, to represent a threat to Washington. It could also be seen as an entrapment in a potential envelopment, and from three sides: by Frémont, advancing out of the Alleghenies to the west, by McDowell from the east and by Banks, if he recrossed the Potomac, from the north. The president sent the necessary orders to McDowell and to Banks on 29 May. To McDowell he wrote, "General Frémont's force should, and probably will, be at or near Strasburg [on the upper North Fork of the Shenandoah] by 12 noon to-morrow. Try to have your force or the advance of it at Front Royal [on the two forks] as soon." Lincoln, in short, was arranging a pincers behind Jackson's back, which would cut him off from the Valley, and from Johnston's army at Richmond, and expose him to defeat in isolation.

Jackson was not to be caught. Acutely sensitive to danger in any case, he was alerted to its correct reality by a succession of reports—that of a loyal Southerner, who had ridden from the Blue Ridge Mountains with word of a move by Frémont, then the transcript of the interrogation of a Northern prisoner who said Shields was marching on Front Royal, finally news of actual contact with Federals near Front Royal. By noon on 30 May, Jackson could no longer ignore the signs that his advanced position just short of Harper's Ferry was overexposed and that prudence required a retreat into the Valley proper.

What followed might have been a rout. The Valley Army, rich with plunder, was encumbered with hundreds of wagons, some its own, some civilian, some captured from the enemy. They occupied eight miles of road. Military caution dictated that they should be abandoned, so that Jackson's men could disengage as quickly as possible. Their commander was set on keeping his plunder, however, and counted on his soldiers' ability to outmarch their pursuers to avoid entrapment. They also retained the capacity to deploy rapidly into battle formation off the line of march. On 1 June, as Frémont staged a thrust to cut the road, Jackson reversed the march of one of his brigades to drive the Union sally back. Some of his troops had marched as much as thirty-five miles in sixteen hours, snatching sleep on wet ground in wet blankets at intervals, but, with skilfully organised bursts of artillery, they succeeded in holding the enemy at bay. Jackson was in frequent conclave with Hotchkiss, who was reconnoitring energetically and measuring off relative distances on the map. He calculated, and so persuaded his commander, that the Valley Army could by quick marching just keep out of danger. On the afternoon of 1 June the Army was beyond Strasburg and still heading south, leaving the vanguards of the armies of Frémont and Banks closing hands on empty space.

During the next two weeks, Jackson would escape from a real or suspected trap several more times. As he headed south from Strasburg, just out of the enemy's reach, his acute sense of danger alerted him to the makings of another. With Frémont, as he believed, hard on his heels and Shields advancing down the westward side of the Massanutten Mountain, he foresaw the two encircling him lower down. That was to overestimate Shields' rate of advance; but, with the barrier of the Massanutten between them, his anxiety was understandable. His solution was to hurry a cavalry force ahead, with orders to pass through the Luray Gap and burn the surviving bridges across the Shenandoah at Luray, thus blocking Shields' way southward.

Jackson's own way south, towards New Market, was impeded by the constant harrying of Union cavalry and by appalling weather, which turned the surface of even the macadamised Valley Turnpike to glue. Men linked arms to keep their footing in the great press of traffic, swollen by the convoy of wagons which Jackson refused to abandon and by the complement of Union prisoners who, sensing how close their own side followed, dragged their feet and had to be bullied onward. The bridge at Rude's Hill, which Ashby had failed to destroy in April, was burnt in the face of the enemy on 3 June.[29] The Valley Army was now running out of room to manoeuvre. Robert E. Lee, who had succeeded the wounded Joseph E. Johnston in command around Richmond, actually contemplated stripping his forces of troops to strengthen Jackson, with a view to his leading an invasion of the Northern states of Maryland and Pennsylvania, as would happen a year later. In the circumstances of 1862, however, such a démarche was impossible. Jackson, like it or not, was still bound to the retreat. His problem was to find the means to continue drawing Frémont and Shields after him, without becoming entangled in a costly battle, and then to disengage on favourable terms.

The trouble was that the earlier cunning of his bridge-burning was now telling against him. He needed routes to fall back on his main base at Staunton and towards the Virginia Central Railroad, which led from it towards Richmond: one was his point of resupply, but also the spot where he could disembarrass himself of his hundreds of wagons, liberating the army for a counterattack if necessary; the other was his line of escape. A key point was Conrad's Store, to which a road ran from the Valley Turnpike, and a way through the Swift Run Gap (in the Blue Ridge) to the railroad. The necessary bridge had been burnt, however, and Jackson's engineers advised him that the Shenandoah, swollen by the exceptional rains, could not be bridged with any safety.

The only way out, therefore, was that followed by the Valley Army the previous month, after its foray into the Alleghenies: a bad track leading to the village of Port Republic, short of Conrad's Store, from which there was a way via Brown's Gap to the railroad at Mechum River Station. It was risky. Shields, moving south from Luray, might catch the Valley Army in column of route and, in its exhausted state, defeat it. Hotchkiss was, however, acting energetically as Jackson's eyes. From a lookout position at the southern tip of Massanutten Mountain he observed Shields encamp his army on the afternoon of 5 June near Conrad's Store. Jackson, having done the distance a month earlier, reckoned that the Union force could not outpace him to Port Republic, now his touchstone of safety.

Not until 7 June, however, did Jackson bring his headquarters into Port Republic, after two days of desperate fighting which had left several regiments shattered and Ashby dead on the field. Union forces were pressing harder than he had anticipated. Pressure was shortly to bring on a battle that threatened to cut off his line of retreat. The Port Republic position was complex. Compressed between the southern tip of Massanutten Mountain and the Blue Ridge, it was also the junction of several key roads and the site of the confluence of three rivers, the South Fork of the Shenandoah, the North River and its tributary, the South River. A surviving bridge at the top of Main Street crossed at the junction of the North River and South Fork, while the South River was fordable at two points, Upper and Lower Fords. Jackson needed to dominate the whole scene of action in order to outface the enemy—Frémont, advancing from the northwest, Shields, advancing from the northeast—and still to preserve his options of retreat southwestwards.

His advantage in intelligence was played out. He was at close quarters with the enemy, who could read the situation map as well as he could. They dominated two sides of the battlefield, including the northern high ground. Unless he could fight them off, they might envelop him to left and right, cut off his line of escape and achieve the Union victory he had staved off for the last four months.

The exit strategy was to fight, and between 8 and 12 June, the Valley Army fought with terrible ferocity. The action of 8 June, conducted in the streets of Port Republic itself, was brought on by an uncharacteristic failure of attention on Jackson's part. Tired himself, from days of marching, and with a tired army, he allowed the need for rest to overcome watchfulness. On what was hoped to be a peaceful Sunday morning, Federal cavalry got into Port Republic and surprised the sleepy Confederates. They were chased out with only difficulty and loss. Meanwhile, Frémont's army, advancing above the South River in strong but unsynchronised support, was defeated at the village of Cross Keys.

Jackson might now have used the time he had won to break contact and retreat in haste to Brown's Gap, the railroad and Richmond. That was what prudence dictated. Instead he decided to engage the enemy again, in the hope of inflicting a conclusive victory but at the risk of falling into a final trap.

The trap almost closed. The jaws were kept apart only by the harshest of fighting on two fronts, at right angles to each other. On the high ground north of the Shenandoah's South Fork, Frémont was held at bay, while on the low ground between the river and the Blue Ridge the bulk of Jackson's Valley Army, recovering from initial disorganisation, eventually formed a strong point and drove Shields back. In several hours of fighting in the early morning of 9 June, the heaviest of the whole Valley campaign, the Confederates eventually drove both Frémont's and Shields' men from the field, at a cost of over 800 killed and wounded to each side. A Confederate survivor later recorded, "I have never seen so many dead and wounded in the same limited space."[30]

The Valley Army was nevertheless the unquestionable victor of these culminating battles, so much so that Frémont and Shields did not merely leave it in possession of the battlefield, traditionally the mark of defeat accepted; each peremptorily withdrew northwards into the Valley, to positions from which neither could resume an offensive. After their defeat at Port Republic "[they] were terrified of the Valley Army. On June 19, with the rebels seventy miles away, Banks fretted that they were upon him."[31] In the face of such moral feebleness, Jackson actually re-entered the Valley, from his positions beyond the South Fork, and rested and refitted while the enemy retreated.

There was even in mid-June a revival of the suggestion that he should march northwards into Pennsylvania, opening an invasion of the North. Lee, now in overall command of the Army of Northern Virginia, organised reinforcements to send him and took no trouble to disguise the disposition from the enemy. It was, in truth, no more than a feint. The Confederacy, at a high crisis of its existence, was not really ready to take war to the enemy. It needed to assure the security of its own capital, not to menace that of its enemy. On 13 June Lee noted on a letter from Jackson, "I think the sooner Jackson can move this way [to Richmond], the better. The first object now is to defeat McClellan. The enemy in the Valley seem at a pause. We may strike them here before they are ready to move up the Valley. They are naturally cautious and we must be secret and quick."[32]

Lee, in short, had decided that he needed Jackson on the Virginia

Peninsula, where the bulk of the Union Army was now deployed. On 16 June, therefore, he sent orders for Jackson to bring the Valley Army to the vicinity of the Confederate capital. Jackson departed on 18 June, riding ahead to meet Lee near Richmond on the afternoon of the 23rd. His army, following behind, crossed the Blue Ridge Mountains and arrived at Richmond shortly afterwards, to take part in the Seven Days Battles, which brought McClellan's Peninsula Campaign to defeat. Richmond was saved and so, for a time, was the Confederacy.

In the second half of 1862 and during 1863 the Confederacy, under Lee's generalship, went over to the offensive, which culminated in the Union's flawed victory at Gettysburg. The progressive destruction of the South's defences followed. What part had Jackson and his intelligence operations played in the postponement of the South's defeat?

Jackson, first of all, had worked on the uncertainties and anxieties of his opponents, at every level from the field commanders to President Lincoln in Washington. He had threatened Frémont and Shields with defeat in open battle in the theatre of the Valley. He had also menaced Lincoln with the danger of an advance across the Potomac to Washington itself. Secondly, within the Valley theatre, he had confronted each of the Northern armies in turn, drawing them deeper into the Valley at the outset of the campaign, when the need to distract Union strength from the Peninsula was paramount, later risking battle when it could not be avoided but almost always on his own terms. He had tried his own army hard—his "foot cavalry" achieved, as its endurance increased, almost unparalleled feats of marching, at times covering as much as seventy miles in a hundred hours—but those who could stick the pace remained able to fight even at the end. Losses from disease and exhaustion ran as high as 30 per cent but battle casualties were surprisingly low, only about 2,000 in forty days of fighting. When it left the Valley, the army, which had admittedly been reinforced, was actually larger than when it had begun the campaign.

Jackson's success was due in large measure to his ability—reinforced by his natural taciturnity and secretiveness—to think faster and more clearly than his opponents and to calculate more moves ahead, making good choices, rejecting bad. That ability rested, however, on his possession of superior knowledge of the Valley's geography and of superior local intelligence, constantly refreshed by the work of a busy intelligence chief, Jedediah Hotchkiss, and a friendly population. The best generals have always valued detailed knowledge of topography, almost above any other sort of intelligence. Jackson was a better general than any of his opponents, and his operations in the Valley, assisted by McClellan's refusal to profit by any of the advantages the North's material superiority gave him, assured the successful defence of Richmond and the Union setbacks of 1862–63 which flowed from McClellan's retreat. The proof of his generalship was demonstrated above all, however, by his exploitation of the secrets of place and passageway in the complexity of the Shenandoah Valley, which he possessed and the enemy did not. He deserved his triumph.

CHAPTER FOUR

Wireless Intelligence

THE USEFULNESS of intelligence had been limited since the beginning of warmaking by the carry of voice, range of vision and speed of message-carriers. No amount of ingenuity could eliminate the delay. The Nelsonian navy had devoted much ingenuity to minimising it. By 1796 a system of fifteen inter-visible semaphore stations had been built between the Admiralty in London and the port of Deal, through which a message could be sent and acknowledged within two minutes; by 1806 the chain had reached Plymouth, transmission and acknowledgement, along a distance of 200 miles, taking three minutes. The system, which depended on each station being able to see the semaphore arms moving on the one before, worked only in daylight, however, failed in fog and shut down earlier in winter than summer.[1]

Mechanical semaphoring, moreover, like flag signalling, could send only such messages as were prearranged by book code; but by the 1840s the American Samuel Morse had devised his code, technically a cipher, which allowed messages to reproduce speech, by allotting each letter of the alphabet a separate identity in the form of short or long symbols, single or combined, soon to be known as "dot" and "dash." When translated into electrical pulses and transmitted along a metallic cable, a method of communication pioneered by the Englishman Charles Wheatstone in 1838, delay in signalling was eliminated. The first successful electric telegraphic Morse code messages were passed in 1844, between Washington and Baltimore.

The electric telegraph rendered mechanical systems almost instantly redundant. The Admiralty sold off its stations, which are remembered today only in such place-names as "Telegraph Hill" or "Semaphore House" in England's southern counties. Yet the electric telegraph did not immediately establish instantaneous communication between commanders and subordinates. For one thing, deficiencies in power supply required frequent retransmission in the early telegraphic links; during the Crimean War, though London was connected to Balaclava by an early undersea cable, messages still took as long as twenty-four hours to arrive at expeditionary headquarters, because of the need to relay at several stages.[2] Secondly, local commanders found interference from above tiresome, and sometimes chose to ignore instructions. Thirdly, the system was inflexible, as it would remain until the invention of a method of electronic communication not dependent on cable, literally "wireless." Commanders who were moving about in the theatre of operations, or otherwise absent from cable head, could not be contacted, nor could they signal in return.

Cable communication, had it then existed, might have benefited Nelson during the Nile campaign, but only when he touched land and only if the cable head had been in the hands of friendly operators. As he was usually at sea, in pursuit of an enemy who chose not to be seen, it is difficult to identify the moments when he might have been helped. Had there been a link between Alexandria and southern Turkey, and had it operated in his favour, he could have saved himself his second transit between Egypt and Sicily and caught the French perhaps a month earlier; but that is a big if. The movement of the French fleet was a model exercise in evasion, likely to have succeeded in any age before the development of wireless, aerial surveillance or radar.

Jackson, operating in the telegraph age, did not much benefit by it. Nor did his enemies. Neither side was helped by the telegraph links which reached into the Shenandoah Valley, at least in the theatre of operations. The Confederates destroyed them in retreat, as they did the railroad track, while their own link to Richmond seems to have had gaps; certainly in the confusing days of order and counterorder, in early June, communication between Jackson and Lee seems to have been carried on largely by letter, at several days' delay.[3] Not until Grant, a master of telegraphic method, took command of the Union armies in 1864 would the telegram come to dominate in the management of armies; and, even then, it remained a strategic, not tactical instrument, again because of the rigidity of the network and the infeasibility of pushing the cable head forward into the maelstrom of battle.[4]

The invention of wireless, requiring only transmitter, power source and receiver to work, and using the atmosphere itself as the carrying medium, in theory ushered in an era of free communication; but at first in theory only. The concept of wireless was first proposed by a British physicist, James Clark Maxwell, in the 1860s; he predicted that electromagnetic waves could be propagated in space and would travel at the speed of light. In 1888 the German scientist Heinrich Hertz published the result of experiments which actually demonstrated electromagnetic wave propagation—and reception—though only over distances of a few feet. By the 1890s, however, following the suggestion by Sir William Crooke that electromagnetic waves could be used for communication, several practical men were building and using "wireless sets." One was a Royal Navy officer, Captain H. B. Jackson; another was an Italian, Guglielmo Marconi.

Jackson's experimental career was blighted by postings to conventional appointments; it may be, in any case, that he was not a true visionary. Marconi was, though, of the entrepreneurial rather than the pure scientific sort. Very early he grasped that the applicability of wireless lay at sea. He was working in the era of the great European naval race, when the adoption of Germany's Naval Law of 1900 had launched it into direct and deadly competition in battleship building with Great Britain. It was also an era of great expansion in maritime trade; the number of steamships plying the ocean had recently exceeded that of sailing ships for the first time. The great majority, nearly 9,000, sailed under the British flag, a principal reason, beside that of preserving control of the seaways within its enormous and worldwide empire, for Britain's determination to maintain its naval supremacy.

Marconi saw that wireless offered the promise of a great improvement in signalling at sea, both between ships and from ship to shore. Ship owners would certainly pay to equip their merchantmen with wireless, if only at first on the larger vessels and passenger liners; but the real scope for commercial exploitation lay with the war fleets. No navy could afford to be left behind in what would as certainly become a race as that over speed, armoured protection and calibre of guns. First, however, wireless actually had to work. Despite relentless experimentation in the mid-1890s, Marconi could not achieve reception at ranges above ten miles. The waves propagated, moreover, could not be tuned, and so blanketed the whole radio spectrum. The receiver could not distinguish between propagated waves and "atmospherics," and if two stations transmitted at once the result was a jumble of unintelligible marks on the tape of the telegraph inker, which was as yet the only mechanism on which signals could be represented.[5]

Between 1897 and 1899, however, Marconi so much improved his apparatus that by 1900 the British Admiralty had decided to adopt wireless as a principal means of communication, had accepted Marconi's hardheaded commercial terms and had purchased fifty sets, forty-two to be installed in ships and eight in shore stations along the south coast, from Dover to the Scilly Islands. The range achieved exceeded fifty miles, and intelligible messages were passed at ten words a minute. Improvement then accelerated at an astonishing pace. In December 1901 Marconi succeeded in signalling between Cornwall in England and Cape Cod in the United States and in January 1902 from Cornwall to the Cunard liner *Philadelphia* over a distance of 1,550 miles.

Wireless worked better at sea than on land for a number of reasons. Marconi realised that, by using high electrical power—readily available in a large ship, with big engines and generators—at low frequency through very large aerials, he could take advantage of the "ground wave" which follows the curvature of the earth; reception was also facilitated by the reflection of waves between the earth's surface and layers in the atmosphere. In the years after his great transatlantic transmission success, he also made other important discoveries, to do particularly with "tuning" transmission in to separate bands of the radio spectrum, which reduced interference, improved reception and allowed stations, whether at sea or on land, to be allotted different wave frequencies. He thereby met naval requirements, particularly British Admiralty requirements, which were for communicating simultaneously with large numbers of ships spread over a worldwide oceanic area.

That also required the building of a worldwide chain of shore radio stations, for even the best of Marconi's wireless sets as yet lacked the power to broadcast globally. Thitherto intercontinental communication had been secured by cables, of which by the end of the nineteenth century Britain owned over 60 per cent; because non-British cables often passed through stations worked by British operators, Britain's dominance of the cable network was even more complete than company prospectuses revealed. Britain had been absolutely ruthless about establishing control of the cable kingdom from the start of the cable era. It had laid the first transatlantic cable in 1858, and though that had been a commercial undertaking, thereafter the government either laid cables itself or subsidised private companies to do so. By 1870 Britain was linked to India by a cable that ran through Lisbon (capital of Britain's oldest and most reliable ally) to Gibraltar (British since 1713), Malta (British since 1800) and Alexandria (effectively British after 1882).

As the British empire expanded, cables followed the flag, down both coasts of Africa, to Australia and New Zealand and across the Pacific, via such remote and tiny places as Fanning Island, 3,450 miles from its Canadian connection at Vancouver and at almost the exact centre of that ocean. The drive to create an "all-red" network (red was the colour by which British colonies were distinguished on the map) became a strategic obsession. To achieve the all-red effect, cables were relaid, or duplicated by sea, until no part of the empire could not be signalled along cables secure from foreign interference. No matter that 140,000 miles of cable, the greatest length by far, belonged to private companies relaying commercial messages. The government exercised the right to priority at any time by prefacing its own signals with "clear the line, clear the line." Moreover, it took the view that the network, private or official, was a national asset. As the French government observed in November 1900, "England owes her influence in the world perhaps more to her cable communications than to her navy. She controls the news and makes it serve her policy and commerce in a marvellous manner."[6] The City of London, in particular, was "the hub of the world's telegraphic and telephone systems," with the result that "most of the industrialised world financed its trade through sterling-denominated bills drawn on London"; not only the industrialised world but also the primary producers, including Egypt (cotton), Argentina (beef), Australia (wool) and Canada (wheat). All that was achieved with a remarkably small reserve of gold, the guarantee of paper transaction, only £24 million in the Bank of England, compared with £95 million of gold and silver in the Bank of France, £40 million in Germany and £142 million in the United States in 1891.[7] Strategic signals may have had priority; but the government was well aware that the Royal Navy ultimately existed to guard the nation's commerce.

For most of the nineteenth century Britain's vast and ever-expanding merchant fleet sailed the seas without thought of foreign interference: as late as 1880, the Royal Navy was equal in size to the next seven navies combined. In the last decade of the century, other navies, the French, the American, the German, the Russian, the Japanese, began to challenge in supremacy, either in particular seas or worldwide. After 1900, when Germany passed the Naval Law inaugurating the creation of a High Seas Fleet, the principal challenge was laid by the Kaiser's navy. Until then it had possessed little better than a coast-defence force. After 1900 it began to lay down battleships which were to be a match for those of the Royal Navy; it had already started to build cruisers, ships designed to scout for the battleships but also to menace an enemy's merchantmen. The architect of German naval expansion was Admiral von Tirpitz, whose mind was fixed on battleship building. His navalist philosophy has been characterised as "risk theory," the posing of a threat to the Royal Navy by a battleship force which, though not big enough to defeat it in home waters, would limit its power by menacing it with the danger of crippling damage in a surprise engagement.

Kaiser Wilhelm II, half English though he was and an honorary admiral in the Royal Navy, espoused Tirpitz's risk theory wholeheartedly. Love-hate of Britain's world pre-eminence informed all his strategic thinking. He longed for parity with, even superiority over, the British battle fleet, while recognising that the costs of his vast army precluded battleship building on a British scale. He also wished to create an overseas empire on the British model. That required the creation of a fleet of cruisers for foreign service. Tirpitz, initially complaisant about cruiser building, became hostile to the project after the passing of the Naval Law in 1900. By then, however, Germany's imperial project was in full swing. It followed the new empire's commercial development. German foreign trading companies had established bridgeheads in east, west and southwest Africa, in New Guinea and in the Pacific islands during the late nineteenth century. By 1914 the German flag flew over German East Africa (today mainland Tanzania), Togo and the Cameroons in west Africa, northern Papua New Guinea and the Pacific Island groups of the Bismarcks, the Marshalls, the Carolines, the Marianas and Samoa. Some of the territories had been annexed outright; others such as the Carolines, remote atolls without apparent value, had been peremptorily purchased from Spain during the era of its imperial collapse following defeat by the United States in 1898. Germany had also joined with other major European powers in annexing coastal enclaves from China after the Sino-Japanese War of 1894. Its prize was the harbour of Tsingtao, on the Yellow Sea, which it rapidly transformed into a well-equipped port for its Far Eastern cruiser fleet.

Germany's overseas possessions were linked to home by cable, in the same way as were Britain's. Acutely aware, however, of its cable network's vulnerability, Germany also early invested in the new technology of wireless and by 1914 had "the most advanced network" in the world.[8] Telefunken, the leading German radio company, was a pioneer of "continuous wave" transmission, which permitted the use of numerous separate channels and so "vastly increased the amount of traffic that could be put 'on the air.'"[9] Telefunken also worked hard to increase the range over which Germany's government stations could broadcast. By 1914 the main transmitter at Nauen, outside Berlin, could reach Kamina in Togoland, 3,000 miles away, while Kamina could communicate with Windhoek in German Southwest Africa, Dar-es-Salaam in German East Africa and Duala in the Cameroons. There were other wireless stations at Yap and Augaur in the Carolines, Nauru in the Marshalls, Apia in Samoa, Rabaul in the Bismarcks and at Tsingtao. The more distant stations, however, were out of range of Berlin and Togoland and could relay only messages sent by cable, of which that between Monrovia in Liberia and Pernambuco in Brazil alone was both German-owned and in what, in 1914, would prove to be neutral territory.

By 1914, however, all major German warships, like those in the British, American, French, Italian and Russian fleets, were also equipped with wireless which, in favourable circumstances, could transmit over 1,000 miles. This was to prove particularly important for Britain's naval operations at the outset of the Great War because, out of a complacency bred by its dominance of the world cable network, it had not kept pace with Germany in building long-range shore-based transmitters. In 1909 it possessed only three long-range stations, at Cleethorpes in Britain, at Gibraltar and in Malta; although a contract was signed with the Marconi Company in 1912 to build an Imperial Wireless Chain, it was not completed until 1915–16, so that at the outbreak of war most of the empire's stations, including that in the remote South Atlantic colony of the Falklands, were low-power and of limited range. Strategic communication was maintained by the cable network which the Committee of Imperial

Defence had assured itself could be interrupted only by a programme of cable-cutting which Britain alone, with its twenty-eight cable ships, twice the number of those of the rest of the world combined, was equipped to undertake.[10]

When war came in 1914, therefore, the naval campaign in distant waters was to be conducted in strangely disparate fashion: a compact German cruiser fleet, only intermittently able to communicate with the home base, was matched against a much more numerous British enemy—supported in places by Japanese, French and Russian naval units—controlled by a mixture of wireless and cable communication which proved almost as erratic. Technically, the means of intelligence and command employable by the admirals on both sides surpassed that available to Nelson and Bonaparte in 1798 by a factor as large as the difference in size between the seas in which they operated, which is to say between the Mediterranean and the Pacific Ocean. In practice, the meeting—and missing—of fleets that took place was almost as haphazard as theirs.

THE BATTLE OF CORONEL

The outbreak of war between Britain and Germany on 4 August 1914 found the main battle fleets concentrated in their home bases. The German High Seas Fleet was in harbour in its North Sea ports, the British Grand Fleet at anchor in the sheltered waters of Scapa Flow, within the Orkney Islands off the north of Scotland. The Grand Fleet exceeded the High Seas Fleet in strength by twenty-one dreadnought battleships and four battlecruisers to thirteen dreadnoughts and four battlecruisers. Both navies also had numbers of obsolete pre-dreadnoughts, the newer with the battle fleets, the older elsewhere; the Germans kept theirs in the Baltic, the British at Portland where they blocked the English Channel. The Grand Fleet's base at Scapa Flow had been chosen so as to deny the Germans exit from the North Sea.[11]

Both fleets also had large numbers of heavy and light cruisers and destroyers, the British considerably more than the Germans. While the latter kept theirs mostly with the High Seas Fleet, the British from the outset based many at Harwich, in the southern North Sea, whence they patrolled aggressively toward the German coast. They were constantly on the alert lest the Germans "came out," though their strategy was to do so only if they were certain of being able to beat a safe retreat. The Royal Navy wanted to destroy the Kaiser's navy; it, by contrast, sought merely to hold the British "at risk." It did so as long as it remained intact and in harbour; but while it stayed there, the naval situation in home waters was a perfect stalemate.

Only in more distant seas was there the possibility of an unforeseen encounter in the old Nelsonian style. The Mediterranean was such an arena of uncertainty, and there, right at the beginning, Germany's *Mittelmeerdivision,* a rather grandiose title for a force of two ships established in 1912, achieved a dramatic success. The battlecruiser *Goeben* and the light cruiser *Breslau* eluding the efforts of the British Mediterranean Fleet to neutralise them, succeeded in gaining Constantinople and joining the Turkish navy. At the beginning of November they sailed into the Black Sea to bombard Russian ports, thus inaugurating Turkey's war with Russia and so with its British and French allies. One outcome of this development would be the Gallipoli campaign; another was the court-martial of Admiral Troubridge "for failure to pursue the chase of His Imperial German Majesty's ship *Goeben,* being an enemy then flying."[12] Troubridge was acquitted but never again employed afloat. It was an appalling humiliation for a descendant of one of Nelson's most trusted captains, and it was to have repercussions that went far beyond the Mediterranean.

Some memory of Troubridge's misjudgement may indeed have influenced another British cruiser admiral at the outset of the Second World War, when a "superior force," the German 11-inch-gun pocket battleship *Graf Spee,* was cornered by one heavy and two light British cruisers, *Exeter, Ajax* and *Achilles,* off Montevideo, Uruguay, at the mouth of the River Plate. The British then did what Troubridge was accused at his court-martial of not doing, so disposing themselves that, by manoeuvre, they nullified the *Graf Spee*'s superiority in range and weight of shell and forced it into flight.

Graf Spee was so christened at her launch in memory of Germany's other leading distant-water admiral of 1914, the commander of the East Asiatic Cruiser Squadron, Maximilian Graf von Spee. A south German aristocrat, a Catholic, von Spee differed in character from the cold east Prussian Protestants who dominated the German army. Sensitive and warm-hearted, he was revered by his officers and men; but they also recognised his fighting spirit and his dedication to the German imperial

idea; in those aspects he was as much a subject of the Kaiser as Hinden-
burg, Ludendorff and von Lettow-Vorbeck, the charismatic German
commander in east Africa, with whose brilliantly evasive campaign in
the bush his own buccaneering exploits at sea were to be intertwined.

In August 1914 Germany's distant cruiser force consisted of eight
ships: five formed the East Asiatic Cruiser Squadron proper, the *Scharnhorst, Gneisenau, Leipzig, Nürnberg* and *Emden*, based at Tsingtao in China.
Scharnhorst and *Gneisenau* were "armoured cruisers," as the type was then
known, mounting a main armament of eight 8.2-inch guns and capable
of twenty knots; both had been launched in 1907, and they were the crack
gunnery ships of the Kaiser's overseas fleet. *Leipzig, Nürnberg* and *Emden*
were light cruisers, launched in 1906–8, effectively unarmoured but
mounting ten 4.1-inch guns and capable of twenty-four knots. Of the
same type were *Königsberg*, based in east Africa, and *Dresden*, which was
cruising off the Atlantic coast of South America. At the moment of the
war's outbreak, *Leipzig* was off the Pacific coast of Mexico, on station to
protect German nationals during the current civil war; *Nürnberg* was en
route to relieve her but was nearer China than her destination. *Emden*
had just left Tsingtao, on news of the heightening tension in Europe, and
was steaming to meet the main squadron in the Central Pacific. Her role
had been to protect and organise the colliers which were to join von
Spee if a cruiser campaign were to begin.

Tirpitz, the creator of the German navy, had disapproved of spending on cruisers. In the memorandum he wrote in June 1897, on which
Imperial Germany's naval programme was based, he wrote that "commerce raiding and transatlantic war against England is so hopeless,
because of the shortage of bases on our side and the superfluity on England's side, that we must ignore this sort of war."[13] He was later to alter
his view but it determined the composition and disposition of the German fleet in 1914, very much to Britain's advantage. It was also objectively
correct. Britain had piecemeal, over 250 years, accumulated a constellation of bases across the world, including—among those which were to
figure importantly in the coming cruiser war—Hong Kong in China,
Singapore in the East Indies, Aden in Arabia, the Cocos and Keeling
group in the Indian Ocean and the Falklands in the South Atlantic.
There were hundreds more, providing cable and often wireless facilities
and also, of even greater significance, coaling stocks.

Oil had just begun to supplant coal as the fuel source in the most modern warships; the turbines in Britain's new fast battleships of the *Queen Elizabeth* class were oil-fired. Most, however, remained dependent on coal. That required a return to base at less than weekly intervals or, even more tiresomely, a rendezvous with a collier and the tedious, difficult business of transferring hundreds of tons of coal from deck to deck, either in an anchorage or, should weather permit, in the open sea. The Royal Navy, with its multiplicity of coaling stations, could spare itself the trouble of coaling at sea. The German cruisers, in the weeks to come, would cruise encumbered by accompanying colliers or be forced to arrange rendezvous with detached colliers in remote inlets.

Despite the difficulties, and despite the disapproval of Tirpitz, Germany's overseas navy was committed to the war against commerce from the moment of the declaration. Orders stated that "in the event of a war against Great Britain, or a coalition including Great Britain, ships abroad are to carry out cruiser warfare unless otherwise ordered. Those vessels which are not suitable for cruiser warfare are to fit out as auxiliary cruisers. The areas of operations are the Atlantic, the Indian Ocean, the Pacific . . . Our ships abroad cannot count in wartime either on reinforcements or large quantities of supplies . . . The aim of cruiser warfare is to damage enemy trade; this must be effected by engaging equal or inferior enemy forces, if necessary."[14] The Kaiser's personal instructions to cruiser commanders went further, urging them to seek an honourable outcome in all circumstances, where honour was understood to mean fighting to the last. Earlier orders, not cancelled, held out the hope, however, of inflicting defeat on potential enemies in foreign waters, citing in 1907 the superiority enjoyed by the Asiatic Squadron over the French, Russian and even British forces in its area. That was optimistic but not unrealistic, as events would show. Germany's colonial navy in the years before the war formed a cohesive unit of high quality, both in personnel and material. The same could not be said of its local opponents, who, though numerically superior, counted on too many old ships manned by second-class crews to be reckoned really formidable.

What was undeniable was the vulnerability of the enemy's trade. Germany, with 2,090 steamships, was the world's second largest mercantile power, a fact often overlooked in discussion of its reasons for building the High Seas Fleet. Its merchant fleet was, however, vastly exceeded in size by the British, with 8,587 steamships; when those of its empire were added, they amounted to 43 per cent of the world's shipping. They sailed every sea along every trade route, carrying not only the majority of the world's trade but also supplies essential to the home country's survival, including two-thirds of its foodstuffs.[15] The Admiralty, moreover, had come to disbelieve in convoy, which had occupied so much of the Royal Navy's energies in the French Wars of 1793–1815. Apart from troop convoys, it had no plans to protect merchant shipping in 1914 and made none until the crisis of the U-boat war in 1917. It trusted to the vastness of the oceans to protect merchantmen, sailing independently, at the war's outbreak and for three years thereafter.

It also expected, of course, that its own distant squadrons would hunt down and destroy enemy commerce raiders should they interfere. Despite the concentration of the navy's most modern warships of all classes in or near the North Sea, enough still remained to assure Britain's worldwide naval presence in the distant waters of its eight historic overseas stations: China, New Zealand, Australia, North America and West Indies, South America, Africa, Mediterranean, and East Indies. Some of the ships on station when war broke out—river gunboats and obsolete cruisers—were unfit for oceanic operations. Those capable to stand in the line of battle included the old battleship *Triumph,* armoured cruisers *Minotaur* and *Hampshire* and light cruisers *Newcastle* and *Yarmouth* on the China station; the modern light cruiser *Glasgow* on the South American station; the old battleship *Swiftsure,* the light cruiser *Dartmouth* and the obsolete light cruiser *Fox* on the East Indies station; and on the Australian and New Zealand stations the modern battlecruiser *Australia* and the light cruisers *Sydney* and *Melbourne* of the Royal Australian Navy and the obsolete light cruisers *Encounter* and *Pioneer.* Australia and New Zealand (which had paid for the battlecruiser *New Zealand,* serving with the Grand Fleet in home waters) had agreed that their navies' ships should come under Admiralty control in war circumstances.

Though numerous—fourteen in all—these warships were of mixed quality. *Australia* could sink any German colonial cruiser at no risk to herself at all, but she was to be tied, at the outset, to convoying the troopships of the Australian and New Zealand expeditionary force; *Dartmouth,* *Sydney* and *Melbourne* were the equal of the modern German light cruisers; the British armoured cruisers, *Minotaur* and *Hampshire,* were obsolescent and not up to the German class. Armoured cruisers had become an anomaly: too weak to fight battlecruisers, too slow to catch light cruisers, capable only of combat with others of their own class. It was to be the Royal Navy's misfortune in the coming cruiser war that its armoured cruisers were the inferior of the German, inferior as both were to the new battlecruiser class.

In the early weeks of the war, Britain was to send reinforcements to the overseas stations. Rear-Admiral Sir Christopher Cradock, commanding on the North America and West Indies stations, received the armoured cruisers *Suffolk, Berwick, Essex* and *Lancaster;* none was to take part in the cruiser war, but their sister ship *Monmouth* was. Detached to South America, she eventually joined the squadron Cradock was to form for anti-cruiser action. So did *Good Hope,* another armoured cruiser detached to him, into which, on 15 August, he shifted his flag. His final reinforcements were the old battleship *Canopus,* launched in 1896—its 12-inch guns were manned by elderly reservists, and its engine-room was supervised by a chief engineer who, as events would reveal, was mentally unfit for service—and the *Otranto,* an armed merchant cruiser. Armed merchant cruisers—to serve, with very mixed results, in both world wars—were liners or fast freighters, fitted with guns and crewed by naval officers and ratings, which admiralties expected to give useful service as convoy escorts or commerce raiders. In favourable circumstances some did; in others they proved deathtraps.

The Germans had also commissioned armed merchant cruisers; indeed, to its merchant fleet of high-speed liners of such companies as Hamburg-Amerika and Norddeutsche-Lloyd belonged some of the fastest passenger ships in the world. At the outbreak most instantly took refuge in neutral ports, particularly in North and South America; but their captains, who frequently belonged, as many of their sailors did, to the German naval reserve, stood ready to join the German raiding force when opportunity offered. When guns and ammunition could be transferred to them, as was to happen, they became effective units in the commerce war.

Other elements in the commerce war were ships of the French, Russian and Japanese navies. France, with bases in Indo-China and the Pacific islands, had several cruisers and destroyers in Asiatic waters, including the *Dupleix* and *Montcalm;* Russia, not a serious Pacific naval power since its defeat in the Russo-Japanese War of 1904–5, nevertheless had one or two units deployed; Japan, which entered the war against Germany on 23 August, altered the balance of naval power in the Pacific to Germany's disadvantage altogether. Japan, whose army had been trained by Germany, had no quarrel with the Kaiser's empire at all. Its declaration of hostilities was narrowly selfish. It correctly anticipated that, by aligning itself with Britain and France, it was likely to acquire possession of the German island chains of the Marianas, Carolines and Bismarcks. It did so; in the short term, the adherence of Japan to the anti-German alliance was also of great importance in limiting the German cruiser threat. In the long term, Japan's annexation of Germany's central and south Pacific islands laid the basis for its successful aggression against the European and American Pacific empires in 1941–42. Rabaul, in particular, Germany's main base in the Papuan archipelago, was to become Japan's principal *place d'armes* in the struggle with the Americans and Australians for New Guinea and the Solomon Islands in 1942–43.

Japan began its war against Germany by laying siege to Tsingtao on 2 September. It was to last until 7 November. The garrison, which knew resistance was hopeless, nevertheless fought with great tenacity. Two local defence gunboats, *S.90* and *Jaguar*, engaged the landing fleet; the defenders manned the redoubts, only gradually giving ground under heavy bombardment. The fortress commander had been cut off from the outside world since 14 August, when the British cable ship *Patrol* had cut the cables to Shanghai and Tschifu.[16] His garrison, mainly naval infantrymen, had also lost any hope of escape with the departure of the East Asiatic Cruiser Squadron some weeks earlier. Admiral Graf von Spee, after an exchange of dinners with his British counterparts, had parted from them on 14 June, on the friendliest terms, to cruise the German Pacific islands with the heavy ships. In the last weeks of July, as news of the heightening crisis in Europe reached him, he persuaded Berlin to cancel an order for *Nürnberg* to return to Tsingtao, calling her instead to meet him at Ponapé, an outlier of the Caroline Islands. There, on 4 August, he learnt that Britain had declared war but also that "Chile is a friendly neutral" and that "Japan will remain neutral."[17]

On this partially correct information, von Spee now decided his immediate course of action. Before leaving Tsingtao, he had instructed von Müller, captain of the *Emden,* that his principal role was to protect the colliers which assured the squadron's mobility. In the event of "strained relations," they were to leave Tsingtao and proceed to Pagan Island, in the Marianas. *Emden* was to seek to rejoin him. In the knowledge that Britain's forces on the China station included a battleship and that HMAS *Australia,* capable of obliterating his whole squadron, was operating to the south, he set course for Pagan on 5 August and there on the 12th was joined by *Emden,* the armed merchant cruiser *Prinz Eitel Friedrich* and the provision ship *Yorck.* They brought four colliers; it was a warning of the dangers surrounding them that four others had been sunk or captured on passage by the British battleship *Triumph* and the armoured cruiser *Minotaur*—aboard which von Spee had dined companionably in June.

The northwestern Pacific was clearly becoming too dangerous a theatre for the East Asiatic Squadron; it was shortly to become even more dangerous when, on 23 August, Japan entered the war. The Imperial Japanese Navy, which had defeated the Russian with spectacular completeness in 1905, now comprised three dreadnoughts and four battlecruisers, all recently built in Japanese yards, seven heavy cruisers and scores of other cruisers and destroyers. Japan was not yet quite a firstclass naval power, as she would become within twenty years. She could, nevertheless, devour Germany's Asiatic fleet. It was time for von Spee to be off. With the British to the west and south, the Japanese to the north and the Australians to the south, he correctly decided to head southeast, towards Chile, where there were many German nationals and businesses, much sympathy for Germany's cause and a labyrinth of uncharted inlets in which a marauding fleet might hide.

Before departing, however, von Spee agreed to a division of his force. Dividing force is a violation of a cardinal military principle. It was one that applied particularly strictly to von Spee. By keeping his ships together, he obliged his enemies to do likewise, which reduced their chances both of finding him and of falling upon German merchantmen plying the ocean. With only four ships—*Scharnhorst, Gneisenau, Nürnberg* and *Emden*—under command, and while awaiting *Leipzig* and *Dresden* to join, logic required he maintain the strongest possible force. Nevertheless, he succumbed to the persuasion of *Emden's* captain. Von Müller argued that, by cruising detached with the squadron's fastest ship into the Indian Ocean, he could spread widespread confusion and do serious damage to Britain's interests, particularly along the coasts closest to its greatest imperial possession, India. On the afternoon of 13 August, von Spee sent von Müller a written order: "You are hereby allocated the *Markommania* [a collier] and will be detached on the task of entering the Indian Ocean and waging cruiser warfare as best you can."[18] Thus began

British & German Naval Bases

Scapa Flow
Rosyth
North Sea
IRELAND
UNITED KINGDOM
ENGLAND
London
Portsmouth
English Channel
ATLANTIC OCEAN
Paris
THE FRONT
DENMARK
Heligoland
Wilhelmshaven
Hamburg
Bremen
GERMANY

28 AUG. 1914
FIRST SEA BATTLE,
BEATTY SINKS
5 GERMAN SHIPS

Baltic Sea
PRUSS
UNITED KINGDOM
Heligoland
GERMANY
FRANCE
ITALY
SPAIN
PORTUGAL
Gibraltar
Mediterranean Sea

CANADA
New York
USA
Bermuda (Br.)
Bahamas (Br.)

60°N
45°N
30°N
Tropic of Cancer

90°W 75°W 60°W 45°W 30°W 15°W 0° 15°E

Canary Isles (Sp.)

Cape Verde Is (Port)

A F R I C A

A T L A N T I C

15°N
0° Equator

St Paul's Rocks
San Pedro (Brazil)
Fernando de Noronha (Brazil)
Pernambuco

TOGO
Gulf of Guinea

KAMERU
(to German

Ascension I. (Br.)

B R A Z I L

O C E A N

15°S

St Helena (Br.)

Rio de Janeiro
Trinidad I.   Martin Vaz (Brazil)
Abrolhos Rocks

Tropic of Capricorn
Antofagasta
PACIFIC
OCEAN

Buenos Aires

GER
SOU
WE.
AFR

30°S

1 NOV. 1914
VON SPEE BEATS
OFF ATTACK BY
CRADOCK, SINKING
THE GOOD HOPE
AND MONMOUTH

X Coronel

C H I L E

45°S

Falkland Islands (Br.)

Stanley Harbour
S.Georgia (Br.)

Cape Horn

X

8 DEC. STURDEE
GIVES CHASE AND
SINKS VON SPEE'S
SCHARNHORST

60°S

90°W 75°W 60°W 45°W

THE BATTLE OF CORONEL 1 NOV. 1914

0   10   20 miles
0 10 20 30 km

Nurnberg
Dresden
Scharnhors
Leipzig

Otranto
Monmouth
Good Hope
Glasgow
Rendezvous point 14.35
Glasgow escapes
Otranto escapes

Monmouth sunk 21.18
Good Hope sunk 19.57

N

Coronel

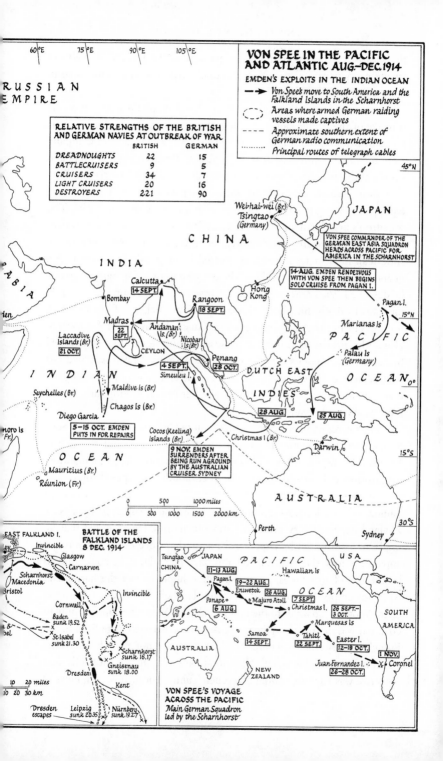

VON SPEE IN THE PACIFIC AND ATLANTIC AUG.–DEC. 1914

EMDEN'S EXPLOITS IN THE INDIAN OCEAN

→ Von Spee's move to South America and the Falkland Islands in the Scharnhorst

⊂⊃ Areas where armed German raiding vessels made captives

--- Approximate southern extent of German radio communication

⋯⋯ Principal routes of telegraph cables

RELATIVE STRENGTHS OF THE BRITISH AND GERMAN NAVIES AT OUTBREAK OF WAR

	BRITISH	GERMAN
DREADNOUGHTS	22	15
BATTLECRUISERS	9	5
CRUISERS	34	7
LIGHT CRUISERS	20	16
DESTROYERS	221	90

RUSSIAN EMPIRE

60°E   75°E   90°E   105°E

45°N

JAPAN

CHINA

Wei-hai-wei (Br.)
Tsingtao (Germany)

VON SPEE COMMANDER OF THE GERMAN EAST ASIA SQUADRON HEADS ACROSS PACIFIC FOR AMERICA IN THE SCHARNHORST

14 AUG. EMDEN RENDEZVOUS WITH VON SPEE THEN BEGINS SOLO CRUISE FROM PAGAN I.

INDIA

Bombay
Calcutta 14 SEPT.
Madras 22 SEPT.
Rangoon 18 SEPT.

Hong Kong

Pagan I.

Marianas Is

15°N

PACIFIC

Palau Is (Germany)

OCEAN 0°

ARABIA

Andaman Is (Br.)
Nicobar Is (Br.)
Laccadive Islands (Br.) 21 OCT.
CEYLON
Penang 28 OCT.
Simeuleu I.
4 SEPT.

DUTCH EAST INDIES

INDIAN

Maldive Is (Br.)
Chagos Is (Br.)
Seychelles (Br.)

28 AUG.   25 AUG.

Diego Garcia
5–15 OCT. EMDEN PUTS IN FOR REPAIRS

Cocos (Keeling) Islands (Br.)
Christmas I (Br.)

Darwin

15°S

9 NOV. EMDEN SURRENDERS AFTER BEING RUN AGROUND BY THE AUSTRALIAN CRUISER SYDNEY

Comoro Is (Fr.)

OCEAN

Mauritius (Br.)
Réunion (Fr.)

AUSTRALIA

0   500   1000 miles
0   500   1000   1500   2000 km

Perth

Sydney

30°S

BATTLE OF THE FALKLAND ISLANDS 8 DEC. 1914

EAST FALKLAND I.
Invincible
Glasgow
Scharnhorst
Macedonia
Carnarvon
Bristol
Cornwall
Invincible
Baden sunk 19.52
St. Isabel sunk 21.30
Scharnhorst sunk 16.17
Gneisenau sunk 18.00
Dresden
Kent
Dresden escapes
Leipzig sunk 20.35
Nürnberg sunk 19.27

0   10   20 miles
0   10   20   30 km

VON SPEE'S VOYAGE ACROSS THE PACIFIC

Main German Squadron led by the Scharnhorst

Tsingtao
CHINA   JAPAN
PACIFIC
USA
Hawaiian Is
11–13 AUG.
Pagan I.
19–22 AUG.
Enuwetok 28 AUG.
Ponape 6 AUG.
Majuro Atoll
Christmas I. 7 SEPT.
26 SEPT.–3 OCT.
OCEAN
SOUTH AMERICA
Marquesas Is
Samoa 14 SEPT.
Tahiti 22 SEPT.
Easter I. 12–18 OCT.
1 NOV. Coronel
Juan Fernandez I. 26–28 OCT.

AUSTRALIA

NEW ZEALAND

the *Emden* epic, as dramatic as any passage in the history of Nelson's frigate captains and a story that was to electrify friend and foe alike in the coming months of naval operations.

It was therefore with only three warships that von Spee set out on his traverse of the Pacific, across 120 degrees of longitude, to offer his challenge to Britain's naval power in the southern oceans; also in company were the armed merchant cruiser *Prinz Eitel Friedrich,* eight colliers and supply ships and the armed merchantman *Cormoran,* formerly the Russian ship *Ryaezan,* captured by *Emden* on passage from Tsingtao and equipped with guns taken from a redundant coastal warship. The squadron's first destination after leaving Pagan on 13 August was Eniwetok, in the Marshalls, nearly forty years later to be the scene of an American nuclear test. Von Spee coaled in the atoll's lagoon between 19 and 22 August, then sailed for Majuro, also in the Marshalls. En route he detached *Prinz Eitel Friedrich* and *Cormoran* to raid commerce; the former was to rejoin him later; the latter, out of coal, was forced to seek internment in the American island of Guam. He also sent *Nürnberg* to Honolulu, already an American possession, with signals to be forwarded by cable to Berlin. His calculation here was sharp. As *Nürnberg* had last been seen by outsiders on the Mexican coast, and news of her joining the main squadron had not been broadcast, her arrival at Honolulu would not reveal his whereabouts.

Nürnberg rejoined von Spee at the remote Christmas Island on 2 September, having meanwhile visited nearby Fanning Island to cut the British cable between Fiji and Honolulu. The action risked giving away his position and, indeed, during the next month, von Spee displayed an uncharacteristic recklessness. At Christmas Island he decided to sail to Samoa, now no longer German, since it had been captured by a New Zealand expeditionary force the previous month. He accepted that he might be confronted by superior force, perhaps even the *Australia,* but apparently took the view that, by approaching at dawn, he could prevail by the use of torpedoes, a very sanguine hope. In fact the harbour at Apia was empty and he sailed away, but left behind a trace of his presence. Another 500 miles to the east he called at Suvarov Island, hoping to coal, but was driven off by heavy seas, so proceeded to Bora Bora in the French Society Islands, where the inhabitants had not as yet heard of the war. They supplied him with fresh food while the ships coaled. His next objective was Papieté, capital of French Tahiti. There, however, news of the war had reached the garrison, which set fire to the coal stocks and put up resistance. There was no wireless station; but, as von Spee drew away, the governor sent a ship to Samoa with a report, which reached the Admiralty on 30 September.

Berlin was out of touch with the squadron since, with the loss of Rabaul in the Bismarcks to the Australians, the rear link to Nauen had been broken. It had therefore decided not to attempt to control von Spee's movements or strategy, but the German naval high command expected him to proceed to South America and perhaps thence, via Cape Horn, into the Atlantic. The British Admiralty, by contrast, was principally concerned by the danger that von Spee might move east, to operate in Australasian waters or the Indian Ocean, where the great "imperial convoys," bringing Australian, New Zealand and Indian soldiers to Europe, were setting sail. Its anxieties were much heightened by the success of von Müller on *Emden,* who, while von Spee made his leisurely way eastward across the Pacific, was cutting a swathe through Britain's merchant shipping in the Indian Ocean.

Against the appearance of von Spee in the South Atlantic, the Admiralty began to dispose ships as early as the first week of September. Local circumstances, particularly the Mexican civil war, had brought about a concentration of ships in the Caribbean during August. Admiral Sir Christopher Cradock, commanding on the North American station, signalled on 3 September, "*Good Hope* [armoured cruiser] . . . visiting St. Paul's Rocks, and will arrive Pernambuco 5th September for orders, *Cornwall* [armoured cruiser] is in wireless touch proceeding south. *Glasgow* [light cruiser] reports proceeding with *Monmouth* [armoured cruiser] and *Otranto* [armed merchant cruiser] to Magellan Straits [Cape Horn], where number of German ships reported, presumably colliers, and where concentration of German cruisers from China, Pacific Ocean and Atlantic Ocean appears positive."[19] Cradock's signal was a remarkably shrewd appreciation by a commander not privy to Admiralty intelligence. Pernambuco, the eastward point of Brazil, abutted the main trade routes from Argentina, whence came much of Britain's beef. St. Paul's Rocks, off Pernambuco, were an obvious coaling area for German commerce raiders; they were to be much used as a refuelling rendezvous by U-boats during the Second World War. South American ports were full of colliers chartered by local German agents to resupply the commerce raiders, as Cradock indicated.

Cradock then became distracted by his inability to locate *Dresden* and another German light cruiser, *Karlsrühe*. About *Karlsrühe* he need not have worried; after disappearing from view among the remoter islands of the West Indies, she was blown up by a spontaneous explosion in her magazines on 4 November, a fact not known in Britain for three months, though speculation as to her whereabouts would continue to complicate Cradock's thinking during September and October. *Dresden* remained a real menace. To guard against her entering the Pacific, Admiral Cradock sent *Glasgow, Monmouth* and *Otranto* to the Magellan Straits, at the tip of South America, in early September. Meanwhile, *Dresden,* having sunk a British collier off the River Plate, itself transferred to the Magellan Straits and then, on advice from the German Admiralty "to operate with the *Leipzig,"* sailed into the Pacific on 18 September. News of *Dresden*'s movements prompted Cradock, disastrously as it would turn out, to take *Good Hope,* his flagship, south to the Magellan Straits also, where he met *Glasgow* and *Monmouth* on 14 September.

Communication between Europe and South American waters was complex. The British Admiralty used its intact cable network to send messages to Cerrito, in Uruguay, whence they were wirelessed onwards to the low-power wireless station in the Falkland Islands; that assured reasonably rapid touch with ships in the South Atlantic. Signalling into the South Pacific was more difficult. The Falklands station could not usually reach the Pacific, because of atmospherics and the barrier of the Andes, so warships had to be sent into port at regular intervals to collect cable telegrams, a tedious procedure entailing many delays. The Germans wirelessed from Nauen, as far as its range would carry, to their consuls, who then communicated by cable with German merchant ships in the port nearest to von Spee's position. South American governments being lax about neutrality regulations, their merchant captains then wirelessed signals onwards and retransmitted those received by the same route homeward.

On 14 September, the Admiralty sent Cradock a long signal that laid the basis for his squadron's and his own destruction:

There is a strong possibility of *Scharnhorst* and *Gneisenau* arriving in the Magellan Straits or on the west coast of South America . . . Leave sufficient force to deal with *Dresden* and *Karlsrühe*. Concentrate a squadron strong enough to meet *Scharnhorst* and *Gneisenau*, making

Falkland Islands your coaling base. *Canopus* is now en route to Albrohos, *Defence* is joining you from the Mediterranean. Until *Defence* joins, keep at least *Canopus* and one "County" class [i.e. *Glasgow* or similar] with your flagship. As soon as you have superior force, search the Magellan Straits with squadron, being ready to return and cover the River Plate or, according to information, search north as far as Valparaiso. Break up German trade and destroy the German cruisers.[20]

This was a strategic rather than tactical directive, and of very wide scope. It committed Cradock to cover both the Atlantic coast of South America, as far north as the River Plate in Uruguay, a merchant shipping focal point, and the Pacific coast as far as the other focal point of Valparaiso in Chile. It instructed him to conduct both commerce warfare and an anti-cruiser campaign. It promised him a ship, *Defence,* which was later to be retained in the Mediterranean; had it come to him, he could not have been outgunned. It represented *Canopus,* an obsolete battleship, as an equivalent, which it was not. It implicitly expected Cradock to produce a victory.

The signal, when sent, disguised the Admiralty's complete ignorance of von Spee's whereabouts. All it knew was that he was somewhere in the southeastern Pacific, between Fanning Island—a fact established by the destruction of that lonely island's wireless and cable station—and Cape Horn, an exercise in location subject to error by a factor of thousands of miles and hundreds of degrees of longitude and latitude. On 16 September there was a correction: "situation changed. *Scharnhorst* and *Gneisenau* off Samoa on 14th September . . . left steering N.W. [back towards the Bismarcks] . . . German trade on west coast of America to be attacked at once . . . Cruisers need not be concentrated. Two cruisers and an armed liner would appear sufficient for Magellan Straits and West Coast. Report what you propose about *Canopus.*"[21]

The report from Samoa was the outcome of von Spee's ill-judged visit two days earlier. It might have resulted in disaster, had the Australian fleet been present. Two weeks later he had transferred to the remote Marquesa Islands, last outpost of the French empire in the Pacific. There he was able to coal again in sheltered waters and load fresh food, from islanders who had not yet heard of the European war. Then he set off to even more remote places, first Easter Island, then Juan Fernandez, Robinson Crusoe's legendary marooning place. At Easter

Island he was joined by *Dresden* and *Leipzig* which, proceeding independently by guesswork, arrived there to meet him during 1–5 October. The first he knew of their approach was by intercepting wireless signals between them.

Meanwhile Cradock, whose communications with the Admiralty were to be increasingly misunderstood, as theirs with him were to be also, was searching for *Dresden* along the Atlantic coast of South America. He was alerted to the fact that he was in the wrong ocean only when, on 25 September, he met a British ship which had been chased by her on 18 September, near Cape Horn. Feeling that he was now on the scent, Cradock immediately led his squadron to the Magellan Straits (the normal means of passage between the two great oceans) and put in at the Chilean port of Punta Arenas, where the British consul confirmed that *Dresden* had indeed been about, using nearby Orange Bay as a base. Finding nothing there, Cradock then reversed course; a complicated toing-and-froing followed, during which he returned in his flagship *Good Hope* to the Falklands, leaving his accompanying armed merchant cruiser *Otranto* behind, but, once arrived, almost immediately sent *Glasgow* and *Monmouth* back to join *Otranto* at Punta Arenas, with orders—in accordance with Admiralty instructions as he understood them—to conduct cruiser warfare on the Pacific coast of Chile. At the Falklands, however, Cradock heard by wireless from *Otranto* that she had overheard German naval wireless signals, which set her off again to Orange Bay, where German sailor scrawls of a "Kilroy was here" sort confirmed *Dresden*'s presence only a few days earlier. Finding no actual German presence, however, he returned once more to the Falklands.

Cradock, who was to be widely blamed for future disaster, was in an unenviable situation. He was acutely aware of ambient danger—the presence of von Spee's big ships, probably in the Pacific but perhaps seeking to break into the Atlantic; the lurking menace of the German light cruisers, preying on British trade; the lack of a British base anywhere in his theatre of responsibility, except the Falklands, which did not offer control of Pacific waters; the penetration of the whole Patagonian region by German settlers and officials, all willing and ready to resupply the Kaiser's ships, shelter their colliers and spy on the Royal Navy; and, as a background to his difficulties, the awful Cape Horn weather which, even in what was the Southern Hemisphere's summer, brought constant gales, sleet, snow and mountainous seas. To cap all, he was oppressed by his difficulties of communication with his masters in London. They in turn, oppressed by fear of a break-out by the High Seas Fleet, were trying to work a worldwide strategy without touching their gold reserve of modern battleships and battlecruisers locked up in northern Scotland, instead hoping that obsolete units left over from the Victorian navy could keep Germany's best cruisers on overseas stations at bay. It did not help the management of British naval strategy that the Admiralty's political chief, Winston Churchill, was currently attempting to direct in person a private war on the north coast of Belgium or that the Royal Navy's professional head, Louis of Battenberg, was under attack by the popular press as a German princeling, an attack which would shortly lead to his removal from office.

In the circumstances, Cradock appears to have tried to straddle two oceans and two incompatible Admiralty demands: to protect British trade in the Atlantic and to destroy the East Asiatic Cruiser Squadron in the Pacific, if that was where it was. Little wonder that his movements in the first days of October appeared confused. However, on his return to the Falklands after his second search of Orange Bay, he received an Admiralty message on 7 October that at last threw light on von Spee's whereabouts and gave him more or less clear instructions.

On 4 October the wireless station at Suva, in British Fiji, had picked up a message from *Scharnhorst* in the German mercantile code, reading, "*Scharnhorst* on the way between the Marquesas and Easter Island."[22] As is now known, the information was correct. The Admiralty anyhow instructed Cradock on 7 October "to be prepared to have to meet them in company . . . *Canopus* should accompany *Glasgow, Monmouth* and *Otranto,* the ships to search and protect trade in combination . . . If you propose *Good Hope* to go, leave *Monmouth* on the east coast."[23]

The question nevertheless remains whether the Admiralty was yet able to read *Scharnhorst*'s code transmissions. A copy of the German mercantile code had indeed been seized in Australian waters early in the war but it did not apparently reach the Admiralty until the end of October.[24] Perhaps the book was already being used locally. More mysterious are Cradock's reactions to the Admiralty's quite clear instructions of 7 October. In his reply on the 8th, he showed that he recognised the likelihood of von Spee's heavy ships being joined by the light cruisers, making a formidable force. He also advised that he had summoned his old slow battleship *Canopus* to join him at the Falklands, where he intended "to concentrate and avoid division of forces." Yet despite his resolve not to divide his forces, he had sent *Glasgow, Monmouth* and *Otranto* into the Pacific, under the feebly limiting instruction "not to go north of Valparaiso until German cruisers are located." He also enquired after the whereabouts of *Defence,* previously promised to him but retained in the Mediterranean; and he was clearly unsettled by the idea that von Spee might go north, pass through the Panama Canal, if the Americans would so permit, and thus either get home to Germany or open up another commerce war in the Gulf of Mexico.

In the last two weeks of October, the Admiralty and Cradock got disastrously, ultimately tragically, at cross-purposes. The Admiralty made new dispositions in the Atlantic designed to backstop Cradock if von Spee evaded him and broke out of the Pacific; they included the deployment of *Defence*—at last—and other cruisers from the African station under Admiral Stoddart to the Brazilian bulge. London was also counting on the disposition of the Japanese fleet in the central and western Pacific to limit von Spee's ability to do harm in that ocean. Over the deployment of strength in what was to prove the critical area—the South Pacific between Valparaiso and Cape Horn—the Admiralty and its admiral on the spot succeeded in misunderstanding each other.

The grit in the works was the condition of *Canopus* and Cradock's misunderstanding of his authority over *Defence. Defence* was an ultimate example of the armoured cruiser idea, bigger, faster, as heavily armoured and more heavily gunned than either *Scharnhorst* or *Gneisnau;* had she joined Cradock, as he believed she would, she would have seen off either of her German equivalents. *Canopus,* though a battleship, was inferior to all three armoured cruisers, British and German alike. She was thinly armoured, and her 12-inch guns barely outranged those of the Germans. Moreover, her timorous chief engineer had persuaded her captain and Cradock that she could not make better than twelve knots, a cripple's speed. Cradock accordingly went on ahead from the Falklands into the Pacific, signalling the Admiralty on 27 October that "*Canopus*'s slow speed" made it "impracticable to find and destroy the enemy squadron. Consequently have ordered *Defence* to join me . . . *Canopus* will be employed on necessary convoying of colliers."[25] Unfortunately, the Admiralty misinterpreted the picture, concluding—by a misunderstanding of the role of *Canopus* or of Cradock's intentions—that von

Spee's squadron was blocked. If he went north he would fall under the guns of the powerful Japanese fleet. If he went south he would eventually run into Cradock's cruisers, which the Admiralty appeared to believe would have *Canopus* in company. It was apparently disbelieved that Cradock would risk an engagement without the support of her 12-inch guns. It therefore concluded that "the situation on the west coast [of South America] is safe" and ordered *Defence*, which had both the speed and guns to defy von Spee, to remain in the Atlantic. Cradock, a sailor in the Elizabethan tradition who was determined not to repeat Milne's mistake during the *Goeben* and *Breslau* episode of letting any German opponent escape, pushed ahead with his collection of weak ships, leaving *Canopus* to limp along 300 miles behind. In the late afternoon of 1 November, the two squadrons made contact off the Chilean port of Coronel.

Wireless had already revealed to them each other's presence. Cradock's progress up the coast had been reported to von Spee by German merchant ships in southern ports, while the British had been picking up distinctive German Telefunken transmissions for some days. While von Spee now knew, however, that Cradock was approaching with several ships, Cradock had been misled by the Germans' clever use of *Leipzig's* wireless alone to believe that only one German cruiser lay in his path.[26] He appears to have thought that the von Spee squadron as a whole was moving northward towards the Galapagos Islands, with a view to traversing the Panama Canal from west to east. In order to verify his supposition, and to send and receive telegrams by cable, he detached *Glasgow*, his fast light cruiser, to Coronel on 31 October, with orders to rejoin next day.[27]

Had *Glasgow* arrived a few hours later, or stayed a little longer, the impending defeat might have been averted. In London, where the veteran Admiral Sir John Fisher had just resumed the post of First Sea Lord, the Admiralty was revising its assessment of the South American situation, had seen the danger that portended, had ordered *Defence* to join Cradock post-haste and had stressed that he should meanwhile not fight without *Canopus. Glasgow* sailed too soon to bring Cradock his fresh instructions. When she rejoined *Good Hope, Monmouth* and *Otranto*, the squadron was receiving strong German wireless signals, apparently transmitted at close range. Since current technology did not permit direction-finding, Cradock decided to form a line of search, with his ships disposed fifteen miles apart—intervals scarcely altered since Nelson's line-of-sight days—and began to look for the transmitting source.

He believed he was seeking a single ship. Ironically, von Spee, who was now nearby, had the same impression. Having left his island coaling base of Más Afuera, part of the Juan Fernandez group, in the Pacific between Coronel and Valparaiso, on 27 October, he had cruised for three days off the coast, awaiting Cradock's arrival, but on receipt of news of *Glasgow*'s visit, moved to cut her off, as he believed, from the main squadron. He, too, was deploying his ships in a line of search when his smoke was sighted by *Glasgow,* which had just taken up station in Cradock's formation. A few minutes later the British ships were seen by the German, and both squadrons moved to form a line of battle.[28]

News of the presence of *Scharnhorst* and *Gneisenau* was wirelessed by *Glasgow* to Cradock over the fifty miles of sea that separated them. The admiral might have decided even then to avoid action. *Monmouth* and *Good Hope* were faster than the German armoured cruisers, *Glasgow* no slower than the German light cruisers. Cradock might have turned away and escaped, but he was encumbered by *Otranto,* which would have had to be sacrificed; and there were other considerations of honour. The Royal Navy always fought. Cradock ordered his ships to form on *Glasgow* and headed towards the Germans.

A South Pacific summer in Cape Horn latitudes can be a bitter season. So it was on 1 November 1914. Though the sky was clear and the sun bright, a Force 6 wind was blowing, seas were breaking over the smaller ships and the air was very cold. Cradock's tactical scheme at the outset of the action was to keep out of range until the sinking sun behind him blinded the German gunners. Von Spee hoped to close the range as soon as the twilight protected his ships but silhouetted the British on the western horizon. At eighteen minutes past six, Cradock wirelessed *Canopus,* 250 miles away, "I am now going to attack the enemy"; it may have been meant as a farewell message.

The Germans, who were waiting for the sun to sink, did not open fire for nearly an hour; meanwhile, the two squadrons slowly converged on a southerly course. About seven o'clock, an officer in *Glasgow* recorded, "We were silhouetted against the afterglow with a clear horizon behind us to show up the splashes from falling shells while the [enemy] ships were smudged into low black shapes scarcely discernible against the background of gathering darkness."[29] The German big guns, twelve in all, outranged all but *Good Hope*'s two 9.2-inch; the British 6-inch guns did not have the range to touch *Scharnhorst* or *Gneisenau,* which carefully kept their distance. "Sharply silhouetted against the red gold evening sky," *Good Hope* and *Monmouth* were hit repeatedly, and an officer on *Glasgow* recorded that "by 1945, by which time it was quite dark, *Good Hope* and *Monmouth* were obviously in distress. *Monmouth* yawed off to starboard burning furiously . . . *Good Hope* . . . was firing only a few of her guns. The fires on board were increasing their brilliance. At 1950 there was a terrible explosion . . . between her mainmast and her after funnel; the gust of flames reached a height of over 300 feet, lighting up a cloud of debris that was flung still higher in the air. She lay between the lines a low black hull lighted by a dull glow. No one . . . actually saw her founder, but she could not have survived many minutes."[30]

Monmouth, though the weaker ship, was still fighting and was able to reply to *Glasgow*'s lamp-signal enquiring "Are you all right?" with the message "I want to get stern to sea. I am making water badly forward." Those were her last words. *Glasgow* observed that "[she was] badly down by the bows, listing to port with the glow of her ignited interior brightening the portholes below her quarterdeck."

At that point *Glasgow*'s captain decided to leave the scene, on the grounds that a warning needed to be taken to *Canopus,* steaming up from the south at her best speed; the radio waves were filled by German jamming. As *Glasgow* fled, seventy-five flashes of shell-fire directed against *Monmouth* were counted before the horizon blanked off observation. That was not the last sight of the sinking ship. Shortly before nine o'clock the light cruiser *Nürnberg* found her "with her flag still flying" and reopened the attack. "The *Monmouth* still kept her flag flying and turned towards the *Nürnberg,* either to ram or to bring her starboard guns to bear. Captain von Schönberg therefore opened fire again . . . The unprotected parts of the *Monmouth*'s hull and also her deck were torn open by the shells. She heeled over further and further and at 2128 she slowly capsized and went down. Von Schönberg subsequently learned that two officers, who had been standing on deck, heard the *Monmouth*'s officers call the men to the guns; [they] were apparently engaged in stopping leaks."[31]

Loss of life in *Good Hope* and *Monmouth* was total: of the 1,600 men aboard, those not killed in the gunnery duel drowned in the darkness of the cold South Pacific. Three Germans were wounded; *Glasgow*, though hit five times, had suffered no casualties at all. Nor had *Otranto*, which had prudently withdrawn from the action, apparently with Cradock's endorsement, early on. She was quite unfit to have taken part. The two survivors escaped southward at best speed to find *Canopus* and, in company, make their way back to the Falklands. The Battle of Coronel, 1 November 1914, was the first British naval defeat since the American war of 1812 and the first defeat of a formation of British ships since the Virginia Capes in 1781. News of it appalled the Royal Navy, the British public, the Admiralty but above all, those in high command, Churchill and Fisher. From the moment they got word of the disaster, they were bent on revenge.

THE SEARCH FOR THE *EMDEN*

Indignation at the defeat at Coronel was enhanced in the homeland by the humiliations currently being suffered at the hands of von Spee's detached commerce raider, von Müller's *Emden*. Since parting company with the East Asiatic Cruiser Squadron at Pagan Island in the Marianas, east of the Philippines, on 13 August, von Müller had slowly made his way westward into the Indian Ocean, where he correctly estimated the richest pickings were to be found. The passages between the islands of the Dutch East Indies were a major shipping route, leading from Calcutta and Singapore to Hong Kong and Shanghai. The Indian Ocean itself was a British lake, always filled with liners and merchantmen and now also with government-chartered ships carrying men and war materials from the ports of the Indian empire to Egypt and Europe.

Von Müller began what was to prove the most sensational commerce-raiding campaign since the eighteenth century at modest tempo. Having been warned out of Dutch imperial but neutral waters by a local coast-defence battleship in late August, he made his way into the Indian Ocean by 5 September, en route avoiding HMS *Hampshire*, a more powerful cruiser, whose presence he detected by wireless interception, but meanwhile capturing a neutral ship, *Pontoporos*, loaded with British government coal. He took her into company, to join his own collier, *Markomannia*. On 10 September he captured, plundered and sank *Indus*, a troop transport which had not yet loaded her passengers. *Indus* had a wireless set, unusual for a ship of only 3,993 tons, but von Müller got control of the bridge before she could send off a warning. On 11 September, *Kabinga* was the victim. Von Müller used her to offload his captives and sent her away, the beginning of a chivalrous practice that would win him international admiration, even from his enemies. On 13 September, *Emden* intercepted *Killin*, loaded with poor-quality coal; she was sunk by gunfire. The same day *Diplomat*, a fine ship carrying tin, was intercepted and sunk. Her loss affected prices on the London commodity market.

Emden's next encounter was with a neutral, the Italian *Loredano*, which was released. As soon as it was out of sight of the Germans, at the entrance to the port of Calcutta, it signalled by semaphore to the British *City of Rangoon*, which had wireless, news of its experience. *City of Rangoon* wirelessed back to the Calcutta authorities, who held up three ships leaving port, and passed the intelligence on. Via the Royal Navy's intelligence officer at Colombo, the naval base in the island of Ceylon, it reached the Admiralty in London on 14 September and was relayed to Admiral Jerram, commanding at the China station, on the night of the 15th–16th. Next day HMS *Hampshire*, which *Emden* had eluded in the Dutch East Indies at the beginning of the month, was sailed from Singapore in pursuit, together with HMS *Yarmouth*; also alerted were HMS *Minotaur* and the Japanese battlecruiser *Ibuki* and cruiser *Chikuma*. All five ships mounted heavier guns than *Emden*, and several could exceed her speed.

None, however, got a smell of her in the course of a concerted search, although *Emden* kept close to the shipping lanes. On 14 September she sank the empty British merchantman *Trebboch* and, soon afterwards, the *Clan Matheson*, which was run down while attempting to escape. The need then was to coal, and von Müller set off towards the Andaman Islands, in the middle of the Bay of Bengal, to load from *Pontoporos*, which was still in company. En route *Emden* intercepted wireless signals announcing his recent sinkings, some from the released *Kabinga*. Near Rangoon he stopped but released a neutral Norwegian ship, *Dovre*, which warned him of the presence of two French cruisers, *Dupleix* and *Montcalm*, and two British armed merchant cruisers.

Von Müller, feeling the pressure of the chase in the upper Bay of Bengal, now decided to steam south to attack the oil storage tanks at the port of Madras. This was a gesture of sheer bravado, tweaking the lion's tail, and it risked an encounter with one of the stronger ships searching for him; but, as he wrote in his after-action report, "I had this shelling in view simply as a demonstration to arouse interest among the Indian population, to disturb English commerce, to diminish English prestige."[32] On the night of 22 September, the *Emden* approached to within 3,000 yards of the harbour, illuminated the six storage tanks of the Burmah Oil Company by searchlight and opened fire. In ten minutes five of the six tanks were hit and 346,000 gallons of fuel destroyed. *Emden* then retreated into the darkness and got clean away.

During the next five weeks, from 23 September to 28 October, *Emden* had an extraordinary run of luck, though luck was combined with cunning and skill. Von Müller, who closely questioned any captive who would talk, and pored over shipping news in captured newspapers, planned his wanderings across the shipping lanes with care and forethought. His intention in late September was to head back to the Dutch East Indies, where local agents would have arranged for him to coal and resupply out of sight of the authorities in that vast archipelago. After the success of his raid on Madras, however, he decided on a sweep of the shipping lanes in the western Indian Ocean, which carried the traffic from British Africa and the Suez Canal to the Bay of Bengal and the China Seas. The remote atolls of the Chagos, Laccadive and Maldive Islands also provided shelter for coaling, and he contrived to keep a collier with him. He was soon to capture several more, among the thirteen ships he took in this period. Most he sank, finding them either in ballast or loaded with cargo he could not use, but on 27 September he took *Buresk*, loaded with 6,600 tons of the best Welsh coal destined for the Royal Navy's China station, and on 19 October the *Exford*, with 5,500 tons. The cargoes of these two ships were sufficient, if he could escape pursuit, to keep him cruising for a whole year. He took them into convoy, having sent *Markomannia* off to the Dutch islands to await his arrival. The ships he did not sink were loaded with captives and sent off to British ports; most cheered the "Gentleman-of-War" on parting company.

Von Müller now decided on another provocative gesture, a descent on Penang, in the British Malay States, which he had been informed was used by enemy warships. So it was; in the early morning of 28 October *Emden* found in the harbour the Russian light cruiser *Zhemchug*, the French light cruiser *d'iberville* and the French destroyers *Fronde*, *Mousquet* and *Pistolet*. *Zhemchug*, whose captain had gone ashore on pleasure, was quite unprepared to defend itself and was overwhelmed by gunfire and torpedo. *D'iberville, Fronde* and *Pistolet* were in dockyard and out of action. *Mousquet* put up a brave fight but was sunk with a few salvoes. *Emden* picked up the survivors, later transferred to a stopped British steamer, and made off into Dutch waters.

By late October, the Admiralty in London was beside itself with rage at *Emden*'s exploits. It was not only that von Müller had turned himself into a hero, almost as much admired by British seafarers as by neutrals and his own countrymen. His depredations were seriously interfering with the strategic as well as mercantile traffic of the empire, besides damaging the prestige of the Royal Navy, mistress of the seas, and of Britain's imperial officials. Shipping clung to port all over the Indian Ocean, afraid to put to sea, while *Königsberg* was also operating independently. The security of the imperial convoys, bringing the Australian, New Zealand and Indian armies to the war in Europe, was severely compromised. The effort to run down the East Asiatic Cruiser Squadron, lost in the wastes of the South Pacific, was hampered by the activity of a single light cruiser, against which dozens of British, French, Russian and Japanese warships were deployed without effect.

The degree of Allied frustration is conveyed by a minute written by Winston Churchill on 1 October:

> Three transports, empty but fitted for carrying cavalry, are delayed in Calcutta through fear of *Emden*. This involves delaying transport of artillery and part of a cavalry division from Bombay ... I am quite at a loss to understand the operations of *Hampshire* ... What has happened to *Yarmouth?* Her operations appear to be entirely disjointed and purposeless ... if the *Königsberg* is caught, the three light cruisers hunting her should turn over to the *Emden* ... It is no use stirring about the oceans with two or three ships. When we have got cruiser sweeps of eight or ten vessels ten or fifteen miles apart there will be some good prospect of utilising information as to the whereabouts of the *Emden* in such a way as to bring her to action ... I wish to point out to you [the First Sea Lord and others] that an indefinite continuance of the *Emden*'s captures will do great damage to the Admiralty reputation.[33]

Churchill's anger was justified, as the search for both *Emden* and the East Asiatic Cruiser Squadron was ill co-ordinated, divided as it was between several Royal Navy stations—the China, the South America, the Australia and the East Indies—and four navies, the British, Japanese, French and Russian; the search was further hampered by the personal and mechanical deficiencies of the French and Russians. Yet Churchill, all the same, was living in the past. His formula for a "cruiser sweep"—disposing eight ships at the limit of visual range and proceeding in line abreast—was no different from Nelson's and covered no larger an area, only about ninety-six miles wide and twenty-four deep, reckoning visual range from the masthead to be about twelve miles. As the Indian Ocean from Sumatra in the Dutch East Indies to the east coast of Africa is over 3,000 miles across, the statistical chance of finding *Emden* was very slim indeed, since von Müller scrupulously observed wireless silence; he used his wireless to listen, not send. That helped him to evade pursuit. Although intercepted signals—the call sign (QDM) of *Hampshire* had become familiar—could not yield a bearing, since the technique of radio direction-finding had still to be discovered, the strength of the signal gave some indication of range and so could be used to manoeuvre away from a pursuer.

Emden's real safety, however, lay in the vastness of the seas, and in late October von Müller decided on a shift to waters he had not yet raided, right across the Indian Ocean in the corner formed by the Horn of Africa and the Arabian peninsula. To get there he proposed to coal first off Sumatra, from the captured *Buresk,* and then in the remote Cocos and Keeling Islands, opposite the Sundai Strait between Sumatra and Java. That move was in the nature of a deception, to suggest to the enemy that he was heading for Australia.

Buresk, when found on 31 October, brought worrying news. Both *Pontoporos* and *Markomannia*, his original colliers, had been taken by HMS *Yarmouth*. Since their crews knew nothing of his new plan, however, von Müller remained confident; he sent another captured collier, *Exford,* to await him at North Keeling Island, whither he slowly followed. Part of his plan was, before coaling, to destroy the wireless and cable station on Direction Island, part of the Cocos and Keeling group, both to cover his movements and to heighten British anxiety. In the early morning of 9 November *Emden*'s steam pinnace and two cutters were loaded with armed sailors, under the command of *Kapitänleutnant* von Mücke, and sent ahead to destroy the wireless mast and cable terminals. *Emden* followed.

The Cocos and Keeling Islands were a romantic anomaly, a private colony owned by the Clunies Ross family under grant from Queen Victoria. Their only importance to the British empire was as a communication point in the wireless and cable system; it was run by technicians of the Eastern Telegraph Extension Company. British telegraph technicians in distant islands saw themselves in 1914, however, as agents of imperial rule, and those on Direction Island were men of fibre. Before von Mücke's party got ashore, but after von Müller had uncharacteristically transmitted a badly timed signal to *Buresk* to join him, they wirelessed, "What code? What ship?" *Emden* at once began jamming, but the station managed to get off two more signals, repeated several times, first "strange ship in entrance," then "SOS, *Emden* here" just as von Mücke arrived.

Von Müller had made a disastrous mistake. The previous day *Emden* had picked up a call sign which was rightly interpreted to be that of an enemy warship; but her operators estimated the sender to be 200 miles distant and departing southward. Von Müller therefore judged her to be steaming to German Southwest Africa, where dissident Boers had raised an anti-British rebellion. The signal's importance was therefore discounted. It was, in fact, of crucial and fatal significance. Direction Island knew that it had come from HMS *Minotaur*, an armoured cruiser, and it was to her, on picking up *Emden*'s transmission to *Buresk*, that it signalled.

Minotaur was not close; she had, however, been sailing in company with the Japanese battlecruiser *Ibuki* and Her Majesty's Australian ships *Melbourne* and *Sydney*, light cruisers like *Emden* but exceeding her in speed and firepower. They were the escort to the first of the imperial convoys to leave Australia and, since *Minotaur*'s departure, were maintaining wireless silence. The convoy commander briskly decided to detach *Sydney*, which was only two hours' steaming from Direction Island, and sent a visual signal. *Sydney* departed at twenty-six knots.

Aboard *Emden* more mistakes were being made. Had von Müller warned von Mücke at once he might have gathered in his landing party and still escaped; it would have been a near thing but possible. Both men, however, wanted the destruction of the wireless and cable station to be done thoroughly, so both tarried. Even when smoke appeared on the horizon, von Müller decided it was from the summoned *Buresk*; a masthead report was of two masts and a single funnel, which fitted. Between

9 and 9:15 a.m., however, the picture altered, ominously: there were several funnels, which could only mean a warship. Von Müller sounded the ship's siren repeatedly, rang the alarm bell, hoisted the international code flag A, meaning that he was weighing anchor. As the pinnace put off, however, *Emden* began to move. The landing party made desperate gestures. *Emden* slowly gathered speed. By 9:17 she had gained the open sea, and her crew were going to action stations.

Sydney, with eight 6-inch guns to *Emden*'s ten 4.1-inch, and a two-knot advantage in speed, was bound to win the coming encounter unless her crew failed the test; but they belonged to an even younger navy than the German and were determined to prevail. Moreover, *Emden* had left all ten of her principal gunlayers ashore. Fire was opened at 9:40, by *Emden*; she hit with her third salvo. Thereafter, *Sydney*'s heavier weight of shell began to count. At ten o'clock she caused major damage; during the next hour she shot *Emden* to pieces. The Germans stood resolutely to their guns, expending 1,500 shells, but by 11:15 most of her armament was knocked out, and von Müller drove the wreck ashore on North Keeling Island. The survivors, spared the awful fate of Cradock's men in the freezing seas of the southeast Pacific, remained aboard until made prisoner by *Sydney* next day.

That was not quite the end of the *Emden* saga. Before *Sydney* could return to Direction Island, von Mücke had commandeered a trading schooner he found in the harbour, embarked his landing party and sailed off to Sumatra. There he found a German collier, working in the local trade, which he appropriated from her co-operative captain. In her he and his *Emden* men crossed the Indian Ocean to Yemen, in south Arabia, a possession of Germany's Turkish ally. Leaving his ship, he commandeered some native craft and sailed up the Red Sea, then transferred to camels, fought a battle against the Bedouin and won, got on to the Hejaz railway—to be destroyed by Lawrence of Arabia during the Arab Revolt—and so arrived to a hero's welcome at Constantinople, the Turkish capital, on 23 May 1915. Most of the party he had brought from Direction Island survived to tell their extraordinary tale.

THE BATTLE OF THE FALKLANDS

Wireless, in the end, had proved *Emden*'s downfall. Her captain had observed all the correct rules of keeping silence. It was his urge to add to his ship's laurels by attacking the communication outpost at Direction Island that cast him under the heavier guns of the enemy.

A similar urge would seal the fate of von Spee's squadron. After his spectacular victory over Cradock at Coronel, von Spee enjoyed a brief triumph at Valparaiso. It was an odd setting for a celebration of defeat of British naval power, despite the presence of a large German colony, for the seafront of the Chilean port was, then as now, dominated by a monument to Chile's principal naval hero, commander of her fleet in the war of independence against Spain, the British admiral Cochrane.

Moreover, strong though German influence was in Chile, the republic's government was anxious to preserve its credentials of neutrality. Von Spee was told, on his arrival, that he would be held to the legal limitation of twenty-four hours for a visit, by not more than three ships. Von Spee took *Scharnhorst, Gneisnau* and *Nürnberg* into harbour, detaching *Dresden* and *Leipzig* to the island of Más Afuera, so far offshore that he correctly calculated he could there breach neutrality regulations with impunity. While in Valparaiso, where thirty-two German merchantmen were sheltering, he received cabled instructions from Berlin. These warned him that enemy warships were operating all over the Central Pacific, West Indies and South Atlantic and advised him to concentrate his ships and attempt to "break through for home."[34]

The Berlin telegram, and news received from local Germans, persuaded him that he had no choice but to leave the South Pacific. British and Australian ships barred the way westward into the Indian Ocean; powerful Japanese squadrons were gathering in the Central Pacific islands; British and French forces blocked the exit from the Panama Canal into the Caribbean. Though there were even stronger enemy concentrations in and at the head of the Atlantic, his only chance of escape lay in the hope of evading them, perhaps covered by bad weather, in a dash up the South Atlantic towards northern waters. He was encouraged in that view by a further message from Berlin, brought to him on 18 November from Valparaiso, which suggested that units of the High Seas Fleet might be sailed to escort him into the North Sea; this message was disingenuous to the point of dishonesty, for the German Admiralty had already learnt, by painful experiment, how closely the Royal Navy controlled the channels which von Spee would have to negotiate.

Von Spee seems understandably to have been plagued by doubt in the days after Coronel. Committed to a break-out into the South

Atlantic round Cape Horn, he dawdled on his way southward. He coaled as he could from colliers despatched by German agents into the maze of bays and fiords that penetrate the Chilean coast above Cape Horn. As he meandered southward, he gathered news of the collapse of the Kaiser's overseas empire. Von Spee knew that the Pacific possessions, in New Guinea, Samoa, the Bismarcks, the Marianas, the Carolines, had already fallen into the hands of the Australians, the New Zealanders, the Japanese. Now he learnt that German Africa was falling away also. Perhaps he retained a hope that the Boer rebellion in German Southwest Africa would distract British naval strength; it would affect his judgement about the Royal Navy's deployment of force into the South Atlantic.

On 6 December, when at Picton Island near Cape Horn, von Spee decided to make a descent on the British colony of the Falkland Islands in the South Atlantic. He gave as his reason to his captains that the squadron could destroy the coal stocks there, and the wireless station, and that intelligence gave no indication of British warships being in the vicinity; he believed that those available had gone to South Africa. He also hoped to capture the governor in retaliation for the New Zealanders' capture of the governor of German Samoa.

Governors apart, a matter of pique, von Spee's arguments for attacking the Falklands suggest a failure of judgement; perhaps he had been too long at sea, too long in the loneliness of command. The attack was only likely to attract attention to his whereabouts, without doing damage to the enemy. It was not a rational decision. It was to result in the destruction of the East Asiatic Cruiser Squadron, in circumstances horribly equivalent to those of its victory over Admiral Cradock, his ships and men.

Coronel had outraged the British people and the Royal Navy. As soon as the news of the defeat was received, Winston Churchill, political head of the Admiralty as First Lord, and Admiral Fisher, its professional chief as First Sea Lord, had agreed that there must be revenge. Admiral Stoddart, nominally commander of the 5th Cruiser Squadron but effectively acting as senior naval officer in South American waters, was ordered to position a collection of cruisers astride the trade routes off Brazil on 4 November. On the same day another and exceptional order was issued. Churchill had first thought of detaching one of the precious battlecruisers from the Grand Fleet in Scapa Flow, to be supported by the armoured cruiser *Defence*, which Admiralty dithering had earlier denied Cradock. Now Fisher, First Sea Lord again since 1 November, demonstrated his legendary dynamism. He persuaded Churchill that the situation in the far south required making doubly sure and that two battlecruisers should be sent, not one. *Invincible* and *Inflexible* were directed to sail at once, to coal at Channel ports and then to proceed to the South Atlantic. They were first to coal again in Portugal, then proceed to Albrohos Rocks, off Brazil, where they would rendezvous with Stoddart's cruisers *Carnarvon, Cornwall, Kent* and *Glasgow; Glasgow*, the sole survivor of the Battle of Coronel, was currently at Rio de Janeiro, repairing damage. Stoddart's squadron also included the armed liners *Macedonia* and *Orama.* Once assembled, the ships would proceed south, under the command of Admiral Sir Doveton Sturdee, who was bringing the battlecruisers.

Sturdee was fiercely disliked by Fisher, who had allowed him to go only to get him out of the Admiralty, where he had been serving as Chief of Staff. He was, nevertheless, a good choice, a complete professional, a devotee of tactical theory and a man of powerful character. He had grasped, moreover, the cardinal importance of maintaining wireless silence. Alerted to the disturbing fact, as he steamed southward, that French wireless stations in west Africa were transmitting Allied warships' callsigns, he instructed the operators aboard *Invincible* and *Inflexible* that "the utmost harm may be done by indiscreet use of wireless. The key is never to be pressed unless absolutely necessary." In practice, he had less success in controlling wireless insecurity than he may have realised. By stopping to coal at the Portuguese port of St. Vincent, he revealed his big ships' presence in the Atlantic, and the news was duly passed on by operators of the Western Telegraph Company to their colleagues in South America. German agents thus learnt of the arrival of Sturdee's ships at Albrohos Rocks on 24 November; by an inexplicable oversight, however, the news was not communicated to Berlin, and so it did not reach von Spee, then still off southern Chile, where he would have been given it by local German officials. Even worse, though the German consul in Buenos Aires also got word of Sturdee's movements on 24 November, he did not telegraph it across the Andes to Valparaiso but sent the news by steamer, to Punta Arenas, where it would take a week to arrive and which the German squadron did not in practice visit.

Von Spee's bad luck was compounded by bad judgement. Instead of making best speed into the Atlantic, on his chosen homeward journey, he tarried around and off Cape Horn, loading coal he did not really need; his decision to attack the Falklands might have been taken several days earlier, in which case he would not have found Sturdee's avenging battlecruisers awaiting him. It was further bad luck for von Spee that Sturdee, too, had tarried on his voyage south, coaling in a leisurely way at Albrohos Rocks and engaging in target practice against a towed target, which fouled one of *Invincible's* propellers with its wire; a diver had to be sent down to clear the obstruction, causing further delay. As a result it was not until 7 December that the squadron arrived at Port Stanley, the Falklands harbour, when von Spee might have come and gone as much as a week earlier. That it did so without the Germans having any inkling of its proximity was due to Sturdee's one substantial effort to preserve intelligence security, his order that any wireless messages were to be transmitted by *Bristol* or *Glasgow,* whose presence in the area was known to the enemy.[35]

Glasgow, since escaping from the disaster of Coronel, had already been once to the Falklands, in company with the doddering *Canopus,* left her there, gone on to Rio de Janeiro to dock thanks to Brazilian complaisance, and was now on a return journey. Once arrived, in company with *Invincible* and *Inflexible,* and the other cruisers *Carnarvon, Kent, Cornwall* and *Bristol, Glasgow's* captain and crew found the situation at sleepy Port Stanley transformed. Under prodding from the Admiralty, via the local wireless station, the colony had been put into a state of defence. *Canopus* had been beached, in a mud berth that allowed her 12-inch guns to command the entrance and its approaches, her marines had been sent ashore to stiffen the local militia, her light guns had been dismounted to provide dockside firepower, and the mouth of the harbour had been closed by electrically controlled mines.

After 7 December, when Sturdee's ships entered the anchorage, it was therefore impossible for Port Stanley to be taken, the governor to be kidnapped, the coal stacks to be burnt or the wireless station to be destroyed. Those dangers, by then, were for the British secondary considerations. The question was whether von Spee could be caught.

Von Spee was working to a different agenda. His decision to attack the Falklands had also been influenced by the calculation that he could re-coal on a major scale from Port Stanley's stocks and that, by firing the residue and causing other destruction, he could deprive the Royal Navy of its most important base in the South Atlantic, including its communi-

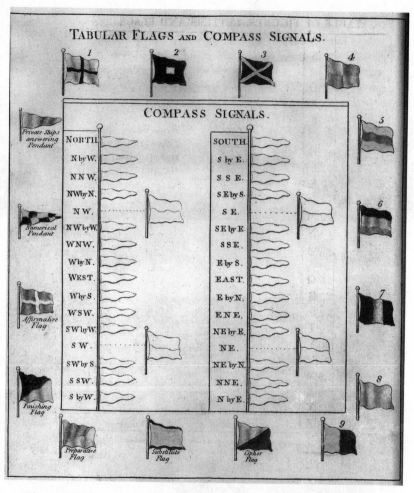

The second illustrated page of Home Popham's *Telegraphic Signals or Marine Vocabulary*, 1803. According to the flags' position, any of the thirty-two points of the compass could be indicated; many thousands of other signals could also be made by his flag system.

The Battle of Aboukir, 25 July 1799. Napoleon's victory over the Turkish rulers of Egypt, near the site of Nelson's destruction of his fleet the previous year.

Thomas Troubridge, who became one of Nelson's most trusted captains during the Nile campaign and founded a naval dynasty.

Vice Admiral Lord Nelson in 1803, five years after the Nile victory, wearing his British and Turkish decorations for the victory.

General Thomas Jackson (1824–63), known from the stand of his brigade at the 1st Battle of Bull Run (Manassas), 1861, as "Stonewall"; after Robert E. Lee, he was the most famous of Confederate commanders.

Some Union generals of the Civil War: Banks, Hooker, Hunter, Ord, Frémont, Sigel, Butler, Burnside. Banks and Frémont opposed Stonewall Jackson in the Shenandoah Valley. Burnside replaced McClellan after his failure in 1862.

George Everest, Surveyor-General of India, supervises the erection of a "trig" point in 1835, during the Great Trigonometrical Survey, 1800–41. His theodolite assistant stands to the right.

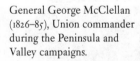 Union supply column in the
Shenandoah Valley during the 1862
campaign.

General George McClellan
(1826–85), Union commander
during the Peninsula and
Valley campaigns.

(*Above*) SS *Great Eastern,* built by Isambard Kingdom Brunel, 1854–58. When launched she was the largest ship in the world (688 feet, 18,914 tons). A failure as a passenger vessel, she was converted to cable-laying and is seen loading cable from an old man-of-war hulk. She laid four transatlantic cables and one from Aden to Bombay. (*Below*) Admiral Graf von Spee (1861–1914) (*left*) Commander of the German East Asiatic Squadron, at Valparaiso after the Battle of Coronel, 1914.

Australian sailors bringing back German prisoners to HMAS *Sydney*, 9 November 1914. The wrecked *Emden* is ashore on Direction Island.

Rear Admiral Sir Christopher Craddock (1862–1914), tragic commander of the British fleet at the Battle of the Coronel.

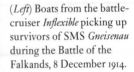

(*Above*) *Kent, Inflexible, Glasgow* and *Invincible* leaving Port Stanley, 9:45 a.m., 8 December 1914, at the outset of the Battle of the Falkands.

(*Left*) Boats from the battle-cruiser *Inflexible* picking up survivors of SMS *Gneisenau* during the Battle of the Falkands, 8 December 1914.

Admiral Wilhelm Canaris (1887–1945), head of the Abwehr during the Second World War. In 1914 he was serving aboard *Dresden* and was interned after her sinking in Chilean waters, 14 March 1915.

cation centre. From the intelligence available, he discounted the possibility of a superior British force being present; the last wireless report he received, on the night of 6 December, from the collier *Amasie,* said that the harbour was empty except for *Canopus.* The report was then correct; that it was falsified by Sturdee's arrival within the next twenty-four hours is a perfect demonstration of the primacy of real-time intelligence.

Real time benefited Sturdee perhaps undeservedly. Had he cracked on from Albrohos Rocks instead of shepherding his flock of cruisers southward in line of search abreast, he would have arrived at Port Stanley in ample time to have been observed and for von Spee to have been warned off. Having escaped that bad outcome, he now risked another by persisting in his lack of hurry inside Port Stanley harbour, a strange characteristic in a man so forceful. By breakfast on 8 December, only *Carnarvon* and *Glasgow* had fully refuelled; the two battlecruisers still had colliers alongside; *Kent* had not begun to replenish, while *Cornwall* and *Bristol* both had engines open for maintenance. It was to an unprepared squadron that at four minutes before eight *Glasgow* hoisted the flags signifying "enemy in sight."

The alert had first come from Sapper Hill, one of the heights surrounding Port Stanley that were to be assaulted by the soldiers of the British Task Force sixty-six years later in 1982.[36] Von Spee, whose surety of touch had been so complete at the outset of his war cruise, had now added a final and fatal addition to his cumulative list of misjudgements; instead of sending forward his light cruisers, whose speed would have permitted escape from the trap, and a warning to the rest of the squadron, he used *Gneisenau* as his lead ship, with the faster *Nürnberg* as companion. As a result, two ships jointly unable to defend themselves or to outrun pursuit ran headlong into disaster.

Von Spee, trundling along astern in *Scharnhorst,* had ordered his squadron to clear for action as early as 5:30 that morning. By 8:30 the captain of *Gneisenau,* then well ahead, made out smoke rising over Port Stanley but concluded that the coal stocks were being fired, as they had been by the French on the squadron's descent on Tahiti three months earlier. He also got the colony's wireless mast in view. Not until 9:00 a.m. did he learn from an officer stationed in the foretop that other masts were visible, tripod ships' masts in Port Stanley harbour. Tripod masts meant only one thing: British big-gun ships.

Maercker, the captain of *Gneisenau*, had always doubted von Spee's belief that any big British ships in southern waters were bound for Africa, where German troops and Boer rebels were waging colonial warfare. His disbelief was now to be confirmed. First, as *Gneisenau* hove into range, *Canopus* opened fire from its mud berth with its ancient 12-inch guns. At over 11,000 yards, fragments of shell hit *Gneisenau*'s after-funnel. She and *Nürnberg* had already turned away and begun to work up speed; but as *Nürnberg* loyally stuck by the bigger ship, both were limited to 20 knots. They were soon under pursuit by *Kent*, a 23-knot armoured cruiser, then by the light cruiser *Glasgow*, 25 knots, and *Carnarvon*. *Cornwall*, a sister ship, was the last to leave, soon working up to 22 knots. Before her had gone the battlecruisers, both capable of 28 knots.

Von Spee, on hearing the report of tripod masts, had apparently concluded that British battleships were present, perhaps *Iron Duke* and *Orion* class, with a best speed of 20 to 21 knots; he may, even after his misjudged arrival, have calculated that he could get away. *Gneisenau*'s hasty withdrawal had conferred half an hour's, perhaps a whole hour's, start, and there was always the chance, in sub-Arctic latitudes, of running into concealing fog. The East Asiatic Cruiser Squadron stretched out, *Scharnhorst* and *Gneisenau* both striving to exceed 20 knots.

Maercker had actually wanted to fight, and there was logic in that urge. The German armoured cruisers might have inflicted disabling damage if they had attacked right at the outset. The anchorage was crowded, it was apparent that few of the British ships had steam up, most were lightly armoured or not armoured at all and even *Inflexible* and *Invincible*, the first and therefore oldest battlecruisers in the fleet, were not much better protected than *Good Hope* and *Monmouth* had been. Von Spee, however, had answered his request to press forward with the signal "Do not accept action, head east at full speed."[37] By 9:45, when the battlecruisers cleared Port Stanley, the German squadron was heading for the horizon.

It was a sunny day, as the evening of Coronel had been, but time favoured von Spee even less than it had Cradock. In the sub-Arctic summer, eight hours of daylight promised. At 10:20 Sturdee hoisted the Nelsonian signal "General Chase"; it was indeed older than Nelson. By 10:50 Sturdee, conscious of having time in hand and not wishing to scatter his squadron, slackened the battlecruisers' speed so that the slower ships could keep up. They nevertheless continued to overhaul the enemy, and it was clear that the battlecruisers' big guns would soon tell. At 12:50, having meanwhile sent the crews to lunch, Sturdee ordered, "Engage the enemy."

Inflexible and *Invincible* opened fire at an estimated range of 16,500 yards; *Glasgow*, the only ship able to keep pace with them, did not have that reach. When the range came down to 15,500 yards, von Spee ordered his light cruisers to disengage—as Cradock had so ordered *Glasgow* at Coronel—and they turned away and made for South America. They were followed by *Kent, Cornwall* and *Glasgow; Carnarvon* was now managing to keep up with the battlecruisers.

Between 1:20 and 2 p.m., they were engaged very effectively by the Germans, whose 8.2-inch guns actually outranged the British 12-inch, though they could do little damage. Von Spee's ships also benefited from the funnel smoke that impaired British range-taking. By clever manoeuvering, von Spee was for a period able to close to a distance at which his secondary armament could hit. His bravado alarmed Sturdee, who could not risk damage to his ships when they were so far from dockyard, and he bore away. Not until after three o'clock, by which time his turret crews and gunnery direction officers had begun to get the measure of the enemy, did he shorten the range again. Then his heavier weight of shell began to tell, and by 4 p.m. *Scharnhorst* had suffered so many 12-inch hits that she was clearly soon to sink. Her upper works were torn and twisted, and fires raged within her hull. Von Spee turned towards the enemy to attempt a final response with torpedoes, but at 4:17, with the sea lapping her deck, *Scharnhorst* turned over and sank.

There were no survivors, as there had been none from *Monmouth* and *Good Hope*. With *Gneisenau* still energetically in action, the British could not pause to lower boats, and they swept on to leave whoever had not succumbed to fire and explosion to drown in the icy seas. Among the victims were von Spee and his sons.

Maerker, in *Gneisenau*, now had to defend himself against three enemies, the two battlecruisers and *Carnarvon*. His plight was hopeless, but he refused the call to surrender. The new German navy was trying to win a reputation for doggedness to equal that of the old mistress of the seas. At 6 p.m., as 200 survivors of the original 850 cheered the Kaiser, the sea overwhelmed the fireswept deck on which they stood, and *Gneisenau* turned over. The British ships rescued 190; Maerker was not one of them.

Kent, Cornwall and *Glasgow*, the only ship to have fought in both the

South American battles of 1914, rapidly overhauled the German light cruisers that von Spee had ordered to save themselves. *Leipzig* was sunk by *Glasgow* and *Cornwall;* her flag still flew, and only eighteen of her crew survived. *Nürnberg* was sunk by *Kent*, twelve of her crew were picked up but only seven survived exposure to the freezing Atlantic. Of more than 2,000 sailors in *Scharnhorst, Gneisenau, Nürnberg* and *Leipzig*, almost all had been killed or died in action. The balance sheet with Coronel was nearly exact.

Dresden, for the moment, eluded her pursuers and was to remain beyond their reach for another three months. At first she hid in the maze of inlets that penetrate the Chilean coast above Cape Horn. Her captain waited in vain for colliers to resupply her. Eventually he went north into the Pacific, hunted by the British cruisers, which he evaded or outran. In early March 1915, however, the intelligence department of the Admiralty was sent an intercepted German telegram by an agent in Chile which, when decoded, revealed that *Dresden* was awaiting coaling off Coronel. *Glasgow* and *Kent*, co-ordinating their movements by wireless, eventually found *Dresden* at Más a Tierra—with Más Afuera it forms the island group of Juan Fernandez—on 14 March and closed. The German ship had only eighty tons of coal left and was at anchor, without hope of escape; a last wireless message from Berlin, relayed via Chile, had given Captain Lüdecke permission to seek internment. The British did not wait for the Chilean authorities to intervene. *Glasgow* opened fire and, though it was returned, inflicted in a few minutes sufficient damage to force Lüdecke to raise a white flag. As firing ceased, he sent a boat to parley for surrender, his object being to gain sufficient time to scuttle. By a bizarre coincidence, the officer he chose was Lieutenant Canaris, who, during the Second World War, was to direct the Abwehr, Nazi Germany's military intelligence service. Luce, the captain of *Glasgow*, refused to negotiate, but Canaris won enough delay to allow flooding and explosive charges to send *Dresden* to the bottom. He and the surviving members of her crew were subsequently taken into internment by the Chilean navy.

The German cruiser campaign in distant waters was over—very nearly. *Königsberg*, which had never belonged to the East Asiatic Squadron, was to survive until July 1915, holed up in the swampy delta of the Rufiji River in German East Africa, where it would eventually be destroyed by the gunfire of two shallow-draught monitors, *Severn* and *Mersey*, directed by the observation of aircraft, the whole force having been sent from England at great expense and difficulty earlier in the year.

The cruiser campaign had never threatened to undermine Britain's control of the seas. It had not even seriously damaged British maritime trade. The total of ships sunk by *Emden* and *Karlsrühe,* the most effective raiders, was thirty-two, gross tonnage 143,630; that was to be set against a total of nineteen million tons of British shipping plying the seas. Two of Germany's armed merchant cruisers, the liners *Kronprinz Wilhelm* and *Prinz Eitel Friedrich,* had done almost as well, sinking 93,946 gross tons. The East Asian Cruiser Squadron proper had sunk no merchantmen at all.

Yet the German cruisers had caused serious alarm to the Admiralty and forced the diversion of very large numbers of warships to distant waters, away from the crucial areas of naval confrontation, the North Sea and the Mediterranean. Of the situation on 12 November, when *Inflexible* and *Invincible* were voyaging southward towards the Falklands, Winston Churchill lamented, "the strain on British naval resources in the outer seas was now at its maximum—a total of 102 ships of all classes. We actually could not lay hands on another vessel." Admittedly, the total included many units that pre-dated the naval revolution, unfit to fight in home waters; but to it must be added French and Russian ships, and Japanese ships, if they were tied to the Pacific. If the maximum German commitment of cruisers to distant waters is reckoned at eight, the strategic return on the ratio was considerable.

Several British and foreign warships had been sunk, *Monmouth, Good Hope, Zhemchug, Mousquet, Zelée* and others damaged. The imperial convoys bringing Australian, New Zealand, Canadian and Indian troops to Europe had been delayed in sailing, and very many British and friendly neutral ships, carrying essential supplies, had been confined to port for fear of capture or sinking, in places as far apart as San Francisco, Rangoon and Calcutta. *Kreuzerkrieg* (cruiser warfare) could not be dismissed as a failure.

Yet it had failed in the end. The prestige of the Royal Navy, dented by Coronel, had been completely restored by the victory of the Falklands, while that of the young German navy had, after the brilliant episode of Coronel and the dashing exploits of the *Emden,* been peremptorily deflated. The Kaiser's navy ended 1914 as it had begun: a service with a reputation to make.

Why had *Kreuzerkrieg* failed? The persistent need to coal, which limited the cruisers' freedom of action, and the drag of accompanying colliers, was one reason; yet *Emden* had coaled only eight times and was never short. Indeed, shortage of ammunition, rather than of coal, may be thought the German captains' real difficulty; after Coronel, von Spee's magazines were half empty and, even had he managed to escape from the Falklands, he would have had insufficient ammunition to fight his way through to home if engaged by British ships in the Western Approaches or the North Sea. The failure to position ammunition ships, as colliers were positioned, may be thought a cardinal error by the German Admiralty.

In the last resort, however, cruiser warfare failed because the Germans could not conceal the movements of their ships. A steady stream of clues as to their whereabouts were picked up, often with great rapidity, sometimes in real time, and circulated with efficiency by the British between the Admiralty, local commands and pursuing naval units on the worldwide wireless and cable network. No delay, such as that which had afflicted Nelson, impeded the chase or obliged a return to base to pick up a lost scent—as after Nelson's first visit to Alexandria.

There were failures. At the outset von Spee concealed his movements with great skill by observing wireless silence and listening to the transmissions of Allied ships that did not keep quiet; *Emden* was particularly skilful at evading HMS *Hampshire* in the Bay of Bengal by steering away from her call sign (QMD), made possible by listening for a weakening of the signal, an anticipation of direction-finding, which current technology did not yet permit. Von Spee was equally skilful, in the days before Coronel, by using a sole ship, *Leipzig*, to transmit and relay messages, thus disguising the size of his force.[38]

Cradock failed to detect the deceptions. On the other hand, von Müller brought about his own downfall by his foolhardy decision to attack the Cocos and Keeling Islands, a quite unnecessary act, which ran him straight into the line of sight of the wireless station and provoked the transmission of perhaps the earliest ever piece of real-time intelligence of the electronic age, "strange ship in entrance." It brought *Sydney*, with its superior 6-inch guns, down to the harbour in less than two hours.

Cradock was also incautious in the preliminaries to Coronel, his signals between ships revealing to the enemy the presence of his squadron, information amplified by messages from German agents ashore. His incaution was more than replicated, however, by von Spee, who chose to put his trust in inaccurate reports of the emptiness of Port Stanley harbour and then, for the sole purpose of destroying its not very important wireless station, steamed his squadron into a position dominated by the big guns of the British battlecruisers, which had arrived undetected, thanks to scrupulous observation of wireless silence, and from which the inferior speed and firepower of his ships allowed him no escape.

Strategically, the First World War, as a naval war, was to be dominated by the new invention of wireless. Coronel and the Falklands, unlike any other naval battles of 1914–18 though they were, belong to an emerging pattern. Before 1914 fleets at war operated in their search for each other as they had always done, working by line of sight and visual signal. After 1914, intelligence gathered by line of sight could be transmitted to infinite distance at the speed of light. Navies would take time to understand and implement the potentialities of the new technology. Yet it had altered for ever the nature of war at sea. Cradock and von Spee were victims of a failure to understand the new world, Sturdee a perhaps undeserving beneficiary. Less than thirty years after his victory, a new electronic dimension, radar, would almost eliminate the importance of line of sight. The Nelsonian world would have gone for ever.

———◄o►———

Crete: Foreknowledge No Help

THE WIRELESS WAR of August to December 1914, in the far Atlantic and South Pacific, was the most dramatic intelligence episode of the Great War. Historians of the Eastern Front were later to suggest that Germany's crushing victory over the Russians at Tannenberg, in east Prussia, in 1914 was brought about by Russian wireless laxity; Rennenkampf and Samsonov, the commanders of the invading Russian First and Second Armies, were accused of transmitting to each other the positions they intended to reach next day *en clair* (without encoding or enciphering their messages). More detailed research suggests that the Germans were guilty of equal laxity and that the cause in both armies was not carelessness but a lack of trained cipher clerks.[1]

The course of the campaign of 1914 in the West is not held to have been affected by intelligence failures, since few important messages were sent by wireless; the French, using the Eiffel Tower in Paris as a transmitter, jammed German wireless comprehensively but without discernible effect. During the years of static warfare that followed, neither wireless messaging nor interference played any significant part, since the available equipment was ill-adapted to trench conditions and most communication, both strategic and tactical, was conducted by hand-carried paper, as was traditional, or by telegraph or telephone. Some overhearing, by erratic earth-conduction, was found to be possible, but its use was short-term and tactical at best.

Wireless interception by navies was of greater significance, although both the British Grand Fleet and the German High Seas Fleet became scrupulous at observing wireless silence. The British, ever on the alert for warnings of the Germans "coming out" into the North Sea, garnered every message that they could. On the only occasion, however, when advance warning might have made a difference, the *superbitas* of the Royal Navy's traditional officer class robbed the fleet of advantage. The Chief of the Operations Division, Rear-Admiral Thomas Jackson, visited the naval intelligence division in the Admiralty, known as OB 40 (Old Admiralty Building Room 40) on 31 May 1916 to ask where its direction-finders (direction-finding had improved since 1914) placed the German signal DK, call sign of the German High Seas Fleet's flagship. He was told, correctly, that the location was Wilhelmshaven and departed without explaining his reason for asking. Jackson was the sort of sea-going naval officer who did not share his thoughts with the non-combat-ant intelligence staff, composed as it was of such lesser beings as naval schoolmasters, university linguists and academic mathematicians. Had he explained why he wanted to know where DK was, he would have been told that the German flagship left its call sign at home when pro-ceeding to sea, to disguise its movements, and adopted another. On the basis of his half-clever question, Jackson therefore telegraphed Jellicoe, commander of the Grand Fleet at Scapa Flow, to assure him that the High Seas Fleet was still in harbour. As a result, Jellicoe eventually heard that the Germans were "out" from Beatty, commander of his battlecruis-ers, which had been sailed south on other information. He was then at sea himself, but making less than best speed in order to conserve fuel, so that he was late meeting the enemy battleships off Jutland, late fighting the battle and late cutting off their retreat. Admiral Jackson's disinclina-tion to take the codebreakers into his confidence robbed the Grand Fleet of a major opportunity to scupper the German navy for good.[2]

Jackson was exceptionally arrogant. A Royal Naval Volunteer Reserve lieutenant, W. F. Clarke, belonging to the OB 40 staff, recorded that he "displayed supreme contempt for [our] work. He never came into the room during [my] time there except on two or three occasions, on one of which he came to complain that one of the locked boxes in which the information was sent him had cut his hand, and on another to say, at a time when the Germans had introduced a new codebook, 'Thank God I shan't have any more of that damned stuff.'"[3] There were many like Jackson, however, if not so bad, and it would take nearly a generation to pass before operations officers would begin to accept that most "raw" intelligence was only as good as the interpretation put on it, often best supplied by the intelligence officers who gathered it on a day-to-day basis.

OB 40 also had its admirers, and rightly so. It began to supply crucial information almost from the start, including forewarning of the raid on the English east coast towns of Scarborough, Hartlepool and Whitby on 16 December 1914.[4] Had it not been for a visual signalling error committed by Beatty's flag lieutenant—who was to repeat his mistake on three later occasions, at Jutland with disastrous effect—the Scarborough raid might have resulted in the destruction of the German battlecruiser force.[5] OB 40 had only then been in existence since 8 November and had been brought into being because of an intelligence windfall. In late October the Russians had delivered to the British a copy of the main German naval codebook (SKM) and a collection of square-ruled charts, used to denote sea areas. They had been recovered from *Magdeburg,* a German light cruiser lost in the Baltic on 26 August. OB 40 subsequently acquired the codebook used for communication between German merchant and naval ships (HVB), found in a German steamer interned in Australia early in the war. Finally, it got possession of a codebook used by German senior officers (VB), allegedly dredged up in the nets of a British trawler off Holland on 30 November, at a spot where four German torpedo boats had been sunk on 17 October.[6]

It was with this material, and intercepts collected via a hastily established chain of coastal listening stations, that OB 40 went to work. They were aided by the Germans' very free use of wireless—forced upon them in part by the dragging up of their oceanic cables by the British cable ship *Telconia* on 4 August 1914—but above all by the nature of the means the Germans used to disguise their signals.

Secret writing takes two forms, known to cryptologists respectively as codes and ciphers. Cipher is a method hiding meaning by altering the form language takes, either by "transposition" or "substitution." Transposition, a technique so ancient that there is no record of its origins, works by changing the order of letters; the simplest system, familiar to any cipher-minded schoolboy, is to shift once along in the alphabet, so that A becomes B, B becomes C and so forth. "The cat sat on the mat" is thereby recorded as "UIF DBU TBU PO UIF NBU"; the result is unlikely to baffle an interceptor for any length of time. There are many ways of complicating a message in transposition cipher; one of the simplest is to run the letter group together—UIFDBUTBUPOUIFNBU—to disguise the word length, but it provides little protection. Another, more sophisticated, is to shift two or three or ten letters along in the alphabet; while straight transposition underlies the ciphering, however, the iron laws of "frequency analysis" will yield the solver a way in. The law of frequency reveals that, in English, E is the most commonly written letter, followed by A and so on. Frequency tables, known to all cryptologists, provide a ready means of decipherment. Frequencies are different in other languages—Z, rare in English, is common in Polish—but the tables cannot be defeated.

Not, that is, unless complexities are introduced. Cryptographers—those who write ciphers or codes—have devised many complexities. Perhaps the best known, and most difficult, is the alphabetical grid, which arranges the twenty-six letters of the Roman alphabet (reduced to twenty-five by combining the letters I and J) in a square five letters wide and five deep, and numbers the columns. If A is the first letter in the top left-hand corner, it is rendered as figures 11, and so on to Z as 55. At its most elaborate, known as a Vigenère, after its sixteenth-century French inventor, the square is twenty-six by twenty-six, presenting a frequency problem of great difficulty. It is not insurmountable, though it was long thought to be so.[7]

Further complications may be devised, particularly when cryptographers begin to use figures rather than letters in transposition. There developed, during the seventeenth century, a strange halfway house between transposition and its cipher alternative, substitution, in which, for example, King Louis XIV's principal cryptographers, the Rossignols father and son, rendered whole words into mathematical figures. The technique had been anticipated by numbering common French "digraphs," e.g., QU, OU, DE, but the Rossignol system, known as the Great Cipher, defeated everyone; it was only unlocked, long after the meaning of the messages in which it was written had ceased to have importance, at the end of the nineteenth century.

By then, however, cryptologists were on the brink of instituting a new cipher system altogether, employing full-scale mathematical "substitution" for letters. Mathematical substitution appeared to promise true impenetrability since by addition or subtraction, a numerical message could be so varied that a cryptanalyst—who attacks secret writing—

would simply be defeated by time; but as long as the intended recipient possessed the "key" to understanding the chosen mathematics, it could be read at the other end.

Keys were the problem: how to ensure that senders and recipients possessed the same set, how to deny keys to the enemy? The simplest solution was to write the keys in a book, logically arranged, which could be owned by all legitimate parties. Codebooks were widely in use during the eighteenth century, if only to disguise the more important words in a message, for example, the proper names of people, places, ships and so forth, the rest being left in plain language. Major Benjamin Tallmadge, George Washington's chief of intelligence after 1778, devised a codebook out of *Entick's Spelling Dictionary* by taking from it the most frequently used words, numbering them in alphabetical or numerical order and adding random words for those not listed. He also chose sixteen numbers for key individuals and thirty-six others for cities or places. Tallmadge kept the original, sent another copy elsewhere and the third to George Washington. It cannot have disguised much from the British if a letter of 15 August 1779 is a fair sample: "Dqpeu [Jonas] beyocpu [Hawking] agreeable to 28 [an appointment] met 723 [Culper Jun.] not far from 727 [New York] and received a 356 [letter]."[8]

This amateurish example discloses the principal weakness of a codebook: that, by collecting the words used, from messages intercepted, parts of the book can be reconstructed; if enough material passes through the hands of the enemy, it can be reconstructed in its entirety.

An apparent safeguard is to avoid the use of the alphabet altogether and employ only figures, singly or in groups. By the beginning of the nineteenth century the British were doing just that. A message from William Drummond, British emissary to Denmark, to the Foreign Secretary, Lord Grenville, in 1801, on the eve of the Battle of Copenhagen reads, in its last sentence, "3749 2253 529 2360 1268 2201 3356," which stands for "Count Bernstorff does not even affect to conceal his alarm and inquietude."[9] The protection, however, is not as great as seems, even though, in the original, there is no indication of sentence length, nor do the groups betray word length. But by painstaking accumulation, again, watching for repetitions, and by guesswork, the codebook can be reconstructed and meanings deduced.

From the use of all-figure codes, it was but a short step to a much more secure system, technically known as "super-encipherment." It employed two—or more—keys: the codebook itself and a system of figure alteration, by addition or subtraction. Since the groups thus altered did not coincide with those in the book, and did not obviously repeat themselves, retrieval of meaning became much more difficult. Yet not impossible: there was an underlying logic, supplied by the second key, which might be established by mathematical analysis. German diplomatic telegrams were, during the First World War, commonly super-enciphered. Oddly the most famous, the Zimmermann telegram, was not. Broken by Room 40, its contents—which encouraged Mexico to attack the United States—prompted President Wilson's declaration of war on Germany.

Many other methods of complexifying secret writing had been devised by the beginning of the twentieth century, many of them variations on the Vigenère square. The most ingenious was invented by an American army officer, Major Joseph Mauborgne, in 1918. It came to be known as the "one-time pad" and was indeed unbreakable. A Vigenère square was constructed in two copies, one held by the sender, one the receiver. It gave the key to the enciphered message; once used by both parties, it was destroyed. One-time pads protected messages absolutely because the coincidence between cipher and plain text was entirely random and the absence of repetition, assured by destruction, forbade all chance of frequency analysis, or any other method favoured by crypt-analysts.

The one-time pad suffered, however, from a disabling defect. To be useful, pads had to be distributed on a very large scale and had to be identified, so that sender and recipient knew that they were working on the same document. The generation of random numbers on a large scale is not a simple matter either; deliberate attempts to randomise will inadvertently follow patterns, the more conspicuously the greater the pace and volume of output, while large-scale distribution in real time poses insurmountable logistic difficulties. Any persuasive solution to the problem of randomising and distribution was, therefore, likely to be enthusiastically welcomed in military circles everywhere.

THE ENIGMA MACHINE

It was to be supplied by a German inventor, Arthur Scherbius, who in 1918 set up a small engineering business to produce and market inventions. One of the ideas he took up was for a machine that would encipher—but also decipher—automatically. Cipher machines were not a new idea; simple versions had long existed. One, indeed, had been invented by Thomas Jefferson, polymath and third president of the United States. It consisted of thirty-six discs, separately rotatable about an axle, on the rims of each one of which were engraved the letters of the alphabet in random sequence. The sender enciphered by turning the discs to produce a plain text (which could not, of course, be of more than thirty-six letters, though it might be of fewer). He then sent another row of letters to the intended recipient. The recipient turned his discs to replicate the message, which was a garble, and then examined all the other rows of letters. One was the plain text. Security was provided by arranging the discs in a different sequence on the axle, the change of order being preordained and known only to sender and recipient. Had the order of discs not been variable, messages would have succumbed quite quickly to frequency analysis; as it was variable, giving thirty-six possible orders, a number having forty-eight digits $(36 \times 35 \times 34 \times 33 \ldots)$, messages were effectively irretrievable in the pre-computer age.[10] In the period 1910–20, several attempts were made to mechanise the rotating disc principle though none achieved commercial success. Nor, at first, did the Scherbius disc machine, when offered for sale under the trade name Enigma in 1923. In the later twenties, however, Scherbius managed to interest the German armed forces in Enigma. Models were bought and adapted and in 1928 the German army began to use it for all secret communication susceptible to interception, which effectively meant radio messages. It was also taken into service by the German navy.

What particularly attracted the German forces to Enigma was a feature unique to the Scherbius system, the "reflector" disc, which equipped the machine to work both to encrypt and decrypt; a message encrypted on one Enigma machine would, when entered in encrypted form on another Enigma machine set up in the same way, yield the plain text automatically. That feature eliminated the need to decrypt by separate process, a tedious and time-consuming business. In that sense, the Enigma was an early example of the "on line" machine (though it was emphatically *not* a computer, but an electromechanical switching system).

Enigma had other characteristics making it attractive to military signal services: compactness and portability. Outwardly it resembled a portable typewriter of the period, with a typewriter keyboard, in the military version originally arranged alphabetically rather than in the QWERTY order, and a strong carrying case; there were no numerals, all numbers having to be spelled out. It was normally powered by dry-cell batteries.

The chief virtue of Enigma, however, lay in its ability to multiply possible encryptions by an order of magnitude so large as to defy decryption by an outsider in any practical dimension of time; estimates of how long it would take mathematicians to break an Enigma encryption by brute calculation vary, but the Germans themselves believed that the lifetimes of thousands, perhaps millions of mathematicians, working without sleep, would not suffice to decrypt a single message. Enigma was supposed to have complicated the making of the "key," which is the heart of the cipher system of secret writing, to a degree lying beyond the power of human intelligence to produce a solution.

The purpose of the key is to disguise letter frequency and to multiply, to as near infinity as possible, the number of mathematical attempts necessary to establish a frequency table. The Vigenère square was one method of lengthening a key; there are many others, including the use of a common text, such as a word possessed by both users. The principle, however, remains constant: to make the key so long, consistent with convenient decipherment, as to defy mathematical process. Total mystification, unless by the one-time pad, is never to be achieved; the key has a logic and is therefore retrievable by reason; the object is so to overload the powers of reason as to defeat it in real time, indeed in any sort of human time at all.

Enigma appeared to do exactly that. Its electromechanical switching process was entirely logical; but unless the steps by which it worked were understood, and unless the basis on which the switching was started was known, then indeed the mathematics of its decryption became insurmountable.

The steps and the starting were separate from each other: the first was intrinsic to the machine, though variable within definable limits, the second was illimitable, in theory at least, being chosen by human decision.

The intrinsic characteristics of Enigma could produce five variables, most dependent on its discs: (1) the internal wiring of the discs, (2) the choice of discs, (3) the arrangement in order, from right to left, of the chosen discs, (4) the alteration of the disc rims, (5) the "plugging" (*steckerung*) of the discs from one to another.

Each Enigma disc, which was removable, had two faces, with fifty-two contact points for the letters of the alphabet; the right-hand face had twenty-six transmitter points, the left-hand face twenty-six receptor points; the interior of the disc wired the transmitter and receptor points together in a secret way.

When the key on the typewriter keyboard was depressed it sent an electric pulse through the right-hand (fixed) disc to the right-hand face of the first rotor. By internal wiring, that rotor transformed the impulse from, say, A to B on the disc's left-hand face (in fact, the course of the wiring; something much more complex was expected by Germany's enemies). The right-hand face of the second disc picked up the pulse and transmitted it by internal wiring to its left-hand face; the third rotor then worked similarly. When the pulse left the third rotor it was picked up by the fourth (fixed) reflector disc and sent back again along the same route as it had been received. With this difference: because each rotor was notched to turn over when it had received twenty-six pulses, the returning signal would find the right-hand rotor in a different position, by one letter, on its first journey, the second letter in a different position (by one letter) on its twenty-sixth journey and the third letter again in a different position, by one letter, on its six hundred and sixty-sixth (26×26) journey.

The eventual destination of the pulse was an electric bulb, each representing a different letter of the alphabet; as each lit up, in sequence, on the receiving machine, the bulbs would reveal the plain-text message. Before the pulse reached a bulb, however, it went through another multiplying process; at the end of the return journey, it moved to a "plug" board, resembling that of a manual telephone exchange, on which six letters were plugged to another six (the number of plugs was later increased): A to E, for example, and G to T and so on; the pluggings were altered according to instructions for monthly, weekly, daily and eventually twice-daily use.

Thus the intrinsic complexity of Enigma. It was enlarged by human alterations. In the original version there were only three rotors.[11] Part of the procedure laid down for Enigma's use, in frequently changed instructions, was to alter the order in which the rotors were arranged in their slots. Finally, each of the rotors had on its outer rim a rotatable ring, often described as the "tyre on the wheel," which would be moved to any one of twenty-six alphabetical positions. When setting up the machine for use, the operator moved the rim to a position laid down in instructions. The number of variables with which a cryptanalyst was confronted was therefore as follows:

Disc positions (three discs): $26 \times 26 \times 26$	= 17,576
Disc sequence (ABC, ACB, BCA, BAC, CAB, CBA)	= 6
Plugboard connections	= over 100 billion
Total	= 10,000 billion.[12]

That number does not allow for rotating the outer rims on the three discs, which multiplies it by 17,576.

The task faced by an interceptor of an Enigma-encrypted text may be represented in this way. If he were able to check "one setting every minute [he] would need longer than the age of the universe to check every setting."[13] Even if he had got possession of an Enigma machine, and so had only to proceed through the initial settings of the discs (17,576) to see if the encrypt rendered a plain text, he would still, working day and night, need two weeks to check all the settings, allowing one minute for each.[14] No wonder Scherbius advertised his machine as generating "unbreakable" ciphers and that the Germans believed theirs to be so.

BREAKING ENIGMA

Yet Enigma was to be broken and not long after it had been put into use. Those who achieved the solution were cryptanalysts of the Polish army which, as the defender of the Versailles state most resented by post-war Germany, took a keen and necessary interest in German military encrypted transmissions. What is extraordinary, positively intellectually heroic, about the Polish effort is that it was done initially by the exercise of pure mathematics. As Peter Calvocoressi, an initiate of the British cryptanalytic centre at Bletchley Park, has succinctly put it, "in order to break [a machine] cipher, two things are needed: mathematical theory and mechanical aids."[15] The Poles eventually designed a whole array of mechanical aids—some of which they passed to the British, some of which the British replicated independently, besides inventing others themselves—but their original attack, which allowed them to understand the logic of Enigma, was a work of pure mathematical reasoning. As it was done without any modern computing machinery, but simply by pencil and paper, it must be regarded as one of the most remarkable mathematical exercises known to history.

To do the work the Polish army recruited in the late 1920s a number of young civilian mathematicians from university mathematics faculties, including Henryk Zygalski, Jerzy Rozycki and Marian Rejewski. Marian Rejewski was to prove the most creative; like the others, he came from western, formerly German Poland, and spoke German fluently. In 1932, soon after the German army had adopted, on 1 June, the Enigma machine as its principal encryption instrument, and his own return from postgraduate study at Göttingen, he began to work on intercepted German encrypts in the Polish general staff building in Warsaw. The Poles had already learnt how to break German super-enciphered codes. From 1928 onwards, however, they had been defeated by strange messages which were clearly enciphered and probably, they concluded, the product of a machine system. The young cryptanalysts were set to learn its secrets.

What the Poles were intercepting were five-letter groups which betrayed no frequency. In technical terms, the message was itself the key, a continuous one which did not repeat unless at very long mathematical intervals (once in many millions of times, as we have seen). Yet it must, as Rejewski knew, obey a mathematical rule. He set out to construct the cipher's mathematical basis.

The messages he was given were, we now know, produced in the following manner. After setting up his machine by printed instruction, which prescribed the disc (or rotor) order, the position of the rim and the plugging, the operator chose his own preliminary rotor setting and typed in a three-letter group, which he then repeated; this instructed the recipient how to set up his own machine for that particular transmission (and was to reveal clues to decipherment that were to be of great use, particularly to Bletchley Park). He then typed in the message with his left hand, writing down with his right hand the letters as they appeared illuminated one by one on the lamp board. Next, he passed what he had written to a radio operator, who transmitted it to the receiving station; it was this process which denied Enigma the status of an on-line system, though it would have been easy to achieve had it been linked directly to a transmitter. At the receiving end, the recipient typed in the letters he received and took down those illuminated on his lamp board, which disclosed the decrypted meaning.

Rejewski got only the encrypt. Quite quickly, however, he recognised that the first three letters were separate from the body of the message, and that the second three letters were an encryption of the first three. These two three-letter groups provided, in short, a key to the very much larger key which was the message itself. If the two preliminary three-letter groups could be broken, two results would follow: first, the electro-mechanics of Enigma itself could be reconstructed, in part at least; second, some intercepted messages could be decrypted.

Rejewski devised a set of equations which would allow him to allot real alphabetical values to the first six encrypted letters. He was able to deduce that, in the groups, say, ABC followed by DEF, D would be an encryption of A (via electromechanical permutation), E would be an encryption of B and F would be an encryption of C. He decided to designate the permutations produced by the first (fixed) disc as S, those produced by the rotors as L, M, N and that produced by the reflector as R. As a result he wrote three equations, the first of which he expressed as:

$$AD = SPNP^{-1} \, MLRL^{-1}M^{-1}PN^{-1}p^{-3}NP \, MLRL^{-1} \, M^{-1} \, p \, N^{-1} \, p^{-} \, S^{-1}$$

The other two were equally complex and, he writes, "the first part of our task [was], essentially, to solve this set of equations in which the left sides, and on the right side only the permutation P and its powers are known, while the permutations S, L, M, N, R are unknown. In this form, the set is certainly insoluble."[16]

"Therefore," Rejewski goes on, "we seek to simplify it. The first step is purely formal and consists in replacing the repeated product $MLRL^{-1} ? M^{-1} \ldots$ with the single letter Q. We have thereby temporarily reduced the number of unknowns to three, namely S, N, Q."

Non-mathematicians will be unable to follow Rejewski's subsequent pages of equations. They conclude, however, as follows: "the method described above for [recovering] N could be applied by turns to each rotor, and thus the complete inner structure of the Enigma machine could be reconstructed."[17]

That was the Polish triumph: the penetration of the Enigma secret by pure mathematical reasoning. During the thirties, the Poles also managed to keep abreast of successive German refinements of Enigma, both electromechanical and procedural, and they succeeded in manufacturing duplicates of the Enigma machine. As its transmissions became more difficult to break, they also devised an electromechanical device (the "bombe," apparently so-called after its ticking, which was thought to resemble that of an infernal machine) which tested solutions of encrypts faster than was possible by paper methods. Meanwhile they shared their knowledge with the French cryptanalytic service, France being Poland's principal ally. The French themselves, through a financially corrupt German informant, known as Asché (French pronunciation of HE, the initials of his cover name), were acquiring documents which revealed many of Enigma's operating secrets; Asché, the brother of a general, appears eventually to have been unmasked and to have been shot for treason in 1943.[18] The Poles and the French certainly worked together closely on German ciphers throughout the thirties: latterly the French were also co-operating with the British Government Code and Cipher School (GCCS) located at Bletchley Park. During the period 24–25 July 1939, just before Germany's invasion of Poland, French and British officials visited Warsaw, where the Poles passed them each a reconstructed model of the Enigma machine.

BREAKING ENIGMA AGAIN

By then the Poles were no longer able to read Enigma intercepts, because of mechanical complications—particularly the introduction of two extra discs, increasing the possible number of disc orders from six to sixty—and procedural changes. Nevertheless, they were able to pass to the British reconstructed machines which reproduced the internal wiring of the discs, which to their annoyed consternation was foolishly simple, A being wired to B and so on. They also introduced to the British—who had hit upon the idea themselves—the concept of subjecting intercepts to treatment by punched sheets. Rejewski, besides being a pure mathematician, also had a practical bent and had grasped, from his theoretical understanding of how Enigma worked, that there would be repetitions in the permutations and that those could be identified by representing encrypted letters as perforations in large sheets of paper. Given enough intercepts, and overlaid sheets, their arrangement on a light-table would reveal repetitions, when they occurred, by light shining through. Repetitions would support disc-settings though not prove them; those would have to be established by subsequent work.

The British decryption operation, though eventually far larger than the Polish, proceeded on the whole by a different method: to use Calvocoressi's distinction, it depended more upon mechanical aids than mathematical theory, though many mathematicians worked at Bletchley and it owed its start to the Polish mathematical endeavour. Gordon Welchman, one of the most gifted of the Bletchley mathematicians, who came from a Cambridge fellowship to Bletchley Park right at the beginning of the war, distinguished four periods in its early history: (1) the preparatory period, ending with the making of complete sets of the perforated sheets in early 1940; (2) the period of dependence on the sheets, ending on 10 May 1940, when the Germans ceased to encrypt the second three-letter group which communicated the setting; (3) a subsequent period when the cryptanalysts were largely dependent on exploiting German operators' carelessness in procedure; and (4) from September 1940 when Bletchley began to acquire its own bombes, similar in principle to those devised in the thirties by the Poles.[19]

Welchman divides the development of his own thinking about how to decrypt the Enigma intercepts into ten steps, spread over several months. It was not officially his concern, since he had been set to study German radio call signs. That was necessary but routine work and Welchman's acute mathematical mind began almost involuntarily to engage with the letter groups on the intercepts he was passed. The first three steps he describes had to do with speculation about whether the two three-letter groups in the preliminaries always contained pairs of encrypts of the same letter three positions apart (as the rotor turned). When he decided that they did, he moved to a calculation of probabilities of how often paired letters would appear, establishing a number which he thought manageable (Step 4). He next concluded that the Germans' military Enigma was much less complex than they—and the

British—thought, because the plugboard did not in practice increase the number of permutations to be tested. "With only the 60 wheel [rotor] orders and 17,576 ring [rim] settings to worry about, we are down to a million possibilities. In fact we have reduced the odds against us by a factor of around 200 trillion. This was Step 5 and quite a gain!"[20] Step 6 was a further calculation of probabilities while Step 7 was his independent perception of how perforated sheets could eliminate many unfruitful possibilities. Steps 8, 9 and 10 led him to see how the sheets should be used; "if we could find twelve females [fruitful pairings] on the Red [war] and Blue [training] key for a particular day, we could confidently expect to discover that key after an average of 780 stackings [superimposing the perforated sheets on the light-table] ... so in great excitement I hurried to tell ... Dilly about it. Dilly was furious."[21]

Dilly Knox, son of the Bishop of Manchester and brother of E.V. (Evoë), editor of *Punch,* and Ronnie, a famous Roman Catholic convert priest, had been a fellow of King's College, Cambridge, was a veteran of Room 40 and had spent all his subsequent life as a government cryptanalyst. On the establishment of the Government Code and Cipher School at Bletchley Park in August 1939 he became principal assistant to Commander Alastair Denniston, another Room 40 veteran who was now GCCS's head.[22] Eccentric and solitary, he was quite unsuited to the task. "Neither an organisation man nor a technical man," in Welchman's words, Knox belonged to an earlier age of code-breaking when puzzles were dissolved by flashes of inspiration rather than rigorous analysis. He had tried his hand at Enigma but had concluded that "there were simply too many unknown factors that had to be solved simultaneously. Although Knox had worked out a mathematical procedure for recovering the daily settings, it depended on first knowing the internal rotor wirings, and there just seemed to be no way of isolating that part of the equation."[23] In short, what Rejewski had achieved, Knox could not. He was simply not a good enough mathematician. No wonder that, with his introverted temperament, he flew into a rage when his clever young subordinate Welchman arrived to claim that he could see a way through the thickets that had defeated him.

Had Welchman been easily put upon, things might have rested there, Enigma might have taken months longer to crack and the Battle of Britain and the Battle of the Atlantic proved even harder to survive. Fortunately Welchman was not to be browbeaten. Though told to go back and get on with his compilation of call signs, he went to the Deputy Director, Commander Edward Travis, but, sensibly, not simply to complain but to present a plan of organisation and action. Welchman had perceived the first lesson in winning bureaucratic battles: present an alternative scheme. He expressed his fear that, once the Phoney War went hot, Bletchley would be overwhelmed by a volume of vital radio traffic it would be unable to read. To cope with the oncoming rush, he proposed dividing Bletchley Park's growing staff into five sections working in shifts twenty-four hours a day: a Registration Room to do traffic analysis; an intercept Control Room to direct the listening stations to the most promising senders; a Machine Room to co-ordinate the work of the first two; a Sheet-stacking Room, under the Machine Room's control; a Decoding Room to deal with any messages that yielded to decryption. Welchman also proposed increasing the number of listening centres, to include one operated by the air force which would listen out for Luftwaffe messages; the principal listening station, in an old fort at Chatham, was, though very efficient, operated by the army.[24]

Travis not only accepted Welchman's scheme but persuaded Denniston to instigate it, so that Bletchley, just in time, was already operating effectively when the storm broke on 10 May 1940. There was another fortuitous event. Bletchley already knew about the bombe, from the Poles. It now acquired bombes of its own. The original design was the work of Alan Turing, another Cambridge mathematics don who had been recruited at the same time as Welchman. Turing was Welchman's intellectual superior. Indeed, he was one of the foremost mathematicians in the world, who, as a visiting fellow at Princeton in 1936, had written the theory of the digital computer, a universal calculating machine which did not yet exist; computers bear the alternative name of "Turing machines."[25] Turing's design for a bombe was being developed by the British Tabulating Machine Company, whose products were largely punched-card devices. Turing's was electromechanical, of much greater speed and power, but Welchman proposed an alteration in the design which allowed possible, but wrong, Enigma settings to be eliminated much faster.

The bombes could not, of course, test every possible Enigma setting, which would have required the calculating speed of a large modern computer. Welchman, but also Turing and others, had realised that parts of many Enigma intercepts were formal and repeated: the full name and rank of the addressee, for example, or the title of the originating headquarters. These might be guessed and came to be called "cribs" (an English public-school term for an illegal guide to a Latin or Greek translation). The bombe method depended upon guessing a crib and testing letter substitutions, by repetitive mathematical process, within the cycle of 17,576 positions in which the rotors could be set. It proved very fruitful.

The crib method, however, would not have worked unless the German operators had revealed clues to the setting by carelessness, laziness or error. "The machine would have been impregnable if it had been used properly," in Welchman's opinion.[26] It was used properly by the operators of a number of branches of the German armed forces and government. Three naval Enigma keys, including the important key code-named Barracuda by Bletchley and used for high-level signals during fleet operations, were never broken; Pink, the high-level Luftwaffe key, was broken only after a year of use and thereafter only rarely; Green, the German army home administration key, was broken only thirteen times during the war and then with some prisoner of war help ("such was the security of Enigma when properly used"); Shark, the Atlantic U-boat key, proved unbreakable between February and December 1942, a crucial period in the Battle of the Atlantic; the Gestapo key, used from 1939 to 1945, was not broken at all.[27]

The pattern of breaks was not random. The Gestapo seems, not unnaturally, to have taken great care; the German army and navy, which had long-established signal branches, made use of well-trained and experienced operators; the weakness lay most obviously with the Luftwaffe, a new service founded only in 1935. Its operators were probably younger and less experienced. A Luftwaffe key was the first to be broken by Bletchley, which thereafter broke almost all Luftwaffe keys intercepted, sometimes on the first day they were identified.

Bletchley called two forms of mistake made by German operators—each was the product of laziness—the "Herivel tip" and "Sillies" (or "Cillies"). John Herivel's tip was the result of a brainwave. He guessed that an operator, after setting the rotor rims, would place them in their slots with the selected letters uppermost. These letters might then form the first three letters of the encrypt, thus revealing the rotor setting which, for the first few years of the war, remained unchanged throughout the day. The "Herivel tip" often yielded a result, which greatly shortened decryption.[28]

"Sillies" were another form of laziness, suggested to a careless operator by the arrangement of the keyboard. Required to choose three letters for his first group, he might, instead of tapping at random, run his finger down a diagonal and then the alternative diagonal, on a QWERTZ keyboard (the later German arrangement) producing QAY and WSX. Because this was a silly thing to do, the crib offered was called by the Bletchley cryptanalysts "a silly," hence "sillies." Other sillies were short German girls' names, perhaps that of the operator's sweetheart, EVA or KAT. Some of the laziest sillies really were silly, ABC or DDD, though their use was quickly stamped out by higher authority; while the habit lasted it provided, nevertheless, numbers of breaks.

The first Enigma key broken by Bletchley was Red, so-called because Welchman used a red pencil to distinguish it from others when he was—as his original task prescribed—identifying call signs. The Luftwaffe's general-purpose key, it was first cracked on 6 January 1940, five months before the opening of the Battle of France, and was thereafter consistently broken until the end of the war, soon quite quickly on the day of use itself and then in real time; that is, as quickly as it was decrypted by its intended German recipients.[29] The Red intercepts and decrypts were of vital importance during the Battle of Britain and the Blitz that followed it. The British listening stations, their listeners straining against static and interference and often having to indicate garbled or indistinct groups in their intercept reports—since the international Morse letters were easily confused with each other, particularly U and V (dot dot dash and dot dot dot dash), so frequently occurring in German—were eventually picking up transmissions from as far away as Russia and North Africa, the signals weakening as the onward march of the Wehrmacht increased distances. In April 1941 the listening stations were struggling to hear faint Morse transmissions from mainland Greece.

THE GERMAN AIRBORNE DESCENT ON CRETE

Hitler had not initially intended to invade Greece. After his great triumph in the West, and following the refusal of the British to accept defeat and make peace, his thoughts turned to invading Russia, a plan long laid. Before the strike—Operation Barbarossa—he decided that it was necessary to lay the diplomatic ground by persuading or coercing the Soviet Union's south-eastern European neighbours, Hungary, Romania, Bulgaria and Yugoslavia, to join his Tripartite Pact alliance. He already controlled most of Russia's borderlands, since, following the incorporation of Austria into the Reich in 1938, he had occupied Czechoslovakia in 1938 and conquered Poland in 1939. Hungary, Romania and Bulgaria acceded easily to membership of the Tripartite Pact: Bulgaria was a former German ally, Hungary had been part of the Austro-Hungarian empire, Romania feared Russian power. Yugoslavia proved more difficult. The Regent, Prince Paul, agreed to sign the Pact. The day after its accession, patriotic officers staged a coup and reneged on the treaty. Hitler was enraged. He at once diverted troops deploying for Operation Barbarossa and on 6 April, nine days after the counter-revolution, invaded Yugoslavia from Austria, Hungary, Romania and Bulgaria. He also launched from Bulgaria a simultaneous invasion of Greece, which remained staunchly anti-Nazi and had already allowed Britain to position forces on its territory.

Churchill at once sent troops from North Africa, to which Hitler had already despatched Rommel and the advanced elements of what would become the Africa Corps, to shore up his failing Italian allies in Libya. The British Expeditionary Force met the invading Germans far to the north in Greece, on the Bulgarian border, but were rapidly pushed southwards, the Greek army also retreating southwards on their western flank. On 26 April the surviving British, forced to abandon most of their heavy equipment, were taken off from southern Greece. Some were evacuated directly to North Africa, some, including large numbers of Australian and New Zealand troops, were landed on the Greek island of Crete, where Britain had already established a base.

Crete, the fourth largest island in the Mediterranean, closes the southern exit from the Aegean, with its many archipelagos of smaller islands. Its people are famously warlike. The last of the major Greek populations to win freedom from the Turks, they are celebrated among Greeks for their fighting qualities and fierce spirit of independence. In 1940 the 5th Cretan Division had gone to the mainland to fight the Italians, whom Mussolini had unwisely committed to invade Greece from recently conquered Albania. The Italians had been defeated and repelled. In April 1941, however, the Cretan Division was still far away on the northern Greek border, while the Cretan homeland lay undefended, except by the disorganised collection of British, Australian and New Zealand troops which had come from North Africa or escaped from the débâcle of the British intervention on mainland Greece.

Hitler might have allowed Crete to wither on the vine. It was not essential to his strategy, either against the Soviet Union or in North Africa. On the other hand, it commanded the sea routes of the eastern Mediterranean and was important for that reason to the British, who intended to remain. Suspicious of peripheral strategies, which he correctly regarded as wasteful of force, and all the more so when he was about to invade the Soviet Union, Hitler had opposed earlier suggestions from Göring that the capture of Crete, together with Cyprus and Malta, would provide stepping stones towards the Near and Middle East. Göring persisted, however, and Hitler eventually gave in; part of the reason may have been a desire to compensate his air commander for the secondary role the Luftwaffe was to play in Operation Barbarossa. Göring, for his part, was less interested in strategic outcomes than in tactical participation. He had a full-strength parachute division available, which had not yet been used in an independent operation, and he yearned to show what it could do.

The 7th Parachute Division had come into being by a roundabout route. When in 1935 the militarised units of the German police were incorporated into the army, to swell its expanding numbers, Göring was allowed, as Minister President of Prussia, to retain control of one regiment of the Prussian *Landespolizei,* which he brought into the Luftwaffe as the Hermann Göring Regiment; that nucleus would form what, during the Second World War, would become the formidable Hermann Göring Panzer Division.[30] In 1936, however, part of the regiment was separated to undergo parachute training, in imitation of developments in the Red Army. The army also formed a parachute battalion at the same time, and while neither individually flourished, they were suddenly deemed to be useful when in 1938 Hitler decided to attack Czechoslovakia if he could not browbeat France and Britain into granting concessions over the Czechs' heads. France and Britain were browbeaten; but by then the idea had emerged of forming a complete parachute division for use in special operations. It was put under the command of General Kurt Student, a Great War fighter ace who quickly brought it to a high level of efficiency. Units of the division took part in the invasions of

Denmark and Norway in April 1940 and then in those of Belgium and Holland in May.

In Belgium the glider-borne elements of the division achieved a spectacular success by capturing the fort of Eben Emael, which guarded a key bridge across the River Meuse, at almost no loss. In Holland things went less well. In Rotterdam and Dordrecht paratroopers seized and held two vital bridges successfully. At The Hague, both paratroopers and air-landing troops, flying in by transport aircraft, suffered heavy casualties on the ground. Losses among officers were 40 per cent, among soldiers 28 per cent, while aircraft losses exceeded two-thirds. Though Dutch resistance generally was quickly brought to an end, the airborne setback offered a warning, if heeded, that the new method of making war was beset by danger.

The warning was disregarded. On 24 April 1941, Hitler wrote a Führer Directive, No. 28, which laid down aims and objectives for Operation *Merkur* (Mercury). It began, "As a base for air warfare against Great Britain in the Eastern Mediterranean we must prepare to *occupy the island of Crete* . . . Command of this operation is entrusted to Commander-in-Chief Air Force who will employ for the purpose, primarily, the airborne forces and the air forces stationed in the Mediterranean area. *The Army* . . . will make available in Greece suitable reinforcements . . . which can be moved to Crete by sea."[31]

Hitler had originally proposed that, if a mission for the airborne troops were sought (the army had by now trained its 22nd Division as an air-landing division), the objective should be Malta. It was a much better plan than that laid down in Directive 28 but Student and, more important, Hitler's operations officer, General Jodl, were against it. They argued that Malta's small size and compact shape would allow the British defenders to concentrate quickly and decisively against airborne invaders; Crete's long and narrow shape, by contrast, would in their judgement force the defenders to disperse, waste their efforts and so predispose the outcome in favour of the offensive. Hitler had concurred. Once he had written Directive 28 the die was cast.

During early May the 7th Parachute (formally *Flieger*) Division left its training areas in north Germany and began to move by train, a thirteen-day journey, to southern Greece. One of its regiments, the 2nd, had gone ahead to Bulgaria on 26 March and had taken part in the seizure of the Corinth canal. The division had an unusual organisation. Its three parachute regiments were composed, as was normal, of three battalions, but they were small, only 550 men each; there was also an engineer battalion trained, by German custom, to fight as infantry. In addition, however, the division also contained a fourth regiment, the Assault (*Sturm*) Regiment, of four battalions of troops trained to land and assault by glider. There was no divisional artillery and few support services. The parachutists, who were loaded in thirteen-man groups into the slow but steady Junkers 52 aircraft, dropped from low altitude (400 feet) on parachutes opened by static line. They carried only a pistol, their rifles and machine guns being dropped separately in canisters which had to be recovered later. The glider troops emplaned their rifles and heavy weapons with them but had to take their chance of surviving a hard landing on unprepared ground.[32]

Supporting the 7th Division was the 5th Mountain Division, chosen to substitute for the 22nd Air-Landing Division which it had been decided to keep in Romania for use in the Barbarossa operation. The 5th Mountain Division had suffered heavy casualties in Greece and had been reinforced by the 141st Mountain Regiment from the 6th Division. All, in the 85th, 95th and 100th Mountain Regiments, were elite troops, originally belonging to the Austrian army, incorporated into the Wehrmacht at the *Anschluss* in 1938; two of the 100th Regiment's soldiers, Kurz and Hinterstoisser (of the Hinterstoisser traverse), had died in the celebrated failure to scale the North Face of the Eiger in that year. The Mountain Division was scheduled to follow the glider and parachute troops by force-landing in Junkers 52 aircraft on the Cretan airfields after they had been captured in the airborne assault.

The British defenders of Crete, whose outriders had arrived far ahead of the despatch of ground troops to the mainland in April, had quickly recognised the danger of a German airborne landing; Brigadier Tidbury, appointed commander of British troops on Crete on 3 November 1940, identified the four parachute dropping zones (DZ) the Germans would use in May as early as that December.[33] All were either close to the three small airfields at Maleme, Rethymno and Heraklion or on the narrow north coastal plain near Canea, the capital. The geography of Crete confines any military operations to the north; the island, though 160 miles long, west to east, is only 40 miles at its widest, and steep mountain ridges, cut by rocky defiles, bar easy access to the south. It is a harsh landscape, though dotted with olive groves and the occasional little fields, and the people, hardy and frugal in their daily habits, were fiercely independent. Disorder always seethed in the highlands, as did internecine conflict.

Had the 5th Cretan Division not been far away on the mainland in 1940, the Germans would not have been able to capture the island. "If only the Division were here" was a phrase heard on every Cretan's lips throughout the battle. Ten thousand trained young Cretans would certainly have overcome the invaders, warriorlike though they were themselves. As it was, the Greek defenders consisted mainly of non-Cretan refugees from the débâcle on the mainland, and some locals too old or young for regular military service, about 9,000 altogether, hastily formed into eight regiments; often lacking uniforms, many were to be shot by the Germans as unlawful irregulars. The British garrison, positioned before the German campaign in the Balkans began, consisted of 14th Infantry Brigade, containing three regular pre-war battalions, 1st Welch, 2nd Black Watch and 2nd York and Lancaster; they were to be joined later from Egypt by the 2nd Leicesters and the 2nd Argyll and Sutherland Highlanders. In the aftermath of the withdrawal from Greece large numbers of Australian and New Zealand troops also arrived on the island, largely lacking their heavy weapons and badly disorganised by the ordeal of the retreat from the northern Greek frontier. They belonged to the 2nd New Zealand Division and the 6th Australian Division; the British units which had escaped from Greece were a miscellaneous collection of regular cavalry, yeomanry, Territorial infantry, Royal Marines and artillery, with few tanks or guns. The Royal Air Force had only five aircraft. The best of the survivors, who numbered 27,000 in all, were the New Zealanders, famously competent soldiers under the command of General Bernard Freyberg, a New Zealand Victoria Cross winner of the Great War. He would assume command of the whole of Creforce on his arrival from the mainland.[34]

As the units became available, during their chaotic arrival from Greece in early May, Freyberg dispersed them as follows: the 2nd New Zealand Division, of nine battalions, around Maleme airfield and to the western end of the island, together with three Greek regiments, the British 3rd Hussars (seven tanks) and the 2nd Royal Tank Regiment (two tanks); around Suda, the main northern port, the marines, four Australian battalions, the 7th Royal Tank Regiment (two tanks), two Greek regiments and a force of Cretan gendarmerie; at the eastern end of the island, around Heraklion, four regular British infantry battalions, Black Watch, Leicesters, York and Lancaster, Argylls, an Australian battalion, ten tanks of the 2nd Royal Tank Regiment and 3rd Hussars, some artillery and two Greek regiments.

Freyberg had arrived in Crete from Greece only on 29 April and did not expect to stay. He was anxious to go on to Egypt and reconstitute the New Zealand Expeditionary Corps. Churchill, however, had decided that he must command in Crete, which he had determined to hold. Freyberg was a favourite of his. He admired brave men inordinately and Freyberg, whom he knew of old, was exceptionally brave. His body bore the marks of twenty-seven wounds. Even before winning the Victoria Cross on the Somme, he had gained a wide reputation in the army for swimming the Hellespont to place guiding-lights on shore before the Gallipoli landings. Freyberg also had the common touch. Ordinary soldiers, British and Australian alike, admired him; to his own New Zealanders he was, of course, a national hero. Physically very large, outgoing in manner and quite without pomposity, Freyberg was a soldiers' general. They knew his mind. When he said, "Go for them with the bayonet," they knew he would do exactly that if he got the chance.

He set up his headquarters in a quarry above Suda Bay, near Canea, as soon as appointed commander of Creforce. In a cave in the quarry his special intelligence officer, Captain Sandover, decrypted the Enigma intercepts—code-named OL (Orange Leonard) after a mythical agent—showed them to Freyberg and then burnt them.[35] Sandover was a member of an exiguous staff. General Weston, the Royal Marine over whose head Freyberg had been appointed, resented his supersession and kept his own subordinates by him. As a result, Freyberg had to scratch to find functionaries. There was, anyhow, a notable shortage of trained staff officers, signallers and even wireless sets available to him. On Crete, an island requiring for efficient military purposes good local and telegraphic communications, but deficient in both, Creforce was hampered from the start.

Yet, simultaneously, it enjoyed almost an embarrassment of intelligence riches. Because Operation *Merkur* was confided to the Luftwaffe, in descending order of responsibility to the IV Air Fleet, the VII and XI Air Corps and the 7th Parachute Division, and because Bletchley, though still struggling with the German army and navy Enigma transmissions, could read the Luftwaffe traffic in real time, very exact warning of the German plans was sent to Creforce well before the operation began.

Warnings of the coming airborne descent on Crete were sent to Freyberg, routed via Cairo, as early as 1 May. The first extensive description of the operation was sent on 5 May. It stated that German preparations would be complete on the 17th and that landings by the 7th *Fliegerdivission* (parachutists) and the corps troops of XI *Fliegerkorps* (glider) would be diverted against Maleme, Candia (Heraklion) and Retimo (Rethymno). Bomber and fighter units would then attack Maleme and Candia. Other army units were allotted, apparently to be carried by sea transport. On 7 May an Enigma decrypt clarified the previous signal, suggesting that "three mountain regiments more likely than third mountain regiment." As we now know, the reference was to the decision to attach a regiment of 6th Mountain Division to 5th Mountain Division, all to be air-landed. As the accompanying army division was originally chosen to be 22nd Air-Landing Division, Freyberg's staff concluded that 5th Mountain Division was to come by sea, not by transport aircraft.

The Enigma decrypts correctly conveyed German intentions, which were to attack Crete with a parachute division (7th *Flieger*), the glider troops of XI Air Corps (the Assault Regiment) and an army division, initially the 22nd, for which the 5th Mountain was later substituted, reinforced by a regiment from the 6th Mountain Division, which were to be flown in by transport aircraft. The substitution of 5th for 22nd Division, and the references to sea transport, succeeded in confusing, disastrously, Freyberg's appreciation of the threat he faced.

The crucial summary of the key Enigma decrypts (OL 2/302) was sent to Freyberg's headquarters on 13 May, at 5:45 p.m. The picture of operations it gave was as follows: the operation would be launched on 17 May (later amended to the 20th). On day one the parachute division would seize Maleme, Candia and Retimo. Secondly, arrival of fighters and bombers on Cretan airfields. Thirdly, air-landing (by glider and transport aircraft) of glider troops and army units carried by transport aircraft. Finally, arrival of seaborne units consisting of anti-aircraft batteries as well as more troops and supplies.

In addition, 12th Army will allot three Mountain Regiments as instructed. Further elements consisting of motor-cyclists, armoured units, anti-tank units, anti-aircraft units will also be allotted ... Transport aircraft, of which a sufficient number—about 600—will be allotted for this operation, will be assembled on aerodromes in the Athens area. The first sortie will probably carry parachute troops only. Further sorties will be concerned with the transport of the air landing contingent, equipment and supplies, and will probably include aircraft towing gliders ... the invading force will consist of some 30 to 35,000 men, of which some 12,000 will be the parachute landing contingent, and 10,000 will be transported by sea ... Orders have been issued that Suda Bay is not to be mined, nor will Cretan aerodromes be destroyed, so as not to interfere with the operation intended.[36]

OL 2/302 was an almost comprehensive guide to Operation *Merkur*, one of the most complete pieces of timely intelligence ever to fall into the hands of an enemy. It revealed the timing of the attack, the objectives and the strength and composition of the attacking force. Moreover, as the success of *Merkur* depended on surprise—as all airborne operations must do—the revelation of the operation order to General Freyberg was particularly damaging.

And yet OL 2/302 did not quite tell the whole story. It did not specify which units would land where, an important omission. As we now know, the 3rd Parachute Regiment was to land at the east of the island, the 2nd in the centre and the Assault Regiment to deplane on Maleme airstrip, at the western end, after it had been captured by the 1st Parachute Regiment. This was vital information but it was either not in the raw Enigma intercepts or was omitted from the intercepted version sent to Crete. Bletchley's policy was not to release raw decrypts, on the grounds that they were often incomprehensible, and even Winston Churchill, who initially insisted on seeing the signals just as they were decrypted, was forced to accept that Bletchley knew better.

Had the raw decrypts revealed which units were to land where, Freyberg might have conducted the battle differently. He might have concentrated more of his available strength at Maleme and thus denied the airfield to the enemy, in which case Germany certainly would have lost the Battle of Crete. On the other hand, he might not. Freyberg was not fully let into the Enigma—properly speaking the Ultra—secret. Few commanders were. The Ultra system allowed only very senior officers, usually theatre commanders, in this case General Wavell in Cairo, to know that German signals were being decrypted in real time. They were instructed to tell subordinates that certain intelligence was particularly reliable—"Special" and "Very Special Intelligence"—but to explain its worth by reference to a supposed agent inside enemy headquarters. Small cells of Ultra-cleared officers handled the material in operational zones but were sworn to complete secrecy. Freyberg, not being in on the secret, was merely told the agent story and forbidden to discuss OL material with anyone else. It was, for him, an intellectually unself-confident man, an unsettling restriction. Instead of being able to discuss his concerns with his close subordinates, his normal method, he was forced to bottle up his Ultra knowledge.

Worse, there is no doubt that he misunderstood what he had been told. He was misled by the confusion caused by references to the 22nd Air-Landing Division, the 5th Mountain Division and the attached regiment of the 6th Mountain Division, in signals OL 2167 of 6 May and OL 2168 of 7 May, to believe that the non-parachute element of the force was much larger than it was. He was also misled, by the references to shipping, to believe that he was faced with a seaborne landing as well as an airborne landing, perhaps simultaneously and perhaps with the seaborne element outnumbering the airborne element. He perhaps should be forgiven, as his son has loyally argued in retrospect.[37]

Ralph Bennett, the authoritative historian of the Ultra system and himself a wartime Bletchley analyst, persuasively puts it thus:

[Freyberg] had known nothing of Ultra until Wavell appointed him to command in Crete [on 29 April, exactly three weeks before the battle began], and so he was quite without experience in interpreting it. Yet almost at once he was compelled by events to make operational decisions in the light of it, without the benefit of a second opinion or any advice whatever [Group Captain Beamish, the Ultra intermediary on Crete, was not in the chain of command]. [Moreover] in the whole course of history no island had ever been captured except from the sea. The only evidence that the new airborne arm could overpower ground defences consisted of [the evidence from Eben Emael and associated minor operations]. The first parachute battalions in the British army would not be formed for another six months. Finally, the fact that the Royal Navy's command of the Mediterranean was being seriously challenged for the first time since Nelson's victory over the French in Aboukir Bay in 1798 was in itself enough to reinforce fears of attack by the traditional means . . . in spite of Ultra, [Freyberg's] apprehension of danger from the sea can only be faulted by an abuse of hindsight.[38]

Yet, when all allowances are made, Ultra did warn that the Germans were going to assault Crete with thousands of airborne troops; the garrison, though disorganised by its Greek ordeal, was not at a disadvantage of numbers (42,460 British Commonwealth and Greek troops to 22,040 German).[39] The seaborne landing did not materialise; but Crete was lost. What went wrong?

THE BATTLE OF CRETE

The twentieth of May was a lovely early Mediterranean summer's day. The diary of the New Zealand 22nd Battalion, positioned at Maleme, recorded "Cloudless sky, no wind, extreme visibility: e.g. details on mountains 20 miles to the south-east easily discernible."[40] There were early German air raids on most British positions, as there had been every morning in the previous two weeks. Then calm returned, briefly, until at eight o'clock bombing, heavier than before, was resumed. At Maleme there were numerous casualties. While they were being treated the noise of a new wave of German aircraft broke in. They were the Ju-52 tugs of the gliders of the Assault Regiment, which began to land in the dry bed of the Tavronitis River, running inland just to the west of Maleme airstrip. In the course of a few minutes, about forty crash-landed, bringing, in ten-man groups, the first battalion, commanded by Major Koch, who had led the assault on Eben Emael in Belgium the year before, part of the 3rd Battalion and regimental headquarters. As the gliders crashed in, they came under concentrated fire, from the New Zealand infantrymen dug in on the airfield and on Hill 107 which dominates it.

The Maleme battlefield, when visited, is, like most battlefields, much smaller than maps suggest beforehand. The airstrip appears to lie under the lip of Hill 107 (today the graveyard of the German invaders); the sea is clearly visible beyond; only the valley of the Tavronitis is, by a trick of topography, hidden from view. Lieutenant-Colonel Leslie Andrew,

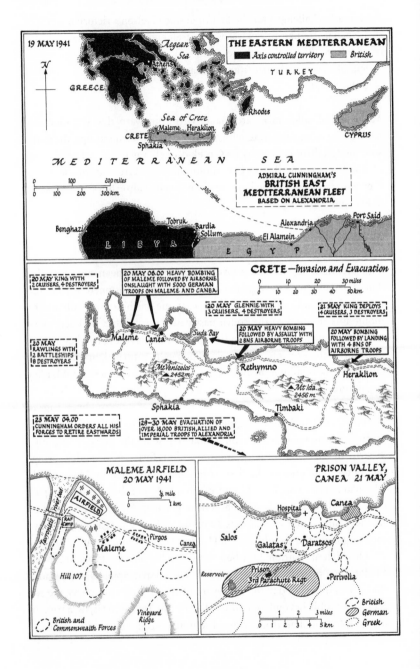

19 MAY 1941

THE EASTERN MEDITERRANEAN
■ Axis controlled territory ░ British

Aegean Sea

TURKEY

GREECE

Athens

Rhodes

Sea of Crete

Maleme Heraklion

CRETE

Sphakia

CYPRUS

M E D I T E R R A N E A N   S E A

0   100   200 miles
0   100   200   300 km

350 miles

ADMIRAL CUNNINGHAM'S
**BRITISH EAST
MEDITERRANEAN FLEET**
BASED ON ALEXANDRIA

Benghazi   Tobruk Bardia   Alexandria   Port Said
Sollum   El Alamein

L I B Y A   Sollum   E G Y P T

CRETE—Invasion and Evacuation

0   10   20   30 miles
0   10   20   30   40   50 km

20 MAY KING WITH
2 CRUISERS, 4 DESTROYERS

20 MAY 06.00 HEAVY BOMBING
OF MALEME FOLLOWED BY AIRBORNE
ONSLAUGHT WITH 5000 GERMAN
TROOPS ON MALEME AND CANEA

20 MAY GLENNIE WITH
3 CRUISERS, 4 DESTROYERS

21 MAY KING DEPLOYS
4 CRUISERS, 3 DESTROYERS

20 MAY
RAWLINGS WITH
2 BATTLESHIPS
8 DESTROYERS

Maleme   Canea   Suda Bay

20 MAY HEAVY BOMBING
FOLLOWED BY ASSAULT WITH
2 BNS AIRBORNE TROOPS

20 MAY BOMBING
FOLLOWED BY LANDING
WITH 4 BNS OF
AIRBORNE TROOPS

Mt Venizelos
2452 m

Rethymno

Heraklion

Mt Ida
2456 m

Sphakia   Timbaki

23 MAY 04.00
CUNNINGHAM ORDERS ALL HIS
FORCES TO RETIRE EASTWARDS

29–30 MAY EVACUATION OF
OVER 16,000 BRITISH, ALLIED AND
IMPERIAL TROOPS TO ALEXANDRIA

**MALEME AIRFIELD
20 MAY 1941**

0   ½ mile
0   1 km

AIRFIELD

Tavronitis river bed

RAF
Camp

Pirgos   Canea

Maleme

Hill 107

Vineyard
Ridge

British and
Commonwealth Forces

**PRISON VALLEY,
CANEA 21 MAY**

Hospital   Canea

Salos

Galatas   Daratsos

Reservoir   Prison
3rd Parachute Regt

Perivolia

0   1   2   3 miles
0   1   2   3   4   5 km

░ British
▨ German
⋯ Greek

commanding 22nd NZ Battalion, and a Victoria Cross winner of the Great War, had allowed for faults in his line of sight by positioning his D Company on the bank of the river bed; A and B Companies were on Hill 107 and its slopes; C Company was actually on the airstrip. He had two tanks, two static 4-inch guns and several Bofors anti-aircraft guns. Together with the rifles and light and heavy machine-guns of his infantry, he deployed formidable firepower.

It took a heavy toll of Battalions II and IV of the Assault Regiment, which dropped by parachute, as did part of Battalion III. Casualties were particularly high among officers; the second battalion was to lose sixteen killed and seven wounded, most in the early part of the fighting. Yet casualties were high all around Maleme. The parachutists had been told that the garrison of Crete was small, only 12,000, an underestimate by nearly three-quarters, that resistance would be light and that the Cretans would show them a friendly welcome. All three predictions proved untrue. The Cretan civilians turned out with any sort of weapon to hand the instant the landings began, an act of collective courage for which they were to play a terrible price in massacre and individual reprisal as soon as the Germans got the opportunity. The British, New Zealand and Australian defenders fought with ferocity, indeed downright glee during the parachuting phase. Parachutists were shot by soldiers sitting down to breakfast, sometimes keeping score as if on a morning's duck shooting. The sense of invulnerability that parachuting confers was quickly dispelled as bodies went limp at the end of the harness. Some of those who were not immediately hit felt outrage at the advantage being taken; others raised their hands in surrender in midair, to no good purpose. Where trees stood they were soon draped with parachutes from which dead bodies hung.

Parachutists who landed unharmed—and most of Battalions I and II of 3rd Parachute Regiment did so, together with the Parachute Engineer Battalion, between Maleme and Canea—had only to find their weapon canisters to be ready for action. Those groups quickly consolidated and started fighting as formed units. Around Maleme itself, even after they recovered from the chaos of their opposed descent, the survivors of Battalion III and from the glider parties remained desperately hard pressed. The New Zealanders, including those of 23rd and 21st Battalions who held the ground to the east of Maleme, were dug in and full of fight. As for the Germans, "even those who landed unwounded and unseen in a vineyard or field of barley could not fight effectively until they found their weapons. And if a container had fallen in the open, retrieving it was like a murderous game of grandmother's footsteps."[41] Quite quickly the battalion was almost destroyed. The commander, his adjutant, three of his four company commanders and 400 of his 500-odd soldiers were killed outright or died of untended wounds among the olive trees and scrub of Hill 107.

General Freyberg, eating breakfast in his headquarters in the quarry near Canea, had greeted the arrival of the Germans at eight o'clock with the comment "They're dead on time!," his only known public acknowledgement of his access to Ultra intelligence.[42] "His attitude," wrote the future Lord Woodhouse, later to be a leader of the Special Operations Executive in Greece, "was that he had clearly made all the necessary dispositions on the basis of his information, and that there was now nothing more for him to do except leave his subordinates to fight the battle."[43]

Freyberg's dispositions, despite his continuing misapprehension of the danger from the sea, did indeed prove effective in the central sector around Rethymno and at Heraklion in the east. Two Australian battalions, 2/11th and 2/1st, defended Rethymno airfield, supported by two Greek regiments. The Australians were well dug in and had good fields of fire, there being little vegetation in the area. They also enjoyed the advantage of ample warning because of delays in Athens, the German parachutists did not arrive until the afternoon, several hours after the descents at Maleme and Canea. The two battalions, I and III of 2nd Parachute Regiment, were also flown in along the coastline, the planes and jumpers presenting excellent targets in the last moment of their approach; some of the aircraft actually flew below the positions of the Australians hidden on the coastal hills. When the Australians opened fire they caused carnage. Several aircraft were brought down, others dropped their parachutists into the sea, where they were instantly taken to the bottom by the weight of their equipment. Those who survived found little cover from either fire or view. They were shot in large numbers, many falling to bands of Cretan irregulars.

The key to the success of the defence was the quality of the two battalion commanders, Campbell and Sandover, who kept their men in hand, organised effective fire and led counterattacks to mop up any remaining resistance. The 2nd Parachute Regiment was decisively defeated at Rethymno. It suffered very heavy casualties and its commanding officer, Colonel Sturm, was taken prisoner by Sandover on the morning of 21 May.

The 1st Parachute Regiment, dropping at Heraklion, suffered even worse ill-fortune. Its Battalions I and III fell among the best-trained defending units on the island, the 2nd Leicesters, 2nd Black Watch and 2nd York and Lancaster. Their soldiers were pre-war regulars, who knew their business. They were also supported by more than a dozen light antiaircraft guns which held their fire during the preparatory German air raids and whose positions were therefore not detected. When the troop-carrying Junkers 52s appeared, even later than at Rethymno, some as late as seven o'clock in the evening, fifteen were shot down in the two hours the parachute runs lasted. The parachutists who got clear were shot in large numbers by the British, as they hung in their harness, as they touched down or as they scrambled to seek cover and their weapon containers on the ground. Whole companies were destroyed—one had only five survivors; Battalion III of 1st Parachute regiment lost 300 killed, and 100 wounded out of a strength of 550. Among the casualties at Heraklion were three brothers, members of the illustrious family of Blücher, Wellington's fellow commander at Waterloo, serving as a lieutenant, corporal and private.[44]

By the second day of the battle, 21 May, the advantage had swung decisively Freyberg's way at Heraklion and Rethymno. Both airfields remained in British hands and, though there were parties of Germans still fighting in the countryside and within the Venetian walls of Heraklion, they were simply hanging on. It was only a matter of time before they would be overrun or forced to surrender—unless, that is, the battle went against the British elsewhere on the island. And it had already begun to do so.

Creforce lacked wireless sets, so that intercommunication between Freyberg's sectors was at best intermittent, often nonexistent. Signalling to Wavell in Cairo on the night of 20 May, he reported: "We have been hard pressed. I believe that so far we hold the aerodrome at Maleme, Heraklion and Retimo and the two harbours. The margin by which we hold them is a bare one and it would be wrong for me to paint an optimistic picture. The fighting has been heavy and large numbers of Germans have been killed ... The scale of air attack upon us has been severe. Everybody here realises the vital issue and we will fight it out." Freyberg actually believed that the tide had turned. What he did not know was that the second sentence of his signal was crucially incorrect. Maleme airfield was about to be abandoned by the defenders under cover of darkness. The Germans would use it to fly in the infantry of 5th Mountain Division, thus turning the balance decisively their way. The Battle of Crete was about to be lost.

Not for want of courage. Both Andrew, the Victoria Cross winner commanding 22nd New Zealand Battalion, and Hargest, his superior commanding 5 New Zealand Brigade, were brave men and experienced veterans of the Great War; their soldiers were brave and experienced also. The unexpected nature of airborne warfare had unnerved them, however, while their means of intercommunication was erratic at best and Hargest in particular shared Freyberg's anxiety about a landing from the sea. Andrew made one concerted effort to drive the Germans away from the airfield in late afternoon, when he sent the two Matilda tanks he had under command forward. Neither was in proper working order and one soon turned back. The other, which might have swept the airfield clear, so frightened were the parachutists of tanks, inexplicably drove past and descended into the Tavronitis River bed, where it soon grounded.

Soon after dark fell on 20 May Andrew came to the disastrous conclusion that his forward companies had been overrun and that his best procedure was to draw his other two companies back on to Hargest's other battalions to the east, perhaps to launch a counterattack in daylight the following morning. In one of their few wireless contacts, Hargest appeared to agree with him, or at least to accept the decision of the man on the spot. Both were quite wrong. The two companies Andrew thought cut off were, though battered, holding their ground and still dominated the enemy, who were now exhausted, often to the point of falling asleep where they lay. Hargest had plentiful reserves, including a whole uncommitted battalion, but declined to organise a full-scale reinforcement of the airfield or Hill 107. During the night, as the New Zealanders in the forward positions learnt haphazard that they had been abandoned, they left their positions, and made their way eastward. The vital ground was falling by default.

In Athens the German senior commanders were concluding, on the night of 20–21 May, that the battle was lost. Student realised he faced the destruction not only of his division but of his reputation and career. He hastily convened a conference to make a new plan. Surplus parachutists would be formed into a battle group under Colonel Ramcke to land directly around the airfield, while Captain Kleye, a particularly daring pilot, was to attempt to land on the airfield at first light, to bring badly needed ammunition but also to test the defences.[45]

Kleye made a successful touch-down and take-off on the morning of 21 May. On his return to Athens, every available soldier was set to preparing the Junkers 52 fleet for the renewed assault. The effort took all day, during which the New Zealanders, partially reorganised by Brigadier Hargest, pushed forward to retake the ground abandoned the night before. They advanced under heavy air attack and against the fire of the surviving Germans, hidden in vineyards and olive groves. The 28th Battalion, composed of Maoris, New Zealand's native warrior race, actually got back onto the airfield but, finding themselves unsupported, turned about. There were other successes as the New Zealanders probed forward. Then, in the late afternoon, the Ramcke parachute group fell in 5 Brigade's area and the New Zealanders were forced to renew the business, as on the day before, of shooting parachutists as they fell out of the sky and mopping up parties that managed to land unscathed.

Ramcke's descent would probably have merely added to the parachute catastrophe had not, simultaneously, the 5th Mountain Division begun to arrive in strength on Maleme airfield. It was not a tidy arrival. The New Zealanders within range opened a devastating fire, thickened by shells from captured Italian field guns fired by British gunners. Twenty-two Junkers 52s were hit on or before reaching the ground, a heavy loss to a transport fleet already severely depleted on the preceding day of action. The Germans, however, were ruthless, using captured British Bren gun-carriers to push wrecks off the runway and turning aircraft round in seventy seconds. On 21 May a battalion of 100th Mountain Regiment was flown in; by the 24th the whole division had landed, bringing the numbers transported by the troop-carrier fleet to nearly 14,000. During the arrival of the mountain division, the New Zealanders, reinforced by the 2/7th Australian Battalion and the 1st Welch Regiment, continued to battle on against the airheads around Maleme and Canea, often with success. It was during this phase of the fighting that Lieutenant Charles Upham, of the 20th New Zealand Battalion, won the Victoria Cross; he was to win another again later in the war, the only infantry soldier, and one of only three men, ever to be awarded two VCs.

Despite the bravery of the defenders of western Crete, and their willingness to return to the fray in an increasingly confused battle, by 22 May the decision was already out of their hands. Although appalling losses had been inflicted on the parachute and glider troops on the day of landing and in the immediate aftermath, the loss of Maleme airfield by 21 May was a decisive setback. Thereafter the Germans, who enjoyed complete air superiority, could reinforce the island at will, while Creforce, unsupported by air and scarcely by sea, began to wither away. Eventually about 20,000 survivors of the fighting, some escaping in formed groups, others straggling, were taken off by the navy from the southern port of Sphakia, having made a hard escape across the White Mountains; others were embarked in units from the north coast. Many remained, to take up resistance with the Cretans, who refused to submit to the occupation; in the end they had to be restrained from attacks on the Germans by the British liaison officers sent to their guerrilla bands, to avert the appalling reprisals inflicted on the inland villages.

British, Australian and New Zealand losses during the fighting totalled nearly 3,500 killed; about 12,000 were taken prisoner. There were also nearly 2,000 losses among sailors of the Royal Navy, fighting to destroy or turn back the seaborne invasion, the prospect of which had so alarmed General Freyberg. The toll among the Germans, though no more than equal, was felt more heavily. Casualty figures rarely agree; estimates of the number of Germans killed in the Battle of Crete, from 20 May to 1 June, vary between 3,352, the number commemorated in the cemetery on Hill 107, and 3,994, Anthony Beevor's calculation, which included aircrew. The gruesome point about the German casualties is that a huge proportion were from the 7th Parachute Division and fell on a single day, 20 May. As many as 2,000 were killed out of a total strength—currently that of the three parachute regiments and the Assault Regiment—of about 8,000; the number may have been higher, there being no one to count.[46]

Crete was therefore a German disaster. It effectively destroyed one of the finest fighting formations in Hitler's army; he resolved never to risk an airborne operation again and largely stuck to his decision. Yet Crete was also a battle that the British lost. Many of those killed, wounded or captured were also soldiers of the highest quality; it is invidious to discriminate but Field Marshal Erwin Rommel, who was to meet them frequently in the Western Desert at the head of the Africa Corps, reckoned

New Zealanders the best soldiers he ever knew, including in that judgement his own Germans. The Australians, less well disciplined but equally self-reliant, were also soldiers of high quality. So were those of the five British regular battalions, York and Lancaster, Welch, Leicesters, Black Watch and Argylls. Some men were taken off by the navy in formed groups, others were evacuated haphazardly. Creforce did not make an organised escape, and the shame of rout, worse than that of defeat, hung about the survivors who eventually found their way back to Egypt.

The navy suffered as badly as the army. In the effort to oppose the seaborne invasion in which Freyberg's fears were so disastrously invested—in reality nothing more than a few flotillas of Greek fishing boats, unprotected and crammed to the gunwales with defenceless German infantry—Admiral Cunningham sacrificed to the attack of German airpower three cruisers, *Gloucester, Fiji* and *Calcutta,* and six destroyers, *Juno, Greyhound, Kelly, Kashmir, Imperial* and *Hereward.* Four battleships were damaged, six cruisers and seven destroyers. Many of the warships had escaping soldiers embarked, and casualties aboard were heavy. The naval battle of Crete, like the land battle, was a defeat for British power, and the loss of life among the naval crews and the embarked soldiers and RAF ground crew was grievous.

How did the defeat come about? Ralph Bennett, the best qualified chronicler of Bletchley and its work, being both an initiate and a professional historian, wrote that "Crete [exemplifies] the truth that both force as well as foreknowledge is needed to win battles." His judgement may be restated. The British possessed ample force to win the Battle of Crete, had it been correctly deployed and committed during the course of the action. They enjoyed almost complete foreknowledge: more information about the objectives of each element of the attacking force, revealing the central importance the Germans attached to Maleme, might have altered Freyberg's assessment of priorities. Given his obsession about the danger of a seaborne landing, however, even that is doubtful.

What the events of 20–21 May on Crete reveal is that a defending force, uncertain of how to respond exactly to impending danger, however well informed it may be of the general risk, is at a disadvantage against an enemy who has his aim clearly in mind. Freyberg knew when the Germans would appear; he knew their objectives, the three airfields; and he knew their strength, a whole airborne division, reinforced by mountain troops. He was unclear, however, about the balance between the attacking elements, having confused the airborne with that of the far less menacing seaborne invasion. The Germans, by contrast, knew exactly what they intended to do: seize the three airfields by an advanced guard of parachutists and glider infantry and then reinforce success. At two of their objectives, Heraklion and Rethymno, they suffered costly defeats. At Maleme they achieved initially partial success, which they completed by taking an operational risk that Freyberg and his subordinates were not prepared to match.

There were a succession of turning points. Had Andrew not withdrawn his two less engaged companies from Hill 107 during the night of 20–21 May, they would have been able to support next day the two companies still engaged around Maleme airfield, which were holding their own. The decision, taken by Andrew alone, though without protest by his superior, was the first and most important mistake. It was not fatal. Well into the next day the struggle for Maleme continued and had not been lost by the British. It was only at 7:15 on the morning of 21 May that German headquarters in Athens got the news that "Group West [the Assault Regiment and 3rd Parachute Regiment] has taken the south-east corner of the airfield and the height 1 km to the south [Hill 107]."[47] Until then, the airborne commanders thought that the battle was lost; it was only on news of the capture of Hill 107, and of Kleye's successful touchdown at Maleme, that they decided to risk the mass descent on the airfield by the air-landed 5th Mountain Division. The mountaineers did not begin to arrive until late afternoon. In the meantime the New Zealanders had mounted several counterattacks and almost regained the ground lost; they had also slaughtered the second wave of parachutists dropped east of Maleme.

The counterattacks of 21 May, which could have saved the situation, failed for a variety of reasons. German close air-support, strafing by Messerschmitt 109 fighters, dive-bombing by Stukas, was both terrifying and deadly; Freyberg did not commit enough troops; he failed to do so because of his continuing concern about a sea landing, an anxiety heightened by an Ultra message received at about four o'clock on the afternoon of 21 May, "among operations planned for Twenty-first May is air landing two mountain battalions and attack Canea. Landing from echelon small ships depending on situation at sea."[48] This message stayed Freyberg's hand. When Maleme airfield finally fell to the incoming waves of mountain soldiers, he still retained at least three uncommitted battalions near Canea.

The Germans were lucky to win the Battle of Crete and the British need not have lost. It is certainly difficult to explain how they came to lose, with the German plans plain before them for several days in advance. Freyberg had force as well as foreknowledge. How did he come to misapply it?

Much has been made of his over-concern at the threat from the sea. Nothing is said of the quality of Ultra intelligence supplied to him. The solution to the mystery of his bad decision-making may lie there. He was not, it should be remembered, supplied with the raw decrypts because it had been decided at Bletchley not to release them, not even to the Prime Minister. That decision was certainly correct. The decrypts were often literally enigmatic or full of German military technical terms and abbreviations; most were "random scraps of secret correspondence which had to be read without the background or context which would have explained their sometimes impenetrable allusions."[49] Yet, as Bennett concedes, Bletchley's Hut 3, which dealt with the interpretation (*not the decrypting*) of all German army and air force decrypts, was full of German scholars who could make quick and accurate translations of the decrypts, but as yet (in 1940–41) it contained scarcely any trained intelligence officers; interestingly, according to Welchman, Hut 6, which did the decrypting, contained scarcely anyone with more than a smattering of German. The result, Bennett concedes, was that "at the beginning of 1941 no one had enough evidence to do more than guess how illuminating the new source of intelligence might become once land fighting was widespread again, or how it could best be used." He goes on, "Nor had anyone much experience of weaving together separate items from it into a pattern intelligible to field commanders and capable of making their battle plans."[50] "It was in fact some time," he says, speaking of his experience in Hut 3, "before the cryptanalysts' achievement was matched by an equal ability on the part of [the interpreters] to convert the product of their skill into precise military intelligence, assess it correctly, and apply it correctly in the field."

Bennett's reflections on the early work of Hut 3—precisely in the Crete period—are of the greatest relevance to the understanding of Freyberg's conduct of the battle, since they disclose the serious shortcomings of the Ultra messages he was sent. What those messages reveal appears to be a complete picture of the impending airborne invasion. What they do not disclose, crucially, is who was going to land where. The objectives—Maleme, Rethymno, Heraklion—are given; so is the strength of the force, 7th Parachute Division, a reinforced 5th Mountain Division. The units of the force are not, however, matched with the target zones. The crucial synthesis of the German operation order, OL 2/302 of 13 May 1941, the work of Bletchley interpreters, not the transcript of the German intercepts themselves, leaves it unspecified how the Assault Regiment and the nine battalions of Parachute Regiments 1, 2 and 3 are to be allotted between targets.

It may be that the intercepts did not so specify. We cannot know. "The translations of the decrypts . . . have not been released [to the Public Records Office]."[51] It is most unlikely, however, that they did not. Military operations orders, in whichever army, conform to a standard pattern, which always specifies, among other points, aim, timing, objectives and units allotted to objectives. It is most unlikely that, among the Enigma intercepts decrypted, there lacked the evidence that Maleme was to be the objective of both the Assault Regiment and 3rd Parachute Regiment in the first wave on 20 May.

Had Freyberg known that, which would have identified Maleme as the main German objective, his mind would have been greatly clarified. He would have been able to leave the garrisons at Rethymno and Heraklion to look after themselves, as they did very successfully, to worry less about the seaborne landing, and to concentrate readily and forcefully at Maleme, with the object of defeating the initial landings and denying the airfield to subsequent arrivals. He had ample force available. The Assault Regiment, even if it had enough gliders to lift all its soldiers, which it did not, numbered at most 2,400; 3rd Parachute Regiment numbered only 1,650. The New Zealand Division, with its associated Australian and British battalions, exceeded the German total of 4,000 men, by an ample margin; the count may be reckoned as seven German battalions to seventeen British, which also had tank and artillery units the Germans lacked.

Freyberg should have been able to organise a victory. If he did not, it was partly because intelligence did not serve him well. The facts were there, but they were fed to him through the "sieve" or "filter," as the process is now known in the intelligence world, of interpretation by young, inexperienced and largely unmilitary officers in Bletchley's

Hut 3, who seem to have been more concerned to provide a smooth narrative on the Oxbridge essay pattern—most were academic linguists—than the sharp assessment of enemy aims and capabilities that a hardened operational intelligence analyst would have composed. Intelligence is only as good as the use made of it. That is the hard lesson of Crete, the "open and shut intelligence example," and the truth of the case remains as clear now as better interpretation would have made it then.

————◄o►————

Midway: The Complete Intelligence Victory?

THE DEFEAT of Admiral Von Spee's East Asiatic Cruiser Squadron in 1914 laid the basis for Japan's conquest of the oceanic perimeter of its Greater East Asia Co-prosperity Sphere in 1941–42, for the island colonies that von Spee abandoned in his retreat from the Pacific were instantly occupied by Japanese naval expeditionary forces. Japan entered the First World War as an ally of Great Britain, in answer to a request for naval assistance in hunting down German armed merchantmen. Japan's motive in responding to the British request was not, however, one of diplomatic goodwill but strict self-interest. Ever since its re-entry into the world in 1854, after centuries of self-imposed sequestration, and particularly since its creation of a modern army and navy in the century's last decades, Japan had sought to become a major Pacific power. Its long-term ambition was to dominate China, but its ruling class recognised that the established great powers, particularly Britain and Russia, which had their own designs on China, would check any attempt at large-scale annexation. The policy of the United States was an even more important obstacle, since it was benevolent and largely disin-terested. While commercial America sought to create and protect mar-kets in China, political and missionary America hoped to make the country democratic and Christian. Since America was a great Pacific power, its attitude was a major factor in determining Japan's strategic policy. By 1908 the Japanese navy was considering the problem of fight-ing the United States navy in Pacific waters; by 1910 it was studying the question of attacking the Philippines, over which the United States had extended a protectorate at the end of the Spanish–American War of 1898.[1]

The Japanese navy recognised that to conduct a naval war against the United States in the Pacific, even as a theoretical exercise, required bases beyond the home islands. By 1914 Japan had already considerably extended its territorial reach. As a result of its victory in the war against China in 1894–95 it had acquired the great offshore island of Formosa (Taiwan) and the nearby archipelago of the Pescadores; it had established an effective protectorate over Korea, which became a Japanese colony in 1910, and secured concessions in the Liaotung Peninsula, the strategic promontory enclosing the Yellow Sea; it had also extracted a "lease" over the productive province of Kwantung. The great powers, wanting for themselves the territory Japan had won, intervened after the China War, forcing it to disgorge the Liaotung Peninsula, in which Britain, Germany and Russia set up their own maritime enclaves, the Russians at the anchorage of Port Arthur.

Japan took its revenge in 1904, opening a war against Russia which resulted in major victories in Manchuria and the destruction of most of Russia's naval power. It had, however, learnt a lesson: that the white imperialists—then including the Americans, temporarily in an imperialist phase—would not allow an Asian state to acquire colonial possessions desired, actually or potentially, by themselves.

Germany's abandonment of its Pacific colonies and Britain's request for naval assistance in August 1914 thus presented Japan with an opportunity not to be missed. Australia and New Zealand, moving briskly to stake their own claims, seized what looked like the best German possessions: New Guinea, Papua, the Solomon Islands and the Bismarck archipelago, including the strategic anchorage of Rabaul, were occupied by Australia, Samoa by New Zealand. Japan, which quickly organised two South Seas squadrons, managed to lay hands on much of Micronesia, including the Mariana, Caroline and Marshall Islands, by early October. The equator became the effective sea frontier between an extended Australasia and a new Japanese empire of the Central Pacific. By November 1914 the Japanese foreign minister had agreed informally with his British counterpart that Japan, at the peace, would retain all German islands north of the equator.

Economically, Micronesia was of trifling worth. Geopolitically, however, when it was mandated to Japan by the League of Nations in 1919, its changed ownership transformed the strategic situation in the North Pacific. Japan, before 1914 a purely regional power, became after 1919 the possessor of a great, potentially offensive bastion, extending almost to the International Date Line and threatening the American possessions of Guam, Wake, Midway and even distant Hawaii in the Central Pacific, as well as the Dutch East Indies, the American-protected Philippines and the constellation of British Central Pacific islands from the Solomons to the Gilberts and beyond. Beyond implied Australia and New Zealand, ultimately dependent on British naval power for their security.

Although the outcome of the First World War made Japan a Pacific oceanic power, both its domestic and external affairs after 1919 and until the late thirties were concerned almost exclusively with China. For centuries, even millennia, China's cultural subordinate, Japan by the twentieth century had determined that its future lay in a reverse subordination, economic but also political and military, of China to its imperial needs. In 1915 Japan had issued a set of "twenty-one demands" which required China to concede rights and privileges to Japan, according it overlord status. The Chinese prevaricated and resisted, as far as they were able. In 1931 they were forced, however, to submit to effective Japanese annexation of Manchuria and then in 1937 to a full-scale Japanese invasion of the south. The nationalist government of Chiang Kai-shek withdrew inland, first to the city of Nanking, then to Chungking. Its capacity to resist was hampered by the attacks of the Chinese Communist Party armies under Mao Tse-tung.

Japan's imperial policy was strengthened and furthered during the 1930s by the rise of an intense nationalist spirit within its military class, particularly in the army. The "Manchuria Incident" of 1931 was largely the work of nationalist officers, in the Manchuria garrison. The "China Incident," so-called by American observers, of 1937 in Shanghai was equally an outburst of ill-discipline by the Japanese occupying troops. By that date, however, the army led the government, which had escaped from the control of constitutional statesmen. By the outbreak of the Second World War, Japan, then in alliance with Germany and Italy, was a totalitarian state, committed to an imperialist programme of territorial expansion directed against China, with which it was in full-scale war, the Asian possessions of the European empires, principally Britain and the Netherlands, and the United States.

The war on the mainland of China consumed most of the strength of the Japanese army, which fielded about twenty-five divisions. Militarily it was far superior to that of the Republic of China, which survived total defeat only by its ability to use space as a means of defence. The Japanese were not able to penetrate far beyond the coastal provinces, though as those contained China's larger cities and main rice-growing areas, they had little strategic reason for mounting deeper offensives.

The Japanese navy was scarcely involved in the China war, which had no maritime dimension. It was, nevertheless, much concerned with the strategic future since Japan's attack on China had provoked the wrath of the United States, manifested in a series of increasingly constrictive trade embargoes. Japan, like Britain, lacked the domestic resources necessary to support an imperial policy. Its home islands did not produce enough food to support its population, which relied heavily on imports of rice, while its industries and infrastructure required large imports of metal ores, scrap and oil. By 1941, after Japan's deployment of troops into French Indo-China, enforced on the defeated Vichy government, an initiative which directly threatened British Malaya, the American oil and metal embargoes were seriously hampering Japan's ability to sustain its manufacturing output. America's intention was to restrain Japan's military ambitions. The effect was to drive Japan towards aggressive war.

The Japanese army and navy operated, to a degree unusual even in the arena of naval and military rivalry, as separate entities. The army, which dominated government, only reluctantly accepted the navy's right to speak on strategy. On the other hand, the navy justifiably argued that since the United States dominated the Pacific, the success of national strategy depended on its plans to overcome American naval power. In 1936 the army and navy had agreed on a statement of Fundamental Principles of National Strategy. The statement committed the army to achieve a strength sufficient to contain the Soviet Union—the old Russian enemy—in the Far East and the navy to acquire both a dominance in the South Seas, meaning the islands and peninsulas possessed by the British and Dutch, and the ability to "secure command of the Western Pacific against the U.S. Navy."[2]

By the summer of 1941 Japan was in a strategic quandary. Though the army had suffered in border conflicts with the Soviet Union in 1936 and 1939, it was in the ascendant in China, controlled Manchuria and had assumed a forward position in French Indo-China. Its leaders recognised, however, that the American embargo policy threatened to terminate its ability to sustain its aggressive strategy within one or two years. The navy lay under an even more exigent threat. American restrictions on the export of aviation fuel promised to end its capacity to conduct carrier operations even sooner. To maintain parity of status with the army, it needed access to a supply of oil outside American control, and the only sources available within its strategic zone lay in the Dutch East Indies and Burma. By the middle of 1941 the Japanese navy was psychologically committed to a war of Pacific conquest.

Although Japan had benefited from the peace settlement of 1919 by its acquisition of the German Pacific islands, it had suffered under the post-war disarmament treaties. The Washington Naval Treaty of 1922, designed to avert another naval arms race equivalent to that between Britain and Germany, widely held to have helped precipitate the First World War, imposed a subordinate naval status on Japan. The United States and Britain, arguing that both their navies had two-ocean commitments, in the Atlantic and the Pacific, succeeded in bringing their wartime Japanese ally to accept that, as a Pacific power alone, it needed only 60 per cent of their naval strength. This 5:5:3 ratio, as it became known, applied to battleships, cruisers, destroyers and aircraft carriers. All signatories—they also included France and Italy—were required as a result to scrap some of their larger naval units and to limit the size of ships planned or under construction.

The Japanese, who bitterly resented what they saw as Anglo-American condescension towards their status as a world naval power, had no recourse but to agree. They proceeded, nevertheless, to exploit any loophole in the treaty that was open to them. The Americans and British did likewise, converting half-constructed dreadnoughts into large aircraft carriers. The Japanese went further. By 1941 they had assembled seven carriers, a larger naval aviation fleet than that possessed by either Britain or the United States. More important, the Japanese carrier air groups were equipped with aircraft respectively superior and greatly superior to those embarked by their American and British counterparts. The torpedo bomber code-named Kate by the Americans was a better aircraft than their Avenger, while the Zero remained unchallengeably the best carrier-borne fighter in the world until the appearance of the American Hellcat in 1942. Japan's naval air arm pilots were also of the first class; those who took part in the attack on Pearl Harbor had 800 hours of flying experience. Yet there were defects in the Japanese naval aviation system: the Zero was essentially a racing sports aircraft, faster both flat out and on the turn than contemporary American fighters, but fragile and combustible; the Japanese flight schools were not adapted to produce pilots en masse, so that any heavy loss of trained aircrew threatened the efficiency of the air groups. After Japan's stunning initial victories, its weaknesses in aircraft design and pilot output would lead quite quickly to a decline in its ability to wage carrier warfare on equal terms with the United States.[3]

By the summer of 1941 the Japanese army and navy were confronting the necessity to go to war and considering how best to deliver the opening stroke. The economic objectives were easily defined: the oil fields of the Dutch East Indies and Burma, the tin mines and rubber plantations of Malaya. The politics of the operation were far more complex. War with Britain could not be avoided, for her colonies were to be directly attacked; but, as her forces were already overstretched in the fight against Hitler, the consequences were manageable. War with the United States could equally not be avoided; the question was how long the outbreak could or should be postponed. Four timetables were considered: to seize the Dutch East Indies first, then the Philippines and Malaya, a sequence that would bring on early a war with the United States, which could not be allowed to retain a base in the Philippines; to attack the Philippines immediately, then the Dutch islands, then Malaya; to begin with Malaya and work backwards towards the Philippines, thus delaying a confrontation with the United States; to attack Malaya and the Philippines together, then the Dutch East Indies.

Since the last sequence was the only one on which the army and navy could agree, it was adopted; but since it entailed provoking war with the United States from the outset, it also required the making of a subordinate plan about how to neutralise American naval power in the Pacific. Since early in the century, the Japanese navy had planned to defeat the American Pacific Fleet by drawing it into Japanese home waters, wearing down its strength meanwhile by attritional attacks as it made its long voyage across the ocean; the logic of the strategy was enhanced after 1919 by the acquisition of the Central Pacific barrier formed by the ex-German islands. An early war with the United States, however, demanded a quicker means of reducing American naval power.

Admiral Isoroku Yamamoto, Commander of the Combined Fleet,

which included the main carrier force, had been considering the problem since early 1941. In principle he opposed making war on the United States, which he knew well as a former English-language student at Harvard and naval attaché in Washington; he did not believe that Japan's small industrial base could ever effectively support a war against the United States' vastly larger economy. His well-known views had made him unpopular both with nationalist politicians and their supporters and within the armed forces; he had been sent to sea in 1939 largely to save him from assassination. The threats were not hollow; in 1936, a group of super-nationalist army officers had killed several moderate politicians, including the finance minister and a former prime minister, occupied central Tokyo and been overcome only after three days of street fighting. Yamamoto undoubtedly had reason on his side, as other naval officers saw. Confronted, however, by the reality of the army-dominated government's determination to solve Japan's economic problems by aggressive measures, Yamamoto stifled his objections and proposed an alternative attack strategy. He suggested using the carrier force to destroy the American Pacific Fleet at its moorings in Pearl Harbor, Hawaii, its Central Pacific base.

Yamamoto's thinking was greatly influenced by the results of the British attack on the Italian battle fleet in the harbour of Taranto on 11 November 1940. Aircraft launched from the carrier *Illustrious* had then sunk three Italian battleships with torpedoes, losing only two out of twenty-one aircraft engaged. While the Japanese naval air arm experimented with adapting torpedoes—the principal Japanese torpedo was faster, of longer range and fitted with a larger warhead than any in use in other navies—to run level in shallow water, Yamamoto and his staff considered how to bring the Combined Fleet undetected to within striking distance of Pearl Harbor. By October 1941 the planners—including Commander Minoru Genda, who would lead a major part of the First Air Fleet in the attack—had outlined a scheme which would take the Combined Fleet from the stormy waters of the northern Kurile Islands, above the isolated American possession of Midway Island and then due south, off frequented shipping lanes, to within 200 miles of Hawaii. The fleet would move in complete radio silence and, if possible, within the leading edge of one of the turbulent weather fronts common in the North Pacific, which impeded visual reconnaissance and interfered with radio transmissions. During October the liner *Taiyo Maru* was sailed down the chosen route without sighting a single ship.

The Japanese and American governments meanwhile continued reasoned diplomatic exchanges. Japan asked for a relaxation of American trade embargoes—a joint army–navy committee estimated in June 1941 that oil reserves were being depleted at a rate a third greater than they were being replaced, a disastrous situation, since it implied inexorably that Japan would run out of oil (stocks plus replacement 33 million barrels, consumption 41 million barrels) in 1942–43.[4] In return Japan would undertake to cease her military intrusion into Southeast Asia and eventually leave French Indo-China. The United States demurred, proposing instead a scheme to settle affairs all over mainland Asia, which would have required Japan to withdraw from China as well as Indo-China, and to leave Manchuria also. The Japanese, as was to be expected, rejected the proposal. On 4 December an imperial conference took the decision to go to war against the United States, beginning with an attack on Pearl Harbor, on the 7th. The Combined Fleet was already en route.

AMERICAN PENETRATION OF JAPANESE CIPHERS

The attack on Pearl Harbor on 7 December has given rise to one of the largest conspiracy theories in history. There are many versions, most alleging American foreknowledge, and scarcely one agrees with another. Allegations of culpable incompetence apart, which deserve attention, the two most important theories allege, first, that the British had foreknowledge of Japanese intentions but chose to conceal what they knew from the United States in order to bring America into the war; second, that President Roosevelt knew independently what the Japanese intended but took no preventative action, since he currently sought a pretext to bring his country into the war on Britain's side. The two theories, in some of their versions, overlap.

The subject is so large that it has stimulated the production of a library of books. Almost the only matter on which they agree is that American cryptanalysts, like their British counterparts at Bletchley, were freely reading Japanese ciphers before December 1941. Exactly what was read and when, what the interpreters made of the decrypts, and how the decrypts influenced the decisions taken by President Roosevelt, his cabinet officers, the chiefs of staff and the commanders in the zone of operations form the substance of the great Pearl Harbor mystery story.

It does not connect with the story of Midway, which began to unroll only six months later. There is, however, this caveat: the cryptanalytic organisation which may have failed before Pearl Harbor was also the organisation that helped to deliver the victory of Midway. It worked in this way.

American cryptanalysis differed from its British equivalent, in organisation, recruitment and ethos. Bletchley was a joint-service civil-military body, in which little distinction of rank was observed, built up on a word-of-mouth basis in the period immediately before the outbreak of war in 1939 and recruited largely among young Oxford and Cambridge dons. Proven mathematical ability was the principal qualification. The atmosphere at Bletchley was creatively amateur, informal and high-spirited; women formed a high proportion of the staff, some in senior positions, and romance flourished. There were many Bletchley marriages.

The American cryptanalytic organisation, by contrast, was sharply divided into naval and military branches, which co-operated uneasily, was highly bureaucratic and almost completely male-dominated. Most of the cryptanalytic personnel were uniformed servicemen and the principal qualification for selection was language skill, particularly in Japanese. Unlike Bletchley, which had a high opinion of itself and cultivated a genial university common room atmosphere, the American intelligence branches were regarded by the rest of the army and navy as backwaters, staffed by officers unsuitable for operational appointments, an opinion of which their members were aware. It is remarkable, in the circumstances, how well they maintained their professional morale. A key indication of the difference between the British and American systems is that, while Bletchley has entered into British national legend and found a popular place in fiction and film, its American equivalents enjoy no such acclaim. Quite wrongly, for what the Americans achieved was equally remarkable, indeed perhaps more so, as the story of Midway indicates.

The origins of the American cryptanalytic service belong, as do those of the British, in the First World War. Major Joseph Mauborgne, head of the army's cipher research section in 1918, was a cryptographer far ahead of his time: he perceived the concept of the random key—one not retrievable by frequency analysis or, indeed, any mathematical or linguistic logic—and devised the one-time pad, still the only intrinsically unbreakable cipher. He would eventually become a general and the U.S. Army Chief Signal Officer.[5] Of the same vintage but of even great importance to the American cryptanalytic effort was a civilian, William Friedman (who coined the term "cryptanalysis"). The son of a Russian Jewish immigrant, who entered the United States at the age of one, Friedman resembled in character both Dilly Knox and Alan Turing. As eccentric as either, he displayed a mathematical ability almost equivalent to Turing's but unfortunately also Knox's psychological fragility. He had suicidal tendencies and just before the outbreak of the Pacific War suffered a nervous collapse brought on by overwork.[6]

Yet Friedman was largely responsible for the most important of America's cryptanalytic successes, the breaking of Purple. In October 1940, the army and navy agreed on a division of labour, not in any spirit of fraternal co-operation but because each lacked the numbers to do much more than concentrate on a single task; in 1938 Friedman had a staff of only eight in the Secret Intelligence Service (SIS), in December 1940 the navy in Washington only thirty-six in its equivalent OP-20-G. There were other personnel at outstations in both the continental United States and the Pacific but most were intercept operators and technicians.[7] The arrangement was that the army would work on foreign diplomatic intercepts on even days of the month, the navy on odd. The navy was meanwhile, naturally, working on Japanese naval intercepts; the army was not particularly interested in foreign army intercepts, since they were too faint to yield text.

The Americans, few though they were in number, had had considerable success in breaking into both Japanese naval and diplomatic traffic in the 1930s, assisted by a succession of night-time burglaries of the Japanese consulate in New York. By 1933 the naval cryptanalysts had solved the main Japanese naval Blue Code, a book code with a cipher additive. When it was replaced in June 1939 by JN-25 (Japanese Naval Code 25, as the Americans denoted it), another book code with a more complex additive, the Americans took time to recover from the setback, but by December 1940, with the help of recently acquired IBM card-sorting machines, they had reconstructed the system of additives, the first thousand groups of the code and two of the keys used to work the system.

They anticipated cracking the system completely in 1941. By then, however, all spare cryptanalytic manpower in Washington had been diverted to a new task: decrypting the Japanese diplomatic traffic enciphered on a new ciphering machine, known to the Americans as Purple. The Purple machine (Type 97 to the Japanese) was designed to achieve the same effect as Enigma—the automatic production of an almost infinitely variable cipher—but differed from it in construction. It was less mechanical, having no rotors, but instead a set of telephonic switches, connected to two typewriters. The first was used to input the text, the second to print out the encipherment for transmission. In between, the switches moved the incoming electrical current to achieve alphabetic substitutions. Because Japanese is a syllabic, not alphabetic, language, however, all texts had first to be written in an alphabetic equivalent; and for an inexplicable reason, equivalent to the Germans' double-encipherment of the operator's chosen indicator at the beginning of a transmission, the Purple machine's switches enciphered vowels and consonants separately; the number of vowel substitutions was considerably smaller than that of consonants, and once that was recognised, a way into Purple was found.[8]

The breaking of Purple—its product was known as Magic, the equivalent of British Ultra—would eventually yield huge intelligence advantages to the Americans, principally through the decipherment of the messages sent by the Japanese ambassador in Berlin, Baron Oshima, to Tokyo throughout the war, which revealed intimate details of Hitler's capabilities and intentions. At the critical moment before Japan's surprise attack on America in December 1941, however, Purple revealed little, while JN-25, the relevant Japanese naval code, was not useful for two reasons: the first was that the Imperial Navy attempted to observe, as far as was possible, radio silence in the period preparatory to the descent on Pearl Harbor; the second was that OP-20-G lacked the staff necessary to deal with the volume of intercepted traffic. The U.S. Navy's Historical Center has now compiled a list of significant messages intercepted—but not decrypted—in the weeks before Pearl Harbor that bear on the issue of foreknowledge. Some, if read in real time, must have alerted a wary admiral to the danger threatening his fleet; in practice, the messages were bundled up and not decrypted and translated until September 1945, a month after the Japanese war had ended.[9]

Pearl Harbor devastated not only the U.S. Navy's Pacific Fleet but also OP-20-G's Pacific outstations. While its Hawaiian station (HYPO) continued to operate, its outpost in the Philippines (CAST) first withdrew into a tunnel in Corregidor, then was evacuated to Australia. American naval intelligence in the Pacific was thus reduced to HYPO, the joint station in Australia and a branch in the British Combined Bureau in Ceylon. The British had had some inter-war success in attacking Japanese naval codes but, in the climate then current, with both the American services and the public adamant about revenge against the Japanese, the prime responsibility for breaking back into Japanese naval traffic lay with OP-20-G. The Commander-in-Chief Pacific, Admiral Chester Nimitz, was on tenterhooks against a renewed Japanese naval offensive; he equally hoped to profit, perhaps from a Japanese mistake but preferably an intelligence coup, and inflict a defeat on the enemy.

THE COURSE OF JAPANESE CONQUEST

There was little sign of stemming the tide of Japanese conquest in the last weeks of December 1941 and the first three months of 1942. Its progress seemed inexorable. Attacks launched out of Thailand quickly collapsed the British defences of northern Malaya. The new battleship *Prince of Wales* and the battlecruiser *Repulse,* operating without air cover in an effort to intercept Japanese coastwise landings, were sunk by bombers launched from French Indo-China on 10 December. Singapore, the great trading city at the tip of the peninsula, was ignominiously surrendered to an inferior Japanese force on 15 February. The British island of Hong Kong and the American islands of Wake and Guam, all indefensible, were captured on 25 December, 23 December and 10 December; the garrisons of Hong Kong and Wake put up a heroic, hopeless resistance. Burma was invaded in January and conquered by May. The invasion of the Dutch East Indies began in January also and was completed by March; there were several attempts by a combined Australian-British-Dutch-American fleet (ABDA) to check the Japanese amphibious campaign, culminating in the Battle of the Java Sea, on 27 February; a miscellaneous collection of Allied ships, bravely commanded by the Dutch Admiral Karel Doorman, but outnumbered and unable to intercommunicate, was overwhelmed. Meanwhile, General Douglas Mac-Arthur was conducting the defence of the Philippines, the attack on which began on 8 December with a devastatingly successful Japanese air raid. The American and Filipino defenders were forced to withdraw as soon as the Japanese landings began, but in January succeeded in establishing a fortified line across the Bataan peninsula. There for the next three months they sustained a heroic defence, inflicting on the Japanese the only setbacks suffered on land during their great campaign of conquest. Eventually, however, shortage of food and supplies forced a surrender in April, and in May of the offshore island of Corregidor.

The defeat in the Java Sea had destroyed the American Asiatic Fleet, originally based in the Philippines, leaving only its submarines to survive. The Pacific Fleet, based at Pearl Harbor, remained operational; but the loss and disablement of its eight battleships had transformed its composition. From being a big-gun capital force, it had become necessarily an aircraft carrier fleet. The three carriers based on Pearl—*Enterprise, Lexington* and *Saratoga*—had been absent on 7 December. The navy's three others, *Wasp, Hornet* and *Ranger*—were elsewhere. It was around these six units that the Pacific Fleet was to concentrate the rest of its strength and design a new offensive strategy that, in a succession of brilliant victories, would halt and then reverse Japan's Pacific onslaught.

In the spring of 1942, the hopes of Japan's strategic optimists seemed to have been wholly realised and the fears of her strategic pessimists altogether disproved. Yamamoto, the only Japanese admiral who knew America well, had predicted that he could "run wild for a year or six months," but thereafter he foresaw only the gathering strength of American industry. The magnitude of Japan's victory seemed to make economic imbalances irrelevant. The Americans, together with their European allies, had been beaten; it was henceforth a question only of where Japan should strike next to capitalise on its success.

There were two schools of thought among Japanese planners, the "Southern" school and the "Central Pacific" school; the Central Pacific school was entirely a naval one, the Southern school also involved the army. The Central Pacific school held that the carrier striking force should resume the attack on Hawaii, to dispose for good of the U.S. fleet's ability to intervene in Pacific grand strategy. The Southern school took a more indirect view, identifying Australia as a potential base for an Anglo-American counteroffensive but also wanting to eliminate British maritime power in the Indian Ocean, thereby weakening Britain's, as well as China's, ability to wage war in Burma and thus laying the basis for a Japanese offensive against Britain's Indian empire itself.

In the early course of the debate, the navy appeared to accept the objections raised against its Central Pacific concept and, during March, deployed two carrier striking forces to the Indian Ocean. One, commanded by Admiral Nagumo, who had led at Pearl Harbor, struck the British base in Ceylon, sank a British carrier, *Hermes,* two large cruisers, *Dorsetshire* and *Cornwall,* and forced the squadron of old R-class battleships to withdraw to east Africa. Meanwhile, the smaller Japanese carrier task force under Admiral Ozawa roamed the Bay of Bengal, sinking 100,000 tons of commercial shipping in five days. Those with a long memory would remember the cruise of the *Emden,* though Ozawa's depredations were far more brutal.

The Americans, meanwhile, were not wholly dormant. On 20 February a task force organised around *Lexington* had attacked Rabaul, the old German base in the Bismarcks, and inflicted heavy losses on the bomber force sent to drive it off; the Japanese lost eighteen aircraft, the Americans two. Then in April an even more daring raid was mounted. President Roosevelt had been pressing for an American attack on the Japanese home islands for some time. The mission seemed infeasible, since it was too dangerous to risk the Pacific Fleet's few aircraft carriers in Japanese home waters, while America's remaining island airfields were too distant to serve as bases for land bombers. In mid-January, however, Captain Francis Low, operations officer to Admiral Ernest King, Chief of Naval Operations, proposed embarking land bombers, which had a range greatly exceeding that of any maritime aircraft, on an aircraft carrier and sailing it to within striking distance of Tokyo.

The idea seemed fantastic, but Colonel James Doolittle, one of the Army Air Force's inter-war bomber pioneers, who was put in charge of planning the mission, determined to dissolve the difficulties. He selected the B-25 medium bomber as the best aircraft available and set sixteen crews training in Florida to learn the technique of very short take-offs. After a month of preparation, the crews watched their B-25s lifted by crane aboard the new aircraft carrier *Hornet* at Alameda Naval Air Station, California, then embarked for the unknown. They had not been told where they were bound. On 13 April, *Hornet* and its escorts met *Enterprise* off Midway, the last surviving American island outpost in the North

Pacific, and set course for Japan. The plan, of which the crews had only just been told, was to fly the bombers off once they were within 500 miles of the Japanese capital, to release their bombs under cover of darkness and then fly on to crash-land in areas of China not occupied by the Japanese.

As they approached take-off point, the Americans discovered that the plan had miscarried. Yamamoto, anticipating just such a revenge raid, had established a line of picket boats 600 to 700 miles eastward of the home islands. American radar, and then visual reconnaissance, identified first one, then a second, then a third picket boat. Admiral William Halsey, in command of the joint task force, recognised that further changes of course would not evade interception. It was decided to fly off Doolittle's force at once, even though they would therefore have to go 650 instead of 500 miles to reach their targets and bomb by day instead of night. In heavy weather, with waves breaking over the bow, all sixteen of Doolittle's bombers took off successfully, reached Tokyo, dropped their bombs and flew on to China; some crews crash-landed, others bailed out; of the eighty who flew, seventy-one survived.[10]

The material effect of the Doolittle raid was insignificant; few citizens of Tokyo were aware that they had been bombed. The psychological effect on the Japanese high command was decisive. Committed to the protection of the body of the Emperor as the supreme value of the warrior creed, Japan's admirals felt dishonoured by the attack. They had failed in an overriding duty. Plans for invading Australia were immediately postponed and all thoughts refocused on the Central Pacific, with the idea of terminating for good the U.S. Pacific Fleet's ability to strike at the home islands. Hawaii was still too distant from the Imperial Navy's centres of power, and too well defended, to be attacked at once. It was seen, however, that its outlier, the tiny island of Midway, could be used as a point of attraction, onto which the surviving American carriers could be drawn by the threat of invasion and then destroyed by the concentration of overwhelming force.

Japan's strategic position in April 1942 was extremely advantageous. Her main aim in resort to war had been to take possession of a perimeter, delineated by the Central and South Pacific island chains, which would eliminate American, British and Dutch naval power in the area of command, isolate China from support, and dominate the long sea route from California to Australia, which the Japanese identified, correctly, as the base the Americans would use to mount a counteroffensive. Much of the perimeter, thanks to the mandating of the ex-German islands to Japan, lay in her possession before the war began. The rest was supplied by the capture of Wake, Guam and the Dutch East Indies; the important land masses inside the perimeter, the Philippines, Malaya and Burma, were acquired by subsidiary offensives.

Despite the overwhelming success of Japan's initial campaign of conquest, however, gaps remained in its strategic perimeter even in April 1942. Only the northern half of New Guinea had been conquered, and the Solomon Islands, beyond the Bismarck chain, remained in dispute. The Americans still potentially enjoyed a way round the tail of the New Guinea "bird" to Australian ports. Even while preparing the Midway operation, therefore, the Japanese decided to complete their conquest of New Guinea by sending a carrier force into the Coral Sea, between the great island and the north coast of Australia, to capture Port Moresby on the south coast and so facilitate the advance of the Japanese ground troops across the island's central spine, the Owen Stanley range.

Japan had already operated in the Coral Sea, bombing Darwin, in Australia's Northern Territories, a week before the Battle of the Java Sea in February 1942. Its plan now was to send a tripartite naval mission, one element to land a force at Port Moresby, a second to capture Tulagi in the Solomons and a third striking force, of two large carriers, *Shokaku* and *Zuikaku*, to cover both operations. American cryptographers had identified Japanese intentions, direction-finders had located the position of the main Japanese units and Admiral Nimitz detached two of his precious carriers, *Lexington* and *Yorktown*, to deal with the intruders.

A very confused encounter, known as the Battle of the Coral Sea, ensued. Japanese aircraft found an American oiler and a destroyer, the *Neosho* and the *Sims*, mistook the first for a carrier, the second for a cruiser, sank *Sims*, damaged *Neosho* and returned to their mother ships exultant. American aircraft meanwhile found the force covering the Port Moresby landing ships, sank the small Japanese carrier *Shoho* and returned equally exultant. On the following day, the main forces found each other, *Lexington* was sunk, *Yorktown* and *Shokaku* were both damaged but withdrew to repair, *Zuikaku* was not touched. The Americans reckoned Coral Sea a victory, since it prevented the capture of Port Moresby. The Japanese, in terms of ship losses, had reason to regard it as a victory themselves.

The Coral Sea left the balance of carriers in the two Pacific navies as follows: Japan, *Zuikaku, Shokaku, Hiryu, Soryu, Kaga, Akagi* and the small carriers *Ryujo* and *Zuiho;* the United States, *Saratoga, Wasp, Ranger, Enterprise, Yorktown* and *Hornet.* Actual numbers were smaller than paper strength. Of the Japanese carriers, *Shokaku* had retired to Truk, in the Carolines, to repair; *Zuikaku* had lost so many of its aircraft at the Coral Sea that it had been withdrawn to refit; *Ryujo* and *Zuiho* were judged too small to take part in a major fleet action. On the American side, *Wasp* and *Ranger* were absent in the Mediterranean, having been generously lent by President Roosevelt to ferry fighters to the besieged island of Malta, while *Yorktown* was in dock in Pearl Harbor. Hit by an 800-pound bomb at the Coral Sea that had penetrated to her fourth deck, killed sixty of her crew and started a serious fire, she was also leaking from multiple splinter holes. The dockyard estimated that she needed a ninety-day repair. Admiral Nimitz announced that he needed her operational in three days. She entered dry dock on 27 May and, after 1,400 men had worked round the clock for two days, was floated out on the morning of the 29th. During the afternoon she embarked new aircraft to replace those lost at the Coral Sea and at 9 a.m. on 30 May she left to join the fleet. Nimitz had organised his carriers into the two task forces: TF 16, under Admiral Raymond Spruance, temporarily replacing Halsey, who was in hospital ashore, consisting of *Hornet* and *Enterprise; Yorktown,* with its escorts, was to form TF 17, under Admiral Jack Fletcher. Their role was to find and destroy the Japanese main fleet.

American cryptanalysis had alerted the high command of the Pacific Fleet to a forthcoming Japanese operation. After the raid into the Indian Ocean against the British, and the frustrated offensive of the Coral Sea, it was obvious that their carriers would strike again. The question was, where? The Americans had no inkling of the effect the Doolittle raid had had on the Japanese sense of honour. There was evidence, however, from intercepted Japanese signals, that the next offensive would be in the Central Pacific. It was in that direction that TF 16 and TF 17 were headed.

MAGIC AND MIDWAY

The first indication that the Japanese Combined Fleet might return to the Central Pacific, after its support of operations around the Philippines, Malaya and the Dutch East Indies in December through February, came on 5 March, not through signals intelligence but because of a minor Japanese bombing attack on Hawaii. The Americans correctly concluded that the attack had been launched from the Marshall Islands, via a refuelling stop at an isolated oceanic anchorage known as French Frigate Shoals. What was important for the future about the 5 March raid was that the American intercept stations were able to identify it with something denoted in Japanese codes as the K operation. The significance of K became more apparent on 6 May, when HYPO, the Hawaiian station of American cryptanalytic intelligence, deduced that K was part of an encrypted geographical designator standing for Hawaii. The American codebreakers were beginning to recognise that the Japanese used three (later two) letter groups to denote geographical objectives, those in the American zone of operations beginning with A (hence AK for Hawaii), those in the British zone beginning with D and those in the Australian zone with R.[11]

This was an important breakthrough but its value was set back by the introduction of new security measures in the Japanese fleet, consisting mainly of a change of call signs between ships but also between shore and ship. The changes greatly complicated the codebreakers' ability to identify the location of individual ships and the composition of fleets.[12] Their difficulties were compounded by the retreat of the main Japanese naval units into radio silence, as before Pearl Harbor. On 13 March the American cryptanalysts broke into the main naval code, JN-25, but that success was shortly negated by the Japanese adoption of a new system of cipher additives to the code groups. Just before the change the Americans got a crucial insight into Japanese intentions via a request from an unidentified ship for a supply of charts, clearly indicating the interest of the Japanese fleet in the Hawaiian group of islands and its western outliers, which included Midway. Nimitz, commanding the U.S. Pacific Fleet, accordingly concluded that he was faced by four possible Japanese operations: an attack on Midway-Hawaii; an attack on the Aleutian Islands, outliers of American Alaska; an attack on other Central Pacific islands; a renewed attack on the islands of New Guinea, any of those operations to be launched between 25 May and 15 June.[13]

As a result of the Japanese adoption of the new cipher additives to their basic code, JN-25, the result being known to the Americans as JN-25B or Baker, the American cryptographers lost their way into Japanese transmissions during May; or should have done. Because, however, of the difficulty the Japanese found in distributing new additive books across their enormous area of conquest, individual transmitters made the mistake, which may be called a classic mistake so often has it betrayed encryptors, of transmitting messages in both the old and the new code, to ensure accuracy of reception. The Americans, able to read the old code, were thereby enabled to read some of the new, so that by late May they had established the outline of their enemy's developing plan.

The composition of the attacking force was the first crucial matter to be discovered, though largely by traffic analysis—identifying individual ship call signs and detecting their location—rather than cryptanalysis. On 17 May, Admiral King was able to publish, to a limited circle, an assessment of the strength available to the enemy for what was now believed—but not confirmed—to be an offensive against Midway and the Aleutians. For Midway, it consisted of four fast battleships, two cruiser divisions, two carrier divisions with a fifth carrier attached, two destroyer squadrons and a landing force; for the Aleutians, a cruiser division, a carrier division, formed of the two old, small carriers *Ryujo* and *Zuiho*, two destroyer squadrons and a landing force.

Next day, 18 May, King was able to narrow the geographical frame. A message was intercepted from Admiral Nagumo, commander of the Combined Fleet, reading, "since we plan to make attacks roughly from the northwest from N minus 2 days until N Day request you furnish us with weather reports three hours prior to the time of take-off on said days." From parallel messages intercepted by the decrypting centres in Melbourne (CAST) and Hawaii it was discovered that the Japanese aircraft for which the weather forecasts were intended would be launched "fifty miles northwest of AF."[14]

AF was one of the indicators used in Japanese encoded messages to indicate geographical locations. So at least some of the American cryptanalysts believed; unfortunately, others thought otherwise. Commander Edwin Layton, the extremely efficient Fleet Intelligence Officer of the Pacific Fleet, took the view that AF was a geographical indicator and so informed Admiral King in Washington; he specified Midway and Hawaii as the targets of the forthcoming Japanese offensive and Saigon and Ominato, in the Japanese home islands, as the strike forces' departure points. This assessment arrived at a moment when the various officers of NEGAT (as OP-20-G was code-named) were deep in argument with Admiral Richard Turner's War Plans Division over what portended in the Pacific. The argument, as so often occurs in bureaucracies, took on a life of its own, separate from the realities of the outside world. OP-20-G had become divided into three sub-branches, OP-20-G1 (Combat Intelligence), OP-20-GZ (Translation) and OP-20-GY (Cryptanalysis). Their counterparts in War Plans began to disagree with the intelligence experts over detail, until a full-scale office war was in progress. Turner, head of War Plans, and Redman, head of Naval Communications, responsible for naval intelligence, eventually came to daggers drawn over the issue of whether the Japanese commander of Fifth Fleet "is to command any force now concentrating in Northern Empire Waters"; Turner, who was senior to Redman, directed him "to assume that Admiral Turner's views are correct." Turner, wrongly, took the view that the Japanese offensive would be a continuation of the Coral Sea campaign, Redman that it was directed at AF, which could not be New Guinea.

This at a time when a huge Japanese fleet was gathering to threaten the United States' last centre of power in the Pacific, Hawaii and its nearby islands, with a potentially devastating attack. What follows cannot be documented, since evidence is lacking. It seems probable, however, that at a local level, HYPO (Hawaii) took steps to resolve the issue unarguably. An undersea cable link, secure from Japanese ears, still operated between Hawaii and Midway, 1,300 miles to the west. The idea of using it deceptively has been attributed to Captain Joseph Rochefort, HYPO's energetic chief. The story goes that on 18 or 19 May, with Admiral Nimitz's permission, a cable message was sent from Pearl Harbor to Midway instructing the garrison of the tiny island to report a water shortage, signalling by radio in plain language. On 22 May CAST at Melbourne reported the interception of a message from Japanese naval intelligence in Tokyo (KIMIHI) reading as follows: "The AF air unit sent the following message to [Pearl Harbor] on [May] 20. 'Refer this unit's report dated 19th, at present time we have only enough water for two weeks. Please supply us immediately.'" CAST appended, "Have requested [Pearl Harbor] to check this message—if authentic it will confirm identity 'AF' as Midway."[15]

The objective of the coming Japanese offensive was now known:

Midway. Subsidiary decrypts identified the Aleutians as a secondary objective; as they were ungarrisoned, that threat could be ignored, even though the islands were sovereign American territory.

Finally, on 25 May, HYPO broke the Japanese navy's date cipher. By applying the decrypts to old intercepts, Rochefort in Hawaii was able to establish that the attack on the Aleutians would begin on 3 June, the offensive against Midway on the 4th. Nimitz, who reposed great faith in his intelligence decryptors and analysts, accordingly called forward his forces to meet the threat. TF 16 (*Hornet* and *Enterprise*) was recalled to Pearl Harbor on 26 May, to prepare for battle. TF 17 *(Yorktown)* was already there, repairing the damage suffered at Coral Sea. He also positioned a submarine screen northwest of Midway to detect the approach of a Japanese strike force.[16]

General MacArthur, commanding in the southwest Pacific, lent important assistance at this stage to Nimitz's countermeasures, by recommending that radio deceptive measures be instigated, to suggest to the Japanese that a carrier group was still in the Coral Sea. Nimitz agreed, and accordingly the cruiser *Salt Lake City* and the seaplane tender *Tangier* steamed due south of New Guinea exchanging radio traffic that simulated carrier transmissions.

It was fortunate, nonetheless, that OP-20-G and its outstations had, by whatever means, established in late May what the Japanese intended, for the intelligence climate then turned against the Americans. The Japanese relapsed into radio silence, as they had done before Pearl Harbor, while their own listening services, on the alert for any American reaction to the despatch of the Midway strike force from the Inland Sea, began to report a significant increase in what it identified as "Urgent" messages from Pearl Harbor. Japanese intelligence also noted the sighting of American patrol aircraft far west of Midway and the interception of messages from an American patrol submarine in the path of the Midway landing fleet. For inexplicable reasons, Admiral Yamamoto withheld the information from the Midway strike force. It may have been that he did not wish the force to break the radio silence imposed on it by requesting clarification; the result, whatever the motive, was that Admiral Nagumo and his carriers steamed on towards Midway in ignorance of a gathering American riposte.[17]

The complexity of the intelligence plot in the first days of June defies easy exposition. Over the enormous expanse of the Pacific Ocean and its surrounding coastlands, the Japanese appeared to be unchallengeably in the ascendant. Indo-China, thanks to the complaisance of the French Vichy government, was under their control. Coastal China was under Japanese occupation. British Malaya and Burma had just fallen into Japanese hands. British India and Ceylon, with the surrounding waters of the Indian Ocean, were under threat of invasion, following vigorous naval attack. The Dutch East Indies had been occupied, together with most of Australian New Guinea and its outlying islands. The Central Pacific islands, mandated to Japan after 1918 or captured from the Americans and British in the first months of the current war, were Japanese oceanic strongholds. Australia itself, whose Northern Territories had already been bombed by the Japanese, was in a state of defence against invasion. The American-protected Philippines had just surrendered. All that remained to Japan's enemies as points of resistance to what seemed its inexorable advance to Pacific domination were the American Hawaiian archipelago and its outlying island of Midway.

THE BATTLE OF MIDWAY

Yet the survival of Midway simplified the coming campaign and battle greatly to American advantage. For unless the Japanese attempted a direct assault upon Hawaii, a most unlikely undertaking, Midway was the only place north of the equator—with the exception of the barren and largely uninhabited Aleutian Islands—that was worth their attention; and as recent intelligence in all its forms, decryption, traffic analysis and visual sighting, put the main body of the Japanese navy in or near the home islands, any operation soon forthcoming must be launched north of the equator. Midway must therefore be the target, a conclusion supported by the work of OP-20-G, HYPO and CAST, which allowed King and Nimitz to concentrate the surviving strength of the Pacific Fleet west of Hawaii with confidence.

Yamamoto's plan was designed to confuse the issue. His fleet was divided into five separate bodies, each committed to a different geographical objective or operational aim. Although the Japanese navy had been created on a Western model, under the guidance of Western, largely British, advisers, its operational methods remained essentially Oriental. Its leaders were well aware of the Western doctrines of singularity of aim and concentration of force. They had failed to rid them-

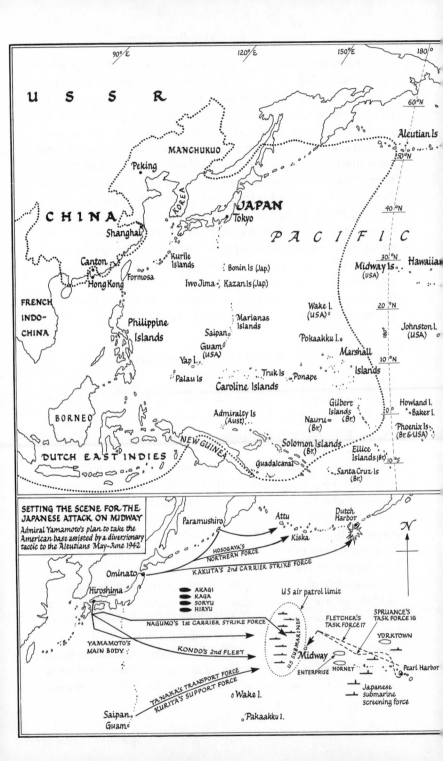

SETTING THE SCENE FOR THE
JAPANESE ATTACK ON MIDWAY
Admiral Yamamoto's plan to take the
American base assisted by a diversionary
tactic to the Aleutians May-June 1942

MIDWAY
THE PACIFIC THEATRE IN 1942
······· Furthest extent of Japanese eastward conquest

| 0 | 1000 | 2000 | 3000 miles |
| 0 | 1000 | 2000 | 3000 | 4000 | 5000 km |

150°W   120°W   90°W   60°W

ALASKA

utch Harbor

UNITED STATES OF AMERICA

San Francisco
(US Naval Base)

San Diego
(US Naval Base)

MEXICO

ands
(USA)

Pearl
Harbor

US OCEAN FORCES

Revilla Gigedo Is
(Mex)

O C E A N

Cocos Is.

Palmyra I. (USA)
Washington I.
Fanning I.
Christmas I.
Jarvis I. (USA)

Galapagos Is
(Ecuador)

Maiden I. (Br.)

SPRUANCE'S TASK FORCE 16
ENTERPRISE, HORNET,
6 cruisers, 9 destroyers

4 JUNE
17.00

HIRYU crippled

HIRYU
scuttled

4 JUNE
06.55

Air strike from ENTERPRISE

FLETCHER'S TASK
FORCE 17 YORKTOWN,
2 cruisers, 5 destroyers

NAGUMO'S 1st CARRIER
TRIKE FORCE
carriers, 2 battleships,
cruisers, 8 destroyers

4 JUNE 08.06

4 JUNE
10.00

HIRYU torpedo aircraft

HIRYU dive bombers

4 JUNE 14.40 YORKTOWN
torpedoed and sunk

SORYU sunk
AKAGI scuttled
KAGA sunk

from YORKTOWN

US air strikes

4 JUNE 07.10

from HORNET

from ENTERPRISE

Aircraft return
to Midway;
11 land 11.35

to Pearl Harbor

BATTLE OF MIDWAY
4 June 1942

N

B 17s

Japanese bombers

Midway

- - - → US air strikes
━ ━ → US fleet movements
──→ Japanese air strikes
━━→ Jap. fleet movements

| 0 | 50 miles |
| 0 | 50 | 100 km |

amoto ordered retirement of the
nants of his fleet at 02.55 5 June

selves, nonetheless, of ancient Asian notions of the value of complexity and diffusion. Yamamoto had therefore sent forward, as his first group, a force of ten submarines to patrol the Midway area; second, he had organised a Midway Occupation Force of transports, embarking the landing party, escorts to protect the transports and two battleships and four cruisers to protect the escorts; third, the Carrier Striking Force, commanded by Nagumo, of four large carriers, *Akagi, Kaga, Hiryu* and *Soryu;* the main body of a light cruiser, a light carrier and three battleships, including the brand-new *Yamato,* 70,000 tons, 18-inch guns, the most powerful battleship in the world, under his own command; and the Northern Area Force, including four battleships and two light carriers. Counting sub-divisions, the complexity was even greater; it has been calculated that there were altogether "sixteen different groups of warships, all working to a complex plan devised by Captain Kuroshima Kamato, Yamamoto's senior operations officer."[18]

The complication was irrelevant. The commanders of America's Pacific Fleet, after the disaster of Pearl Harbor, the devastation of the air force in the Philippines, the defeat of the Java Sea and the drawn battle of the Coral Sea, had no margin of strength to dabble in diversions. Like a gambler with one last throw up his sleeve, Nimitz had to stake all on the appearance of the enemy at Midway. It was not a blind gamble. The American intelligence organisation had counted the cards. By the end of May the chips were stacked. The outcome depended upon how the hands would be played.

The two American carrier task forces departed Hawaii for Midway in the last days of May, *Enterprise* and *Hornet* (Task Force 16) on 28 May, the hastily repaired *Yorktown* (Task Force 17) on the 30th. Japanese counter-intelligence had probably become aware of American movements, because its traffic analysts had reported in late May that 72 out of 180 American messages originating in Pearl Harbor were prefixed "Urgent," while the intelligence detachment on recently captured Wake Island reported that American patrol planes were operating in its area, and the Midway Landing Force, which had left Saipan on 28 May, detected an urgent transmission from the American submarine that appeared to be tracking its progress.[19] So the Japanese knew something; but the Americans knew more: strength, place, time. Most significantly, a HYPO intercept disclosing the departure date of a Japanese oiler, believed to be part of the Midway attack force (MI), suggested, by reason of its known speed, an arrival date near Midway of 30 May.

There was much other imprecise but collateral intelligence confirming the MI fleet's movements and purposes from several intercept sources in the first three days of June. Then at 6:04 a.m. Midway time on 3 June, a Catalina search flying-boat, on patrol from Midway, transmitted the report "Many planes heading Midway from 320 [degrees] distant 150 miles."[20] The pilot was Ensign Jack Reid, who had decided to extend his search time for a few minutes. He said to his co-pilot, "Do you see what I see?" His co-pilot answered, "You're damned right I do." Spread out before them, at their limit of vision, was an enormous assembly of warships. They knew at once that they had sighted part of the Midway attack fleet.

What Reid and his fellow aviator had seen was in fact the leading element of the Midway Occupation Force. On receipt of their sighting report, Captain Cyril Simard, in command at Midway, ordered nine of the fifteen Flying Fortress bombers based on the island's airstrip to the attack. The Fortress, a four-engined aircraft flown by the Army Air Force, operated at high altitude and could strike targets with great precision; but army pilots always found bombing over water difficult, and 3 June was no exception. On return they reported hits on two battleships or heavy cruisers—none were present—and two transports. In practice they had hit nothing. Next morning four Catalina flying-boats, equipped with radar and torpedoes and flying at low level, did better, damaging an oiler. The attack did not, however, deter the advance of the Midway Occupation Force.

It may, however, have confirmed Nagumo's intention to attack Midway itself with his carrier-based aircraft. Chuichi Nagumo was a bluff sea dog, much admired in the Japanese navy for his outspoken manner and fighting reputation. He had been a dashing destroyer captain and was notably unimpressed by American naval power, a matter of disagreement between him and Yamamoto. Yet, although he had commanded the carrier group that attacked Pearl Harbor, he was not a carrier specialist and does not seem fully to have understood naval aviation. Moreover "as a fleet-handler in wartime, he always hesitated, never being quite sure what to do."[21] After Pearl Harbor, when Genda, the leader of the air attack, had urged him to send a second strike, he had prepared to rest on his laurels and withdraw the carrier fleet to a safe distance, though, as we now know, he could have attacked again with impunity. His indecision and wrong decision-making during the coming Midway battle would do much to rob Japan of victory.

At 4:30 on the morning of 4 June, Nagumo's four carriers launched seventy-two Val and Kate bombers, escorted by thirty-six Zero fighters, to strike Midway. The Val was a dual-purpose aircraft, a dive- but also high-level bomber, with a speed of 200 mph and a range of 800 miles, superior to its American equivalent, the Douglas Dauntless. The Kate was a torpedo bomber, also able to drop bombs, with the same speed but a slightly shorter range than the Val, again superior to its American equivalent, the Douglas Devastator. Nagumo's aircraft, led by Lieutenant Joichi Tomonaga, had 276 sea miles to fly to their target, well within the operational radius, but, unlike modern carrier aircraft, they had no radar, so that navigating the return flight to their mother ships was beset by hazard. There were other important differences between then and now. Second World War carriers were "straight deck," unlike the modern angled-deck ships which can park recovered aircraft off the flight path of those landing on. Aircraft landing on did so with the recently recovered aircraft parked in front of them, unless there had been time to "strike down" the incomers, sending them below by elevator to the hangars on the next deck. Landing-on was therefore always a fraught business: there was the danger of an incoming aircraft missing the arrester wire with its hook and crashing into the aircraft park; there was also the possibility of enemy aircraft catching a carrier with much of its air group on deck, waiting to be struck down while landing-on took place.

Real safety for a carrier group lay in not being discovered by the enemy. Thanks to the decrypting activity of Joseph Rochefort and his staff at HYPO on Hawaii, however, the whereabouts of Nagumo's carrier fleet was already known to Nimitz and his two task forces even before it had been sighted by the Midway Catalina on 3 June. He had already predicted that the gathering Japanese attack force would be found 175 miles from Midway on 4 June 1942, bearing 325 degrees, at 7 a.m. local time. "That forecast was the most stunning intelligence *coup* in all naval history."[22]

Tactical intelligence then confirmed the prediction. As Tomonaga's 108 carrier-launched aircraft headed for Midway, they were detected by a radar station on the island at 5:30 a.m., then lost, then identified again by a maritime radar. Midway received a report, "Many bogey aircraft bearing 310 degrees distance 93 [miles]." Midway at once launched all its fighters, six Wildcats, twenty Buffalos, to intercept the marauders.

The Wildcats and Buffalos, flown by U.S. Marine Corps pilots, were obsolete and outclassed as well as outnumbered by the Zeros. Only nine survived. Tomonaga's aircraft, though they did a great deal of superficial damage, also suffered heavy losses to anti-aircraft fire and failed to put the Midway base out of action. As he began his return flight to *Hiryu,* Tomonaga reported to Nagumo that a second strike was required. Meanwhile, more than an hour before Tomonaga's aircraft had attacked, Reid's Catalina had sighted Nagumo's force; Tomonaga's aircraft had not yet left their mother ships. The Catalina, 200 miles outward from Midway to the northwest, first reported in a short cipher group "enemy carriers" at 5:34 a.m.. At 5:45 it signalled in plain language, intercepted by the combat information centre on *Enterprise,* "Many enemy planes heading Midway bearing 320 degrees distant 150 [miles]," an almost exact confirmation of HYPO's prediction. Finally at 6:30 the Catalina, whose crew deserved to be regarded as one of the most efficient reporting units ever to have performed a naval reconnaissance mission, sent the message, "two carriers and battleships bearing 320 degrees distant 180 [miles] course 135 degrees speed 25 [knots]." The only falsity in the Catalina's signal was the report of battleships; the pilot may have miscounted the number of carriers, in fact four, or have mistaken carriers for battleships.

Despite being seen by Nagumo's force, and despite its very low speed, the Catalina got away. Soon after it left, other American aircraft appeared. They belonged to the land-based bomber squadrons on Midway, which Captain Simard had launched before Tomonaga's arrival. It was their absence which had prompted Tomonaga's warning that another attack was needed; he correctly anticipated that Midway was still an offensive base. He apparently did not conclude that the missing aircraft might be on their way to intercept the Japanese mother ships. They were. Soon after seven o'clock the carriers were attacked by six Avenger dive-bombers and four B-20 Marauder medium bombers. The Avengers, though quite fast by contemporary standards, were too few in number to swamp the defence and four were shot down by anti-aircraft fire or fighters. The Marauders, which were equipped with improvised torpedo launchers, pressed their attacks right home but scored no hits;

two were destroyed. Just before eight o'clock a squadron of Marine Corps aircraft from Midway, sixteen Dauntless dive-bombers and eleven obsolete Vindicators, continued the attack: the Dauntless was a robust modern bomber, soon to win a reputation as the best carrier-borne American attack aircraft in the Pacific, but the Midway Marine pilots were unfamiliar with it, and the squadron commander did not attempt to dive-bomb. Six of his Dauntlesses were shot down, two damaged; no hits were scored. Finally, at about 8:10, Midway's fifteen Flying Fortresses appeared overhead at 20,000 feet, dropped a concentrated pattern of heavy bombs on to the carrier group and departed, unscathed, believing they had hit several ships.

Wrongly; none of the Midway aircraft had inflicted damage to Nagumo's ships, though they had killed some of his sailors. They had, however, seriously shaken Nagumo's ability to think clearly. Always impulsive rather than analytical, he now allowed events instead of reason to prompt his responses. Admittedly, he was faced by a dilemma. The point of the Midway operation was not to destroy the island's defences, or even to capture it, but to attract the surviving American carriers into a battle. The advance to Midway was the preliminary to springing a trap. Even though there was as yet no evidence of American carriers in the vicinity, Nagumo's duty, as fleet commander, was to keep his ships prepared to fight a carrier battle if one suddenly erupted. On the other hand, he was also supposed to cover the landing force, which the defenders of Midway, if still active, might defeat. Moreover, they might launch a fourth strike on his ships.

In the circumstances he decided at 7:15 to "break the spot," in American parlance: to alter the arrangements on his four carriers' flight decks from preparation for an anti-ship strike to preparation for a repetition of the attack on Midway. That required the torpedo bombers to be re-armed with bombs, the dive-bombers to be re-armed with similar fragmentation bombs instead of armour-piercing bombs. Time-consuming work, particularly as the aircraft on deck had to be struck below to the hangars. As the work began, Tomonaga's Midway aircraft started to land, together with the Zeros of the Combat Air Patrol, to refuel. While all this complex activity was in progress, Nagumo was given word, at 7:28, of the proximity of American surface ships after all. A seaplane catapulted from the cruiser *Tone*, silent until then, suddenly reported "sight what appears to be ten enemy surface ships in position bearing 10 degrees, speed over 20 knots." *Tone*'s seaplane, because of catapult trouble, had left half an hour late. It was now almost at the limit of its search radius.

The news came at the worst possible moment for Nagumo. His decks were cluttered with aircraft just landed, and strewn with refuelling hoses. Many of his strike aircraft were below in the hangars, exchanging torpedoes for bombs or one sort of bomb for another sort. Yet instead of making the firm and obvious decision to launch an anti-ship strike, Nagumo dithered. He apparently thought he could cover himself by preparing for two missions at the same time. At 7:45 he signalled the fleet, "Prepare to carry out attacks on enemy fleet units. Leave torpedoes on those attack planes which have not as yet changed to bombs."[23] Then, in a brisk afterthought, he radioed *Tone*'s seaplane, "Ascertain ship types and maintain contact."

Perhaps, Nagumo may have thought *Tone*'s seaplane would not have found American carriers. In any case, as if to vindicate his decision to preserve some capacity to resume the attack on Midway, it was at this point that the last attack from the island, by Dauntlesses, Vindicators and Flying Fortresses, was received. Despite its failure, it disturbed the fleet's formation and further discomposed Nagumo's ability to analyse the tactical situation. At 7:58 *Tone*'s seaplane reported that the enemy fleet had changed course from 150 to 180 degrees. Nagumo demanded, "Report ship types." At 8:09 the seaplane answered, "Enemy is composed of five cruisers and five destroyers." Nagumo's anxieties seemed allayed, particularly as at 8:29 the last bomb-splashes raised by Midway's Flying Fortresses collapsed harmlessly into the sea. Refuelling and re-arming were almost complete. The moment of danger appeared to have passed.

Then, at 8:20, just before the Fortresses departed, *Tone*'s maddeningly deliberate seaplane radioed, "The enemy is accompanied by what appears to be a carrier bringing up the rear." Nagumo had made a mistake; quite how serious the next hour and twenty minutes would reveal. *Tone*'s seaplane crew were not wholly to be blamed for their dilatory identification of danger. The fourth of June 1942, in the north-central Pacific, was bright and sunny but the sky was spotted by clouds. The Americans flying from Midway had found the Japanese ships they sighted appearing and disappearing with bewildering rapidity. The clouds broke up their field of vision, denying a panorama. *Tone*'s seaplane had had the same experience.

The consequences of its incomplete reporting, however excused, were disastrous. In the hour and twenty-seven minutes between 7:28 and 8:55, from first sighting of the American task forces and reception of the *Tone* seaplane's last ominous message, "Ten enemy torpedo planes heading towards you," Nagumo might, by better thinking, have put his fleet into a state of defence, prepared his bombers and torpedo aircraft for an anti-ship strike and got them, with a refuelled Combat Air Patrol, off the decks. As it was, though most of his Zeros were refuelled and aloft by the time the crisis came, his other aircraft were either below decks or not yet struck below, while the decks of his four carriers were littered with fuel hoses and loose ordnance.

Spruance, commanding *Enterprise* and *Hornet*, because he could not enjoy the luxury of indecision, had reacted with single-minded positivity to the Midway Catalina's report of Japanese proximity received at 5:34. He had first decided to close the distance to no more than a hundred miles before launching. When he got news of Tomonaga's attack on Midway, he decided to launch earlier, making the calculation that he might thereby catch Tomonaga's aircraft landing or waiting to be refuelled and re-armed. It was an acute judgement. Shortly after six o'clock, and although he committed his pilots to a flight of 175 instead of 100 miles, he decided to advance his launch time from 9 a.m. to 7 a.m. Fletcher, commanding *Yorktown* (Task Force 17), operating north of Task Force 16, decided to hold his hand. He believed that at the Coral Sea he had launched too soon and did not intend to repeat his error.

Spruance's strike force comprised almost equal numbers from *Enterprise* and *Hornet*: sixty-seven Dauntless dive-bombers, twenty-nine Devastator torpedo bombers and twenty Wildcat fighters to fly escort. The earliest aloft were ordered to orbit, while they waited for a full launch, so that the carriers could deliver a concentrated blow. At 7:45, however, concerned that his leading flights might run out of fuel, Spruance ordered them to set off for the Japanese. By 8:06, all were on their way. There were six squadrons in the air, Bombing 6, Scouting 6 (bombers) and Torpedo 6 from *Enterprise*, Bombing 8, Scouting 8 and Torpedo 8 from *Hornet*, with those fighters of Fighting 6 and 8 not flying Combat Air Patrol in company.

Flying off, in 1942, was almost a stunt performance. The pilot worked up to full revolutions, released his brakes and accelerated down the deck, pulling the joystick back as he crossed the bow; engine failure or mishandling dropped him into the sea. All TF 16 aircraft made successful departures, and after formating, sixty-seven Dauntless dive-bombers, twenty-nine Devastator torpedo bombers and twenty Wildcat fighters set course for Nagumo's calculated position.

Circumstances were to prevent their arrival in concentration. At the outset, Spruance decided to send the first four squadrons on ahead, since orbiting wasted precious fuel. Then, as they strung out towards the target, Nagumo, warned by his scouting aircraft of their approach, altered course at 9:05 from northeast to southeast. At 9:20, when *Hornet*'s dive-bombers reached the indicated position, they found the sea empty. Bombing 8's leader therefore decided that Nagumo must be heading towards Midway, turned, and led his squadron due south. The aircraft were running out of fuel, however, and fifteen were forced to land on the island; the rest returned to their own carriers but all the Wildcat fighters fell into the sea with dry tanks.

Hornet's torpedo bombers, led by Lieutenant-Commander John Waldron, had become separated from the dive-bombers but, arriving near the target area, spotted funnel smoke on the horizon and turned to investigate. As they approached the Japanese carriers at sea level, to make their torpedo runs, they were attacked by the combat air patrol of sixty Zeros. In a few minutes all fifteen Devastators were shot down, only one pilot surviving. No hits were scored. Torpedo 8 was shortly followed by Torpedo 6, from *Enterprise,* which had also lost its fifteen escorts. Manoeuvring into a favourable approach, the Devastators attracted the Zeros which had just destroyed Torpedo 8 and were massacred. Only four of fourteen survived and the squadron achieved no hits. Finally, at about ten o'clock, *Yorktown*'s torpedo squadron, VT 3, appeared. It, too, was attacked at sea level by the Japanese combat air patrol, lost seven out of twelve aircraft and achieved nothing.

The destruction of the torpedo bombers was not, however, in vain. Because they descended to sea level to make their dropping runs, they thereby brought down the Japanese combat air patrol from its protective high altitude. When at 10:25, therefore, yet another wave of American aircraft approached, to bomb from 14,000 feet, Nagumo's four carriers lay open to destruction. Their decks were crowded with aircraft waiting to be launched in a retaliatory strike, draped with fuel hoses and littered with torpedoes and bombs. *Akagi* was the first to be hit. Nagumo's chief of staff, Ryunosuke Kusaka, reported "a terrific fire ... bodies all over the place." A bomb from a dive-bomber had hit the midships elevator, penetrated to the hangar deck and set off a torpedo store. Another fell into the aircraft park. Kusaka went on, "There was a huge hole in the flight deck, just behind the amidships elevator. The elevator itself, twisted like molten glass, was drooping into the hangar. Deck plates reeled upwards in grotesque configurations. Planes stood tail up, belching livid flames and jet-black smoke, their torpedoes began to explode, making it impossible to bring the fires under control. The entire hangar area was a blazing inferno and the flames swiftly spread to the bridge."[24]

Akagi's fate had come about by accident, not intelligence activity. The intelligence supplied to TF 16 and TF 17 had, indeed, thus far resulted only in catastrophe. The three torpedo bomber attacks, by eighty-three aircraft, had resulted in the loss of thirty-seven, together with many of their fighter escorts, and no damage to the Japanese at all. *Enterprise*'s dive-bombers had been led to Nagumo's carriers by hazard. The Japanese were not where they were expected to be; they were discovered by chance. Among Nimitz's preparations for the encounter near Midway had been the deployment of a submarine screen. One of the submarines, *Nautilus,* attempting to set up an attack, had been detected by the destroyer *Arashi,* which lingered to drop depth charges, without effect. Working up to speed to rejoin the fleet, it created a vivid white wake. Lieutenant-Commander Clarence McClusky, leading the Dauntless dive-bombers of *Enterprise,* saw the white streak on the surface of the ocean, guessed and turned to follow at 9:55. At 10:20 he sighted *Akagi, Soryu* and *Kaga* steaming north-west in "a circular disposition of roughly eight miles"; *Hiryu* was farther ahead. Their original tight, self-supporting formation had been broken up by the torpedo bomber attacks. McCusky turned to engage, leading his dive-bombers down from 14,000 feet in a seventy-degree dive. Their terminal speed almost exceeded that of the carriers' Zeros; but they, in any case, were flying too low, having driven off the torpedo bombers, to attain a defensive altitude.

One after another, three of the big Japanese carriers succumbed, first *Akagi*, Nagumo's flagship, then *Kaga,* whose parked aircraft, fuel lines, bomb stores and hangars were set alight by 500- and 1,000-pound bombs. Finally, *Soryu* was attacked by dive-bombers from *Yorktown* which, launched late, were attracted to the scene by the smoke of battle and dropped bombs that, among other damage caused, folded the midships elevator back against the bridge.

Between 10:25 on 4 June, when Nagumo was preparing to launch his anti-carrier strike, and 10:30, when *Enterprise*'s Bombing Squadron 6 delivered its attack, Japan's plan to conquer the Pacific was reduced to ruins. Three of its six big carriers had been fatally struck; the fourth was to fall to American naval power before the twenty-four hours were out. *Soryu* was finished off in the early morning of 5 June by torpedoes from *Nautilus,* the submarine whose intervention had inadvertently guided the dive-bombers of Spruance and Fletcher to Nagumo's carriers the day before. At about the same time, *Hiryu,* hit by dive-bombers launched from *Enterprise* on the afternoon of 4 June, succumbed to fatal damage sustained in that raid.

The Japanese defeat was not to be quite unbalanced. *Yorktown,* which had survived grievous damage at the Coral Sea and launched one of the decisive strikes of 4 June, was found by aircraft from the still-surviving *Hiryu* about noon the same day and, despite the desperate efforts of its combat air patrol, hit hard. Abandoned, then reboarded by a damage control party, she was limping towards the safety of Pearl Harbor when a Japanese submarine, one unit of a screen deployed by Yamamoto to entrap Nimitz's fleet in his great Midway scheme, found her proceeding eastward at very low speed, on 5 June, manoeuvred to intercept and fired four torpedoes. Two hit, and after a desperate death struggle, she capsized on the morning of the 6th.

The battle of 4 June 1942—Midway—was nevertheless instantly reckoned, exultantly by the Americans, reluctantly but no less certainly by the Japanese, a dramatic victory for the power of American arms. Beginning with all the advantages, the Imperial Japanese Navy had been reduced in a few hours, indeed minutes, of hectic conflict from dominance to subordination in the struggle for control of the Pacific. The Japanese empire's long-laid plans, to acquire an impregnable strategic holding in the Central and South Pacific and to create a world-class fleet capable of defending it against any counteroffensive, had been reversed in a few hours of violent combat.

Yet the question remains to what extent exactly Midway was an intelligence victory. It was hailed as such by those in the know at the time and, when the facts became public knowledge, in general opinion. OP-20-G, and its outstations on Hawaii and at Melbourne, was credited with identifying, first, Japan's decision to switch the axis of its naval offensive from the South Seas—against Australia—to the Central Pacific, next to identifying Midway as the offensive focus, then to establishing a narrow time bracket for its launching and, finally, to constructing an accurate plot of the Japanese order of battle; on 31 May, Nimitz issued a signal, 13/1221, beginning, "Estimate Midway organisation stop striking force 4 carriers [*Akagi, Kaga, Hiryu, Soryu*], 2 Kirishimas [battleships] 2 Tone class cruisers 12 destroyers . . . ," an almost exact tally of the ships under Nagumo's command. More conventional intelligence—radar contact by Midway's station, visual sighting of Nagumo's fleet by one of Midway's search aircraft—established the Japanese carrier fleet's oceanic position and speed of advance just before its first strikes were launched. That information allowed Captain Simard on Midway to launch his bomber and torpedo strikes and Admiral Fletcher to position the two task forces for their ship-to-ship attacks.

The exactitude of the intelligence available to Nimitz and his subordinate commanders about the Midway attack was indeed extraordinary: enemy objective, timing, strength, direction of approach, launch position, a tick list of "information enemy" requirements; and all the more extraordinary in that most was the product of cryptanalysis. Yet it has to be recognised that, despite the riches cryptanalysis bestowed on the Americans, the result was not preordained, the outcome hung in the balance even after Fletcher had launched his aircraft towards Nagumo's position and that contingencies and chance were critical determinants of the victory.

Spruance risked all by his decision to launch "a full load" from *Hornet* and *Enterprise*, every dive- and torpedo bomber he had. Despite the intelligence the crews had been given, many failed to find the target. Nagumo, so much vilified in the aftermath, made a correct and prudent decision to alter course, based on a reconnaissance sighting of the incoming aircraft, before *Hornet*'s dive-bombers arrived. They found empty sea at the expected point of encounter, were led away in the wrong direction to search for the Japanese, missed the battle and, in the case of their escorting fighters, missed their mother ships on their return flight. TF 16's torpedo bombers, which had become separated from the dive-bombers on the approach, detected Nagumo only by chance, at the extreme limit of vision, and were then devastated by his combat air patrol. By that stage of the battle, Nagumo would have had every reason to believe that he was winning. The events of the next few minutes would have reinforced that belief. *Enterprise*'s torpedo bombers had also missed Nagumo and only spotted his ships at the extreme limit of vision.

They were then overwhelmed by the combat air patrol as they made their attack.

Nagumo, moreover, had been able to land on and re-arm his Zeros—they were operating so close to their mother ships that they did not need refuelling—during this stage of the fighting. By 10:25, the four Japanese carriers, though somewhat dispersed by taking evasive action against TF 16's torpedo bombers, were untouched and were preparing to fly off their own strike planes against the enemy, whose position and distance could be estimated by scouting reports and observation of the Americans' line of approach.

What happened next was the outcome of random factors. The first was that the opening attacks had been delivered by torpedo bombers, which drew the Japanese combat air patrol down to sea level, at a moment when the American dive-bombers were to begin their descent from 14,000 feet. The second, a truly haphazard event, was Bombing 6's sighting of the wake of the destroyer *Arashi*, departing from its depth-charging, unsuccessful, of the U.S. submarine *Nautilus* and leaving a signature that the quick-witted Commander McClusky realised pointed to Nagumo's position. The third, which ante-dated the opening of the engagement, was Nagumo's time-wasting indecision at the very outset.

Poor Nagumo; the bold destroyer commander had not been equipped, by either training or experience, to perform the intricate and rapid calculations of relative speeds in three dimensions that a successful carrier commander needed to make. An outside observer can see in retrospect that, on receipt of the *Tone* seaplane's sighting report of American warships within flying distance of his irreplaceable carriers, he should have cancelled the order for his bombers to prepare a second strike against Midway, as Tomonaga urged, and readied all his strike aircraft for a ship-to-ship attack. It was his inability, after seven o'clock, to make up his mind, despite the promptings by light-signal from his fellow admiral Tamon Yamaguchi, commanding the *Hiryu-Soryu* group, that led to his decks being cluttered by fuel hoses, loose ordnance and re-arming aircraft when, three hours later, Lieutenant-Commander McClusky's Bombing 6 began its dive which culminated, in less than five minutes, in the sinking of three of the four Japanese carriers.

Results in war, in the last resort, are an affair of body, not mind; of physical force, not plans or intelligence. Over the longer run, of course, a power of superior intellectual resource will, if its superiority translates into possession of superior industrial, technical and demographic means, ineluctably overcome a power inferior in those qualities. There are no examples in military history of a state weaker in force than its enemy achieving victory in a protracted conflict. Force tells. Mind, however, is usually also its concomitant. The governing class of the Japanese empire, with less than a third of the population of the United States and a fraction of its industrial capacity, had been deluded to believe that its painfully acquired collection of modern warships and aircraft, even when enhanced by the warrior spirit of its sailors and airmen, could overcome. That had been Admiral Yamamoto's warning. His estimate of "running wild" for a year or six months had been exactly realised. The Japanese had risked all and, at Midway, lost all.

Nevertheless, Midway demonstrates that even possession of the best intelligence does not guarantee victory. Nimitz, Spruance and Fletcher had the enemy's plans, thanks to the relentless intellectual effort of Rochefort and his fellow cryptanalysts, laid clear before them, or as clear as the obscurities of war will ever allow. They had, all the same, nearly lost. A little less intuition by McClusky of Bombing 6, a little more intellectual resolution by Admiral Nagumo and it would have been the carriers of TF 16 and 17, not those of Yamamoto's Mobile Force, which would have been left burning and bereft in the bright waters of the Pacific on 4 June 1942. Japan would still have lost the Pacific War; but how much longer would it have taken the United States to triumph?

————◆○▶————

CHAPTER SEVEN

Intelligence, One Factor Among Many:
The Battle of the Atlantic

THE OFFICIAL HISTORIAN of British intelligence in the Second World War, Professor Sir Harry Hinsley, made muted claims for its importance. It did not, he stated firmly, win the war; but it did shorten it.[1] It did so particularly, he argued, by the part it played in the successful conduct of the Battle of the Atlantic, first by preventing the domination of the U-boats in the last six months of 1941, and again in the winter of 1942–43, and finally by contributing heavily "to the defeat of the U-boats in the Atlantic in April and May 1943 and then to the Allied success in so crippling the U-boat command during the second half of 1943 that it could never return to the convoy routes."[2] These achievements, though put strictly by Professor Hinsley into the context of a far wider and more complex war, are impressive, for it was upon the ability of Britain to survive U-boat attack on its oceanic supply routes that its capacity to wage war depended; and, had Britain not sustained the effort in the seventeen months between the fall of France in June 1940 and Japan's attack on Pearl Harbor in December 1941, Hitler would have completed his conquest of Western Europe, perhaps defeated the Soviet Union and then been able to deny the United States entry to the continent.

Defeat in the Battle of the Atlantic would have been a catastrophe. No one recognised that more clearly than Winston Churchill, who wrote, in his magisterial history of the Second World War, that "the only thing that really frightened me during the war was the U-boat peril . . .

How much would the U-boat warfare reduce our imports and shipping? Would it ever reach the point where our life would be destroyed? Here was no field for gestures or sensations; only the slow, cold drawing of lines on charts, which showed potential strangulation."[3]

Strangulation would have been slow; but had Admiral Dönitz, Hitler's U-boat chief, been given the time, it would have been sure. Dönitz was a U-boat officer of the First World War who, in the inter-war years, when Germany was denied possession of a submarine fleet, had worked out in cold theory how to conduct a campaign of commerce destruction that would destroy an enemy's—meaning Britain's—merchant navy. Dönitz's tool of experimentation was the surface torpedo boat, which the Versailles Treaty allowed Germany to possess. Well before 1936, when Hitler succeeded in extracting from Britain agreement to his rebuilding of a U-boat force, Dönitz had, by trials at sea, designed a scheme of torpedo-boat attack which was to underlie the "wolf pack" tactics of the Battle of the Atlantic. Whether the target was a convoy of merchant ships or a squadron of warships ("itself a convoy") the technique must be to make contact by daylight, with a dispersed patrol line, hang on at the limit of visibility and then, under cover of darkness, deliver the torpedo attack. The surfaced U-boat, Dönitz argued, was a torpedo boat, and the capabilities of the one predicated those of the other.[4]

At the outbreak of the Second World War, Germany once again had a U-boat fleet; but it was small, only fifty-six boats, thirty of which were tiny coastal models. The principal ocean-going submarine was the Type VII, of which there were eighteen in service, 220 feet long, capable of 17 knots on the surface under diesel power, 7½ knots submerged on its electric motors. It mounted a 3.5-inch gun and had four bow and one stern torpedo tube, for which there were nine spare torpedoes. The crew numbered forty-four. During 1939 the larger Type IX was introduced but, though it mounted a heavier gun, had six torpedo tubes and sixteen spare torpedoes and a operational range of 11,500 instead of 8,500 miles, it was reckoned by Dönitz less suitable than the Type VII for convoy battles, being slower to dive and less manoeuvrable. In 1939 there were eight Type IX.[5]

U-boats, in the early days of the war, found their targets by patrolling the regular shipping lanes leading to the United Kingdom, which they reached by going round the north of Scotland; the English Channel was closed from the outset. Boats patrolled independently, scanning the sea by periscope during the day, surfacing at night. To begin with, there was little co-operation between the few boats on station—usually fewer than fifteen—and Dönitz made little attempt to co-ordinate their operations.

The U-boat commanders' target was the British merchant fleet, still the largest in the world by far, with 3,000 ocean-going ships and a carrying capacity of seventeen million tons. It was fully employed, since Britain was dependent on imports for over a third of its food and most of its raw materials, except coal. Annual imports in 1939 totalled fifty-five million tons, to be paid for largely by the export of finished or semi-finished manufactures. Britain, uniquely among the major powers, was a country reliant upon maritime trade, both inward and outward. Interruption to sailings quickly produced shortages, as well as harm to credit. Sinkings threatened permanent damage, since the output of all shipyards in Britain and the empire amounted to only a million tons a year, or about 200 merchantmen or tankers of average size.

Dönitz, a keen student of the economics of maritime trade, was well aware of Britain's vulnerability and, as a result of his experience in the First World War, believed firmly, indeed with passionate conviction, that an expanded U-boat arm, attacking without the restrictions imposed by traditional prize regulations, could end Britain's ability to wage war. Between 1914 and 1918, the German navy sank 4,837 Allied merchant ships, totalling over eleven million tons, most of them British, most sunk by U-boats and most in the period from 1917 onwards. Germany had launched 365 U-boats and lost 178.

On 28 August 1939, as Hitler completed his preparations to attack Poland, Dönitz submitted to Admiral Raeder, the head of the German navy, his proposal for a major expansion of the U-boat fleet. He wanted 300 U-boats, together with some larger submarines to serve as supply ships to the attack boats, which would allow fifty U-boats to be kept on patrol on the shipping routes at any one time; he had elsewhere calculated that if each sank three ships a month, a rate achieved in the Great War, half the British merchant fleet would go down in a year, at a rate vastly outstripping replacement. Britain would starve, as it had nearly done at the end of 1917, and be forced to surrender. He also wanted the U-boat building programme to be placed under the control of a single officer and, at a conference held on 9 September, six days after the war with Britain had begun, proposed himself for the post. "This task now becomes the most important of all, which should be under the direction of an officer with expert knowledge of the theory and practice of U-boat warfare."[6]

Raeder demurred. He recognised Dönitz's talents and dedication but wanted him to remain in day-to-day command of the existing U-boat fleet, though he gave assurance that it would be greatly and rapidly expanded. Raeder was probably right to decide as he did. Dönitz, though physically unprepossessing, humourless and intellectually obsessive, had undoubted qualities of leadership. His U-boat men, who included at the outset some buccaneering captains of exceptional seamanly quality, always looked up to him, craved his approval and served devotedly to the end. Theirs was a horrible life. The U-boat was cramped and smelly, always either too cold or too hot and usually dripping with damp. Food soon went off, clothes were clammy, toilet arrangements foul and the air for much of the time scarcely breathable. U-boat life was characterised by prolonged periods of boredom, particularly as the war drew on and crews had to spend long periods submerged motoring to their patrol stations. Above all, it was extremely dangerous. Of the 40,000 sailors conscripted into the U-boat arm—unlike the submarine services of the U.S. or Royal Navies it was not voluntary—28,000 were killed in action, most of them drowned as the result of attack by escort vessels of the Royal Navy, Royal Canadian Navy and United States Navy or their associated air forces and naval air arms.

ANTI-SUBMARINE WARFARE

During the First World War, 178 of the 365 German U-boats built by Germany had been lost at sea, despite the then Allied lack of any effective means of detecting submerged boats. Acoustic methods were employed, and aircraft and airships attempted to spot U-boats from the air in shallow waters, almost always unsuccessfully. Most U-boats sunk, at least forty-eight, fell prey to mines in mine barriers. Ramming, by warships or merchantmen, accounted for another nineteen, attack by British submarines seventeen. Destruction by depth-charge, the specific anti-submarine weapon, was responsible for only thirty losses.[7]

The depth-charge was a bomb containing usually 40 pounds of high explosive, dropped from a rack over the stern or projected abeam, and activated by a pressure fuse, which could be set to explode at a chosen depth. It created high-pressure waves and, if detonated close enough to a U-boat hull, cracked its plates. Accurate depth-charging was fatal but accuracy was difficult to achieve; throughout the Battle of the Atlantic, but particularly in the early days, damage rather than destruction was a common outcome of depth-charging. After 1942, depth-charging was supplemented by the firing of large numbers of contact bombs, from the Hedgehog and subsequent Squid systems, which, given accurate location, could be deadly. In mid-1943 another weapon appeared, the Mark 24 Mine so-called; in fact, an acoustic torpedo, dropped from an aircraft to home on U-boat propeller sounds. Lethal in most circumstances, it suffered from the disadvantage of being deemed so secret that it could be launched only under special conditions. Aircraft also dropped depth-charges and fired high-explosive rockets against surfaced or submerging U-boats, those motoring to Atlantic patrol lines across the Bay of Biscay providing most of the targets.[8]

Most of these developments lay far in the future at the outset of the U-boat war, when the advantage lay heavily with the Germans. The advantage would have been decisive, had Dönitz been able to deploy the numbers he desired and would eventually achieve. Even so, the advantage was enhanced by the enemy's lack of anti-submarine warships. Britain in 1939 appeared to have a sufficiency of escorts. The Royal Navy deployed 128 destroyers and 35 sloops.[9] Most of the destroyers, however, including the superb Tribal and Javelin classes, were high-speed ships, designed to accompany the battle fleet and lacking the endurance to linger as convoy escorts. Many of the older destroyers had been built during or soon after the Great War and were coming to the end of their lives. The Hunt-class destroyers, designed specifically as escorts, were entering service but were too few in number as yet—only twenty—to tip the balance. The sloops were generally too old to be effective. A whole new generation of escorts—slow but sturdy corvettes, modelled on South Atlantic whalers, and speedier frigates—were in gestation but had not yet reached the fleet. Trawlers and drifters from the fishing fleet had been pressed into service; but they were too small and too slow to make capable escorts. The result was that convoys had too few escorts to be able, when attacked, to defend themselves.

Convoy was adopted by the Admiralty at the outset of the war, in sharp contrast to its policy during the First World War. Then the admirals had resisted it, for wholly mistaken reasons. Convoy was a hallowed

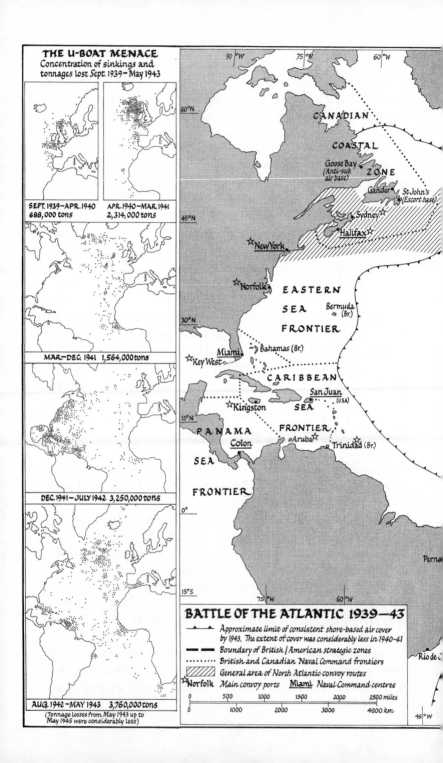

THE U-BOAT MENACE
Concentration of sinkings and
tonnages lost Sept. 1939 – May 1943

SEPT. 1939–APR. 1940
688,000 tons

APR. 1940–MAR. 1941
2,314,000 tons

MAR.–DEC. 1941 1,564,000 tons

DEC. 1941–JULY 1942 3,250,000 tons

AUG. 1942–MAY 1943 3,760,000 tons
(Tonnage Losses from May 1943 up to
May 1945 were considerably less)

CANADIAN

COASTAL

Goose Bay
(Anti-sub
air base)

ZONE

Gander
St John's
(Escort base)

Sydney

Halifax

New York

Norfolk

EASTERN

SEA

Bermuda
(Br.)

FRONTIER

Miami
Bahamas (Br.)

Key West

CARIBBEAN

San Juan
(USA)

Kingston

SEA

PANAMA

FRONTIER

Colon

Aruba

Trinidad (Br.)

SEA

FRONTIER

Perna

Rio de

BATTLE OF THE ATLANTIC 1939—43

Approximate limit of consistent shore-based air cover
by 1943. The extent of cover was considerably less in 1940–41
Boundary of British/American strategic zones
British and Canadian Naval Command frontiers
General area of North Atlantic convoy routes
Norfolk Main convoy ports Miami Naval Command centres

0   500   1000   1500   2000   2500 miles
0   1000   2000   3000   4000 km

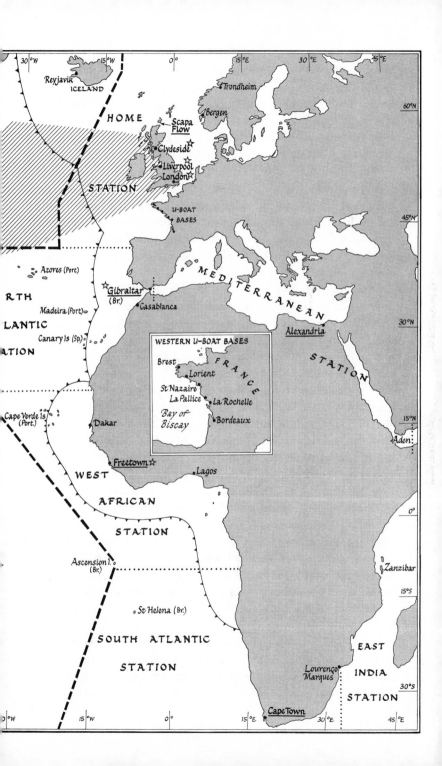

British maritime practice, which had protected British trade from attack by the French fleet and roving corsairs throughout the wars of the French Revolution and empire. At the outbreak of the Great War, however, and until late in 1917, the Admiralty had mistakenly calculated that, given the submarine's ability to mount attacks submerged, the practice of massing merchantmen in convoy merely multiplied the targets available to a predator, which the navy lacked the means to locate. It was therefore better, it was concluded, to let ships sail independently, thereby, apparently, forcing submarines to choose their targets singly and, so it was supposed, with greater difficulty. The Anti-Submarine Division of the Admiralty also shrank from the task of providing escorts for what it believed were 5,000 sailings into and out of British ports each week.

These two objections to convoy were separate and different but were dissolved by analysis of the second. In April 1917, Commander R. G. H. Henderson, RN, dissected the figures for maritime trade and established that only 120–140 arrivals and departures each week were by ocean-going ships, those on which Britain's survival depended; the rest were by coastal and short-sea-crossing vessels which it was not vital to protect. Because of the enormous number of destroyers and other small warships that been built during the war, provision of escorts to convoy the essential merchantmen was not seen to be a difficulty at all. The only problem was to learn the technique of convoy. Once that was mastered, sinkings began to fall. In October 1918, tonnage lost was 178,000 tons against an average of 550,000 tons a month during 1917. Most ships sunk were sailing independently; losses from convoy were under 2 per cent.[10]

The Admiralty's immediate adoption of convoy in September 1939 averted any large-scale toll of sinkings in the first year of the war. There were several ancillary reasons for that, the paucity of U-boat numbers being one and the confinement of the U-boats to German bases far from the shipping routes another. The most spectacular U-boat successes, indeed, were achieved against naval targets, particularly the torpedoing of the British battleship *Royal Oak* inside the protected anchorage of Scapa Flow in October 1939. It owed much to intelligence success. A German captain, who had visited the Orkneys just before the declaration of war, reported that he had heard the defences of the eastern approaches to the anchorage were neglected; aerial photographic reconnaissance confirmed the existence of a gap. Dönitz then briefed the thrusting young U-boat commander, Gunther Prien, about the possibility of making an entrance at slack water under cover of darkness. On 13 October, U-47 found its way through the defences, fired torpedoes which detonated *Royal Oak*'s magazine and sent it to the bottom with most of its crew. Militarily the attack was not significant, for *Royal Oak* was obsolete; its sister R-class battleships would have to be hidden from the Japanese in east African ports after Pearl Harbor, so low was their ability to defend themselves. Nevertheless, the attack was a humiliation for the Royal Navy, besides being an awful warning of the vulnerability of capital ships to unorthodox attack, particularly when at anchor, as Pearl Harbor, Taranto and the Italian attack on Alexandria were subsequently to demonstrate.[11]

The comparative ineffectuality of Dönitz's U-boat campaign of September 1939–July 1940 was to be sharply reversed after the fall of France. In the immediate aftermath, the German navy hurried supplies of torpedoes and other submarine warfare material to the French Bay of Biscay ports—Lorient, Brest, La Pallice, Saint-Nazaire, Bordeaux—which were henceforth to be the bases for its U-boats during the Battle of the Atlantic; the first arrived in the Bay of Biscay, at Lorient, on 7 July. The Biscay ports provided Dönitz's submarine fleet with direct access to Britain's Atlantic trade routes, shortening by hundreds of miles those from Germany's bases and sparing it attack on passage in the constricted waters of the North Sea.

As soon as the Biscay bases were acquired, Dönitz embarked on the realisation of his plan to defeat Britain, and its surviving allies, by destruction of its Atlantic convoys. Advantage seemed on his side. The number of U-boats, which had to survive only one outward passage from German shipyards to the French ports in order to become effective, was increasing. The number of British escorts, and of replacements of British merchantmen lost to attack, was increasing much more slowly. Dönitz's belief in his ability to win the naval—and thereby the European—war, by destruction of the Atlantic shipping trade, seemed ready to be realised.

By a strange reversal, the First World War fears of the Admiralty, that it lacked the escorts necessary to protect convoys, seemed about to be confirmed in a subsequent war twenty years later. In the second half of 1940, the Royal Navy, wholly committed to the concept of convoy as it was, was attempting to protect much larger convoys than it had organised in 1917–18 with far fewer warships. In 1918 a typical oceanic convoy of 16–22 merchantmen was protected by seven destroyers, first-class warships of a speed (over 30 knots) double that of a U-boat on the surface, where U-boats usually attacked. In the winter of 1940, convoys of as many as thirty ships or more might be protected by only one inadequate escort.

An example was Convoy SC7 (convoys were identified by acronyms, usually denoting point of departure, and numbered consecutively; those most used were HX, originating in Halifax, Nova Scotia; later New York, OB, outbound from Britain; CU, Caribbean–United Kingdom; MK, Mediterranean–United Kingdom; SL, Sierra Leone; PQ, Britain–North Russia). SC7 originated in Sydney, Novia Scotia, and consisted of thirty-five ships, all slow, four of them inland freighters from the American Great Lakes. The only escort was the sloop *Scarborough*, built in 1930 with a top speed of 14 knots and so slower than a surfaced U-boat. On the fourth day out, 8 October 1940, the convoy ran into a gale and that night into U-boats. Over the course of the next ten days, SC7, though reinforced by two more sloops and two corvettes, and attended by a Sunderland flying-boat, lost seventeen ships. The horror of the experience scarcely bears thought. For the torpedoed seamen, even if they were able to launch lifeboats or floats, there was no hope of rescue. The convoy could not stop; the escorts' duty was to stay with the merchantmen. The survivors of sunken ships drowned or died of exposure.[12]

SC7 was a ghastly example of the pre-1917 fears of the anti-convoy Admiralty, which thought there were not enough escorts to protect merchantmen, encountering the plans of a commander, Dönitz, who had contrived means to maximise the offensive power of what should have been an inferior weapon, the U-boat. The U-boat was the realisation of an ancient conception, the idea of the invisible weapon. Most of its early forms had been devised to undermine, literally, the power of the British surface fleet, as had the first practical submarine, invented by the Irish-American, J. P. Holland, in 1900. The Holland boat, however, had been intended, like all its ineffectual predecessors, to attack submerged. The genius of Dönitz—he was a sort of evil genius—was to perceive that the submersibility of the U-boat should be used merely to protect it from counterattack, once its presence was detected, and that in offence it should be used on the surface, where it could achieve speeds superior to most of its targets, the merchantmen, and not greatly inferior to those of all but first-class escorts.

The other ingredient of the Dönitz idea was that of the wolfpack (*Rudel*). His time as commander of a U-boat in the Great War had persuaded him that the deployment of single U-boats was wasteful. Better, he convinced himself in the war's aftermath, to mass them in groups which could, first, detect convoys by forming a patrol line—similar to that organised by Nelson with his frigates—and then close for the kill. SC7 had had the misfortune to fall under attack by one of Dönitz's earliest wolfpacks. It overwhelmed the escort. At one stage seven U-boats were operating against four escorts, since *Scarborough* had detached itself to hunt for one of the first predators, U-48, which it did not find; nor did it find its convoy again.

The other ingredient of the wolfpack method was central control from headquarters; after June 1940 from La Pallice. The medium of control was radio, just as it had been during von Spee's cruiser campaign against British shipping in the Pacific and the Indian Ocean in 1914. Radio, as before, overcame the limitations of visual sighting, which so restricted Nelson's ability to command the Mediterranean, because a single visual sighting, transmitted by radio, allowed La Pallice to concentrate a wolfpack against a convoy even if the U-boats constituting it had been scattered across several hundred miles of sea. Pack strategy, plus radio, was a deadly weapon against the convoy system.

All strategies, however, have weaknesses. Radio was the weakness of the pack system. Bletchley, supplied by intercepts from the listening stations, was provided with the material by which Dönitz controlled his U-boats. The difficulty was to break it. By late 1940, Bletchley had had no success against the German naval keys. Unlike those of the recently founded German air force (Luftwaffe), the *Kriegsmarine*'s operators came from a long-established signal service, which had strict procedures and severe schooling. Not only were German naval signallers trained not to make mistakes—for Bletchley the most fruitful source of breaks into the Luftwaffe traffic; the whole German naval signalling system operated on the belief that the enemy was listening. The *Kriegsmarine* therefore strove not only to keep enciphering secure but also to limit the amount of material transmitted, on the sound principle that the smaller the quantity of intercepts, the harder it is for an enemy to find a way into them.

Assurance of secure encipherment was attempted by two principal means: enlarging the number of rotors used in naval Enigma machines and designing certain keys to be used only by officers. Even before the war, naval Enigma operators were issued eight rotors from which to select three; from 1 February 1942 onwards Atlantic and Mediterranean U-boats used four rotors in an adapted machine.[13] The "Officer" keys introduced were versions of the *Heimisch* key, the *Süd* key, and *Triton*, known as Shark at Bletchley, by far the most important since it was the key used in Atlantic U-boat operations from February 1942. Officer keys were regularly broken but usually with some delay.[14]

Limitation of material transmitted was achieved by the devising of "short" signals, a form of code which was enciphered within longer messages or used simply as answers to enquiries from U-boat headquarters at La Pallice (later Berlin). Most short signals, transmitted as "digraphs" (two-letter groups), referred to a chart of the Atlantic and adjoining waters, which was divided into an irregular grid. Bletchley, beginning with some captured material, managed to reconstruct some of the grid by April 1940. In May 1941, as a result of the celebrated capture of U-110, it reconstructed the grid of the whole North Atlantic and most of the Mediterranean. The Germans, who constantly reviewed the security of their signal system, became concerned in mid-1941 that the U-boat position transmissions might have been compromised and introduced a more complex short signal by relating positions at sea to fixed points of reference—Franz, Oscar, Herbert, etc.—arbitrarily chosen and changed at short intervals. When deciphered, a typical Enigma order to a U-boat now read: "If boat is in a fit condition for night attacks occupy as attacking area the northern waters of the 162-mile-squares [of the naval grid] whose central points lie 306 degrees 220 miles and 290 degrees 380 miles respectively from Point Franz. If boat not in a fit position, report by short-signal 'No.' "[15]

Bletchley managed to overcome the difficulty thus presented quite quickly, a vital matter since the position reports provided the data by which the Admiralty rerouted convoys, on passage, away from wolfpack patrol lines. Other short signals used by U-boats at sea were sighting and battle reports and announcements of expected dates of return to port. Most useful of all were the short weather reports, essential to Dönitz's headquarters in positioning U-boats. Bad weather, paradoxically, was welcomed by convoy commodores and escort commanders, since it usu-

(*Left*) An Enigma operator aboard a U-boat.
(*Above*) An Enigma machine team, Army Group Centre, Russia, autumn 1941. The text is clipped inside the lid, the operator is entering the letters, his assistant is taking down what appears on the lamp board (one lamp has just lit), under the three rotors.

(*Left*) German paratroopers collecting their equipment after landing in an olive grove, Crete, May 1941.

(*Below left*) General Bernard Freyberg, VC, British commander in Crete, May 1941.

(*Below right*) General Kurt Student in Crete, June 5, 1941. British prisoners behind him.

(*Left*) USS *Lexington*'s crew abandoning ship after she had been set on fire at the Battle of the Coral Sea, 8 May 1942.

(*Above left*) Admiral Isoroku Yamamoto (1884–1943), Commander of the Combined Fleet at Midway.

(*Above right*) Admiral Chester Nimitz (1885–1966), Commander of the U.S. Pacific Fleet, 1942–45 and mastermind of American strategy at Midway.

(*Left*) Captain Joseph Rochefort, head of the U.S. Navy's cryptological department (HYPO) in Hawaii, 1941–42.

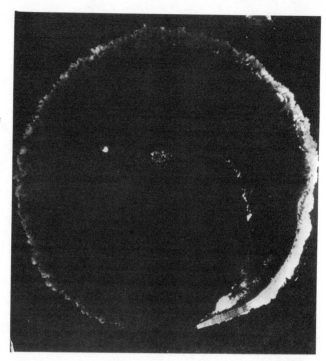

(*Right*) The Japanese carrier *Kaga* circling under attack at Midway, 4 June 1942.

(*Below*) USS *Yorktown* sinking from torpedo damage after the Battle of Midway, 7 June 1942.

(*Above*) Admiral Karl Dönitz (1891–1981) (*right*), Hitler's Commander of U-boats, later Grand Admiral and last Chancellor of Nazi Germany. Here with Grand Admiral Raeder, he stands in front of gridded charts of the North Atlantic.

(*Below*) Taken from USS *Greer* on convoy escort in heavy Atlantic weather. Storms were welcomed by convoy captains because they deterred U-boat attacks, *Greer* was the first American warship to engage a U-boat, on 4 September 1941, under Roosevelt's policy of "armed neutrality."

(*Right*) U.S. Coast Guard Cutter *Spencer* firing a depth charge (upper right) at the outset of the attack on U-175, April 1943. U-175 was blown to the surface and sunk, some of her crew surviving.

(*Below*) Admiral Ernest J. King (1878–1956), Chief of Naval Operations, 1941–45, meeting warrant officers of the U.S. Navy, 1942. Autocratic, arrogant, Anglophobe, King was an outstanding naval commander.

(*Below left*) A convoy conference at Derby House, Liverpool, headquarters of Western Approaches Command during the Battle of the Atlantic. The merchant captains are being briefed by the convoy commodore, a naval officer.

(*Below right*) Captain F. J. "Johnny" Walker (1897–1944), the most successful U-boat killer of the Battle of the Atlantic. His 2nd Escort Group sank twenty U-boats. Here he directs an attack from the bridge of his sloop *Starling* by TBS (Talking Between Ships) radio.

(*Above*) A V-2 rocket on its Meillerwagen transporter-erector at Peenemünde, being raised to the vertical position.
(*Below*) A V-1 flying bomb on its launch ramp in France, 1944.

(*Above*) A V-1 at the end of its flight, about to impact near Drury Lane, London, 1944.

(*Above*) General Walter Dornberger, head of the V-2 programme (*right*) with Dr. Todt (*on his right*), Hitler's armaments minister, at Peene-münde in 1941.

(*Right*) Werner von Braun (1912–1977), designer of the V-2, as an American citizen in 1961.

(*Right*) SAS troopers after the helicopter crash on South Georgia, Falklands campaign, 22 April 1982.

(*Left*) An Argentinian Pucara ground-attack aircraft destroyed by the SAS on Pebble Island, 14 May 1982.

(*Below*) A British anti-tank missile team launches a Milan during the second Gulf War, March 2003

ally prevented U-boats from attacking. The short weather reports became a fruitful source of decrypts because, early in the Battle of the Atlantic, Bletchley found that they were rebroadcast by a meteorological shore station in a code that it could read; later, because the reports were made in three-letter groups, Bletchley discovered that U-boat operators were not using the fourth rotor on their Enigma machines, thus greatly simplifying the mathematics of decryption.[16]

BLETCHLEY AND THE BATTLE OF THE ATLANTIC

Winston Churchill confessed that he would much rather have faced the danger of a German invasion of Britain than had to sustain resistance to the U-boat war. The point is understandable. An invasion would have taken place within the dramatic unities of time, place and action. The U-boat war went on and on, always destructive and formless and apparently endless. As long as Dönitz could find boats and crews to send westward into the waters of the Atlantic, across which Britain's necessities were convoyed, ships would be sunk, sailors drowned, cargoes lost, and the issue of the war suspended in precarious balance.

Yet, despite that perception, the Battle of the Atlantic, like any great battle, can be seen in retrospect to have had chronology and shape. It divides into five broad periods. From September 1939 until July 1940, the combat between Dönitz's U-boat arm and the Royal Navy was not strictly a Battle of the Atlantic, since Germany's lack of forward bases made passage into great waters difficult and largely confined the U-boats to the seas around Britain. There were rarely more than ten U-boats on station, often as few as four, and, though few were lost, only nineteen in the first ten months of the war, little shipping was sunk either. Dönitz's dream of deploying 300 U-boats, to sink 100,000 tons of shipping a month (about twenty ships, given the current average size of ocean-going merchantmen), seemed a fantasy.

Then, following the fall of France and the Franco-German armistice of July 1940, Germany acquired occupation rights over French territory which included the ports of the Atlantic coast. Dönitz at once set up his U-boat command headquarters there, in the chateau of Kerneval near Lorient, and began to bring his flotillas out of the narrow waters of the Baltic and North seas to the Bay of Biscay. Sinkings at first rose but then fell again as the Royal Navy deployed more escorts and Dönitz was obliged to commit a huge proportion of newly built U-boats to training. Throughout the war, the *Kriegsmarine* never skimped on training, subjecting new boats and crews to as much as a year's practice in the Baltic before allowing them to the "front."

The third period of the Battle of the Atlantic began in April 1941, the month after Winston Churchill coined the phrase "Battle of the Atlantic." Dönitz had by then assembled enough trained U-boat crews to begin organising patrol lines and wolfpacks in the central North Atlantic, though his boats had been driven into those deeper waters, away from the sea approaches to the British Isles, by the increasing numbers of Royal Navy escorts and Coastal Command aircraft. Sinkings rose but the Admiralty also had considerable success during the year in routing convoys away from U-boat patrol lines, thanks to Bletchley decrypts. In September 1941, for example, when thirty-two U-boats were on patrol in the North Atlantic, twelve sank no ships at all, and only four sank more than 10,000 tons, or two ships each.

Dönitz's prospects changed suddenly with the inception of the fourth period in January 1942, when he was able to withdraw his U-boats from the central North Atlantic, break up his packs and patrol lines and deploy individual boats, under now often highly experienced skippers, against the coastwise shipping of the United States on its east coast and in the Caribbean. U-boat captains described the next six months as their "Happy Time." Targets were numerous, so were sinkings. In January 26 U-boats operating in American waters sank 400,966 tons of shipping, 71 cargo ships or tankers, for no losses at all. February was worse proportionately: 18 U-boats sank 344,494 tons, 57 ships. In April, after a very bad March when 406,046 tons were sunk, 31 U-boats sank 133 ships of 641,053 gross tons; and so the awful summer went on. By the end of August, when the Americans at last instituted proper anti-submarine measures, 609 ships, of 3,122,456 gross tons, had been sunk, for the loss of 22 U-boats, out of 184 engaged.[17]

The extent of Dönitz's success was due to the refusal of the United States Navy to institute convoy at the outset, in a bizarre repetition of the British Admiralty's policy of 1914–16. Admiral Ernest King, Chief of Naval Operations, formed the view that weakly escorted convoys would merely provide more plentiful targets than individually sailed ships, and so left America's coastwise traffic to its fate. In his defence, it is argued that he used such warships as he could muster in Atlantic waters to escort the convoys taking American troops to Britain, and that not one suffered loss; it was also the case that he was meanwhile waging a life-and-death struggle with the Japanese navy in the Pacific, which consumed almost every serviceable warship his navy possessed. Nevertheless, there was undoubtedly an American anti-convoy prejudice, as evidenced by the U.S. Navy's organisation, as by the Admiralty in the First World War and again at the outbreak of the Second, of U-boat "hunting" groups which, as reason should have taught and experience did teach, found few if any U-boats to attack. By 1941 the Royal Navy was wholly committed to the correct view that, if U-boats were to be found and sunk, they had to be presented with targets to attack that could defend themselves, in short, convoys with strong close escorts.

The conclusion of the American Happy Time confronted Dönitz with the need to risk his U-boats against such targets again. The inception of the period that ensued, the fifth and climactic stage of the Battle of the Atlantic, from September 1942 until May 1943, ushered in a dreadful episode in sea warfare, marked by heavy losses of merchant ships and tragic loss of life, all suffered in North Atlantic weather at its worst. Nevertheless, the crisis can be seen in another light. It was the moment in the maritime conflict between the *Kriegsmarine* and its opponents— the Royal Navy, the Royal Canadian Navy, the United States Navy and their associated air forces—when Dönitz was forced, in the classical terms of sea warfare, to give battle. He had argued throughout his life as a professional naval officer that there was a victory waiting to be won between a surface fleet and its submarine enemy. At the end of 1942 he was challenged to win such a victory—and lost.

The part played by Bletchley Park in that victory, though crucial, was complex. Its struggle was two-sided, not unilateral, for the German navy had its own interception and decryption service—the *Beobachtungs* (Observation) *dienst*, known as the B-dienst—and it had a record of considerable success against the Royal Navy's disguised transmissions. Because of Britain's efficiency in breaking into Germany's codes during the First World War, an undeniable complacency prevailed in the Naval Intelligence Department during the post-war years and well into the Second World War. The British believed that decryption was a one-way traffic even if, at first, they were unable to break Enigma. Because also they had unwisely trumpeted their decoding achievements of 1914–18, they had put the Germans on their mettle so that, long before the outbreak of war in 1939, the B-dienst was breaking the then current Admiralty code, a system of five-digit groups called the Naval Code, which was super-enciphered mathematically. Through carelessness, the more secure Naval Cypher was also betrayed to the enemy. It was a familiar story; a cipher officer used the super-encipherment book of the Naval Cypher to super-encipher messages sent in the Naval Code. As the latter could be read, the former was quickly broken and was read currently and continuously until 20 August 1940.[18]

The English-language department of the B-dienst employed 900 people before the war; the number would rise to 5,000 by 1942. The B-dienst was located at German naval headquarters in Berlin and was led by Wilhelm Tranow, a radio technician who had first been employed to test the security of Enigma. The security of their ciphers was a matter of deep concern to the Germans throughout the war. It was constantly reviewed, as their own was by the British. Both navies remained convinced, nonetheless, that they could not be overheard, the Germans with far better reason. They correctly reasoned that, even were the British able to acquire three of the four elements of the Enigma system—the machine, the setting list, the indicators and the tables of bigrams which designated the grid-squares on the oceanic chart—they would still not be able to read messages. They discounted the possibility of the British acquiring all four and, as the result of a prolonged investigation during 1942, instituted new precautions against the operators' resorting to short-cuts. More to the British disadvantage, they also altered the U-boats' Enigma machines to accept a new, fourth rotor which, in combination with an adapted reflector, multiplied by twenty-six the number of possible keys.[19]

The result was that between 1 February 1942 and the following December, Bletchley lost its way into Enigma altogether, with a calamitous impact on sinkings. The effect was heightened by a sudden German breakthrough into British naval codes. Suspecting, correctly, that both Naval Cypher No. 2, the successor to the Naval Code, and the first Naval Cypher had been penetrated, as the latter had been since September 1941, the Admiralty had introduced in December 1941 the new Naval Cypher No. 3, still a super-enciphered code, not a cipher proper. It worked by the traditional method, the adding of numbers, from a book of number groups, to the groups indexed in the main code ("Cypher") book and was decoded by subtraction. The books were issued to the

Royal, Royal Canadian and United States Navies for use in the passage of convoys across the North Atlantic.

During January 1942 the B-dienst succeeded in reconstructing both the codebook for Naval Cypher No. 3 and the subtraction tables used with it. In consequence, it was able to read 80 per cent of the convoy traffic, often twenty to thirty hours in advance of the movements signalled. This warning allowed ample time for U-boats to be positioned in a convoy's path, since the difference between the speed of an average convoy, seven to eight knots, and that of a surfaced U-boat, at least sixteen knots, meant that patrol lines and wolfpacks moved at twice the speed of their prey. In twenty-four hours, during which a convoy advanced 180 miles, a hunting U-boat could move 360 miles to cut it off, submerging only just before the moment of contact.

U-boat commanders were trained, according to Dönitz's pre-war experiments with torpedo boats, to linger at the limit of visibility, until the fall of dark, on the convoy's projected line of advance, then to surface, if possible within the convoy columns, fire salvoes from bow and stern tubes simultaneously, and to submerge and make their escape as the escorts appeared. Depending upon how much disruption had been caused in the convoy pattern, a second attack might be mounted.

The difficulty for the U-boats was to locate a convoy in the first place. The limit of visibility at best from a conning tower was ten miles; a patrol line of ten U-boats could therefore cover 220 miles of sea. Dönitz attempted to extend the area of sweep by securing the services of 1/KG40, a squadron of long-range Condor aircraft, from the Luftwaffe; but the growing efficiency of British air patrols in 1941–42, which had driven the U-boats into the central Atlantic, also put them beyond the Condors' range; an effort to fly an autogiro on a cable from U-boat conning towers proved as impractical as it was dangerous. A convoy of fifty ships, meanwhile, occupied a front of only 2,400 yards. In the enormous spaces of the Atlantic—at least nine million square miles of operational waters—the area occupied by a convoy and the space covered by a questing U-boat patrol line were both relatively tiny. The one could be missed by the other with the greatest ease and usually was. Between 1 January and 31 May 1943, for example, the height of the Battle of the Atlantic, forty-eight out of eighty-six convoys sailed were not found by U-boats at all.[20]

Bad weather played a part, hiding the convoys from German eyes or forcing U-boats to seek shelter from the elements below periscope depth; so did routine alterations of course and emergency turns if contact were made. Deliberate rerouting of convoys, however, away from patrol lines and wolfpacks, located by Bletchley, was the most productive method. Indeed, it was Hut 8's main task to provide such intelligence, which was Bletchley's principal contribution to winning the Battle of the Atlantic and so, arguably, to assuring that the war would not be lost.

David Kahn, the great historian of cryptography, gives a dramatic account of one such rerouting contest in his book on the U-boat war, *Seizing the Enigma*. It describes the progress of Convoy SC127—so code-named for the point of departure from Sydney, Cape Breton Island at the mouth of the St. Lawrence River in Canada—towards Liverpool in England in April 1943. The direct distance, measured from Halifax, Nova Scotia, was about 4,000 miles. The planned course, including changes of direction, was longer. The course actually sailed, around identified U-boat traps, was longer still.

SC127 comprised more than fifty ships, arranged in thirteen columns, escorted by five Canadian warships. It first steered east, at about 7.5 knots, then turned slightly northeast towards a spot in the ocean designated Point F by the Admiralty's Trade Movements Section. On 16 April, when it departed, both Bletchley and OP-20-G in Washington were decrypting Dönitz's radioed instructions to his U-boats, and their encrypted reports, at a delay of three days. Both tracking organisations knew, however, that he had over sixty boats in the Atlantic (in fact sixty-three) and that twenty-five were located on SC127's path due east of the Gulf of St. Lawrence. They formed a patrol line 650 miles long, aligned north-west–south-east, due south of Cape Farewell, the southernmost point of Greenland.

On 18 April Allied cryptanalysis—at either Bletchley or Washington or both—broke Dönitz's transmission of the 17th, ordering the formation of the line code-named Titmouse. Though it referred to the U-boats involved only by their captains' names, a new security measure, the key was one that had been solved; less clear were the boats' locations, since neither Bletchley nor OP-20-G had comprehensively established the inner encryptions of the grid-square designations. Moreover, the B-dienst was reading correctly Naval Cypher No. 3 and so knew, from Admiralty transmissions to SC127, that it was aware of the existence of Titmouse.

Dönitz nevertheless, with some complacency, decided that SC127 would maintain its current course; he may have done so because he was also tracking the progress of the convoy following SC127, HX234, which had just made a sharply evasive change of course, and perhaps concluded that the British would not order two convoys simultaneously to divert from their planned line of advance, trusting to HX234's alteration to distract Titmouse from SC127. In this he was wrong. On 20 April the Convoy and Routing Section in Washington, the U.S. Navy's equivalent of the Admiralty Trade Division, ordered SC127 to make a radical change of direction, just before it reached the designated Point F due east of Cape Race in Newfoundland. Instead of continuing northeastward it was to turn almost due north, leaving the coast of Newfoundland to port, and so evading by several hundred miles the grid squares in which the twenty-six U-boats of Titmouse were lying in wait for it.

Alerted by Titmouse's failure to find SC127, Dönitz now formed another patrol line, code-named Woodpecker (bird names were currently in fashion at U-boat headquarters), and deployed it to the south, wrongly guessing that SC127 had gone in that direction. SC127 therefore proceeded untouched on its leisurely way. Two factors intervened to spare it. One, unfortunately, was that Titmouse had found HX234 and was beginning to sink ships; the sinkings would be sustained by yet another hastily formed wolfpack, Blackbird. The other was that because of SC127's northward diversion, in one of the worst winters of the twentieth century which brought ice as far south as Newfoundland, it began to run into bergs and floes that forced it to change course yet again, into the Denmark Strait between Greenland and Iceland. The danger from ice also caused the convoy to slow down, thus throwing out German presumptions about its rate of progress towards other U-boat concentrations further east in the central Atlantic.

When on 22 April the convoy resumed the changed course ordered by Convoy and Routing in Washington, the allied cryptanalysts were again reading Dönitz's Enigma transmissions, but now at only one or two days' delay. As a result SC127 was redirected due east, thus sending it outside and to the north of the U-boat trap lying in the middle of the Denmark Strait. On 25 April the convoy passed from the control of the U.S. Commander-in-Chief Fleet (Cominch) to Western Approaches in Liverpool, the Admiralty command fighting the Battle of the Atlantic. Long-range aircraft, based in Iceland, had also appeared on 26 April to fly cover, an intervention that would have forced any U-boat surfacing (none did) to submerge and so lose contact with the convoy. On the last three days of its voyage, 29 April to 1 May, SC127 was continuously escorted by aircraft. On 2 May, having meanwhile detached twenty-two of its ships to Iceland and Scotland, it made port in Liverpool, having lost none of its number in seventeen fraught days at sea.[21]

SC127's escape from attack by the wolfpacks was very largely due to Bletchley's and OP-20-G's ability to overhear Dönitz's instructions to his U-boats. They did not yet enjoy complete dominance of the airwaves, however, nor had the Admiralty wholly overcome the B-dienst's ability to decrypt Anglo-American convoy traffic. Not until June 1943 would changes to Naval Cypher No. 3 absolutely assure the security of the signals used to fight the Battle of the Atlantic and Bletchley—and OP-20-G—achieve mastery of naval Enigma. Nevertheless, the steering of SC127 through the traps laid for it by U-boat headquarters demonstrated that the corner had been turned.

In mid-March 1943, two convoys leaving New York, SC112 and HX229, had fallen into the grip of two wolfpacks, "Robber Baron" and "Attacker-Driver," and been massacred. Out of the two convoys' ninety merchantmen, escorted by twenty warships, twenty-two were sunk, for the loss of only two U-boats. One was destroyed while returning to base far from the scene of action, by a British anti-submarine aircraft in the Bay of Biscay.

The battle around SC112 and HX229—the two became intermixed, as SC127 and HX234 almost did—was the costliest of the war and later reckoned to mark the crisis of the Atlantic battle itself. Any more battles like SC112/HX229 and Dönitz would have won. As it was, the others that followed tilted against him. Later in March the two convoys SC123/HX230, escorted by one of the new small "jeep" carriers, outfaced two wolfpacks, "Seawolf" and "Seadevil," and got home to Britain with the loss of only one merchantman; HX231 and ONS176 did not fare as well, but there was no massacre. In April, 313,000 tons of shipping was lost, but so were fourteen U-boats. In May, when Dönitz sent three wolfpacks, "Woodpecker," "Blackbird" and "Ram," comprising sixty U-boats, against ONS5, sustaining contact for ten days, twelve merchantmen were sunk, but so were eleven U-boats, seven in a single night. At the end of May, in which thirty-four U-boats were sunk, he decided that the exchange rate could no longer be borne. "In May in the Atlantic the sinking of about 10,000 tons [approximately two ships] had to be paid for by the loss of a boat, while not long ago [he meant during the most profitable period in 1942] a loss came only with the sinking of 100,000 tons. Thus losses in May have reached an intolerable level." He accordingly ordered "a temporary shift to areas less endangered by aircraft," by which he meant away from those flown off the escort carriers, from Britain over the Bay of Biscay and from Iceland, Ireland and North America into the former "air gap" in the mid-Atlantic. It was an admission of defeat, effectively total defeat, for, though new weapons and new U-boat technology would allow sinkings to continue, they would never again approach the level of 1942 to early 1943, let alone his projected "war-winning" total of nine million tons of merchant shipping a year. Dönitz had lost.

What part had the cipher battle played in the defeat of the U-boats? It was greatly to the Germans' disadvantage that from the middle of 1943 onwards they lost their ability to read British transmissions. During its best days, in early 1943, the B-dienst was able to give Dönitz advance warning by as much as thirty hours of convoy movements and was breaking up to 10 per cent of the convoy traffic in real time. After an unjustifiably laborious and time-consuming examination of the security of Naval Cypher No. 3, the design of an improved super-encipherment, Naval Cypher No. 5, and its distribution to the British and American fleets, the B-dienst found itself shut out. Nevertheless, until then, its contribution had been formidable. Dönitz, lacking the access to long-range air patrol enjoyed by the British and Americans, never could have achieved the number of convoy interceptions he did by the formation of U-boat patrol lines alone. The ocean was too large, the number of U-boats too few. They needed direction onto the convoy tracks and that, from 1941 to 1943, was provided by the B-dienst.

Even so, most convoys crossed the ocean, east–west or west–east, without interference, even when the Battle of the Atlantic was at its worst. In the five wartime months of 1939, 700 ships were sailed in convoy across the North Atlantic to the British Isles, and only five were sunk; sinkings in inbound convoys were far more damaging to Britain's war effort than in outward convoys for, though the ship was lost in both cases, so in the first case also was the cargo; outward-bound ships were often sailing empty in ballast, at worst carrying export goods to pay for purchases. In 1940, 5,434 ships arrived and 133 were sunk. In 1941, 12,057 ships arrived and 153 were lost. The total of sinkings in convoy, even the weakly escorted convoys of those first three war years, totalled 291, or .02 per cent of those sailed.

In 1943, the year of the biggest convoy battles, when Dönitz had more than 300 U-boats in commission and was organising wolfpacks of as many as 40 boats, 9,097 ships arrived across the North Atlantic and 139 were lost; in 1944, 12,007 arrived and 11 were lost; and in the five war months of 1945, 5,857 arrived and six were lost. Out of a total of 838 convoys and 35,449 ships sailed in 1943–45, 325 ships were lost altogether, a percentage loss of .009.[22]

These figures need adjustment. They do not include sinkings in outward-bound convoys, in convoys sailed elsewhere than the North Atlantic or of ships not in convoy, which were numerous; nor do they include sinkings by aircraft, surface warships or mines. They make no allowance for the fact that many ships crossed the Atlantic several times, exposed to risk on each occasion, as were their crews. They are a record only of transits by ships loaded with war material, including essential foodstuffs, fuel and raw materials, on the crucial route, the North Atlantic lifeline between North America and Britain—in short, on the battleground of the Battle of the Atlantic, the course of which so preoccupied and tormented Churchill during the dark years of the war.

Clay Blair, the most meticulous historian of the U-boat war, has concluded, by analysis of the statistics, that Churchill's fears were exaggerated and that Dönitz, even when he acquired his 300 boats, never threatened Britain with starvation, as the country undoubtedly was so threatened in 1917, nor even came close to inflicting defeat on the Allied anti-submarine forces. Indeed, Blair actively argues that reliance on evasive rerouting of convoys away from U-boat patrol lines, by intelligence derived from Bletchley and later OP-20-G decipherment, positively reduced the rate of U-boat destruction by lessening the frequency of contact between the wolfpacks and the convoy escorts and so effectively prolonged the Atlantic battle. By the end of 1941, he writes, "it was clear ... that the British could not rely so completely on convoy 'evasion' much longer. In addition to strengthening defensive convoy escort, they needed to hurl offensive air and submarine forces at U-boat construction yards, training areas, bases and pens, the Bay of Biscay and elsewhere to kill U-boats at a much higher rate."[23]

By the middle of 1943, a variety of measures adopted by the British,

Americans and Canadians had greatly increased the rate of U-boat destruction both in the vicinity of convoys and on the oceanic approaches. In May 1943, of the forty-nine U-boats which sailed on patrol to the North Atlantic convoy routes, eighteen were lost, a destruction rate of over one in three. Many were commanded by men sailing as captains for the first time; eleven, however, had experienced skippers. Between them, the forty-nine boats sank only two merchantmen.[24]

What factors, as opposed to intelligence derived from decryption of Enigma, so set back Dönitz's dream of victory through convoy destruction? They were many, ranging from a successful effort to increase the size of the merchant fleet supplying Britain, by charter and requisition, to such technological advances as the installation in escorts of high-frequency direction-finding (HF/DF or Huff-Duff) and the adoption of centrimetric radar, but also including improvements in underwater weapons, additions to the number and improvements in the quality and organisation of escorts, the introduction of escort ("jeep") carriers, the commitment of more anti-submarine aircraft, with improved equipment, to the routes leading from U-boat bases to the Atlantic, as well as of Very Long Range (VLR) aircraft to the middle Atlantic "air gap," improvements in underwater detection methods and many other measures.

ENLARGEMENT OF THE MERCHANT FLEET

Britain began the war with a merchant fleet—soon to be christened by Winston Churchill, in one of his flashes of linguistic inspiration, the Merchant Navy—of 3,000 ships, totalling 17.5 million gross registered tons. By the end of 1941, when the United States (with 1,400 ships of 8.5 million gross tons) entered the war, the Merchant Navy had lost 1,124 ships, including neutrals sailing with British war cargoes, totalling 5.3 million gross tons. However, 483 ships had been acquired from German-occupied Norway, Greece and Holland, and another 137 requisitioned as prizes from the enemy, totalling four million gross tons. British yards meanwhile built ships of about two million gross tons, with the result that between September 1939 and December 1941, the month of Pearl Harbor, the Merchant Navy effectively increased in size to 3,600 ships totalling 20.7 million gross tons. In retrospect it seems astonishing that owners, to say nothing of their employed crews, were prepared to acquiesce in the chartering of their ships into the Battle of the Atlantic. Their readiness to take the risk, commercially and personally, can be explained only by what Clay Blair has demonstrated to be the relatively low rate of loss.[25]

In what may be characterised crudely as a personal struggle between Churchill and Dönitz, Churchill can thus be seen to have been outbuilding and outchartering Dönitz in the period September 1939 to December 1941. Measured in terms of tonnage sunk, Dönitz's preferred index of success, Britain was keeping ahead. After 7 December 1941, when the might of United States industry was thrown unequivocally into the balance, the Anglo-American alliance drew unchallengeably away. At the forefront of the tonnage struggle was the dynamic American industrialist Henry Kaiser, a civil engineer who had built the Hoover Dam and other major works during Roosevelt's programme of reconstruction after 1932. Requested to devote his time-and-motion techniques to ship construction, Kaiser designed a standardised merchant ship, the Liberty ship, modelled on a British general-purpose freighter, and built shipyards on the west and east coasts that could turn it out, partly by prefabrication, in as little as four days. The average was forty-two days and, by the Kaiser method, 2,710 were built during the war. The number included the superior Victory ship, which could do eighteen as against eleven knots. Kaiser also built the T2 and superior T3 tankers and numbers of escort carriers. When the output of the Kaiser shipyards is set against U-boat sinkings in the North Atlantic, 1,006 ships in 1942 but only 31 in 1944, it becomes clear that Dönitz's hope of winning a tonnage war was quite misplaced. As had not been the case in the First World War, Germany's enemies could outbuild the U-boat's ability to sink. However hard Dönitz drove his captains, statistics, by mid-1943, had turned decisively against them.[26]

CONVOY COMBAT

A battle conducted on the basis of replacing losses faster than they can be inflicted is, however, a sterile and dispiriting business. The Admiralty rightly strove, from the outset of the U-boat war, to hit back and to give better than it got. While its overriding policy was to assure "the safe and timely arrival of the convoy," a policy pursued from the war's first day to the last, and its first requirement of the escorts was that they should drive attacking U-boats out of contact, it also sought their destruction. At first, given the paucity of escorts, convoy protection was all that could be attempted. Moreover, escort crews and captains were inexperienced, weapons inadequate and means of detection primitive. Britain entered the war with about 180 destroyers, 60 surviving from the Great War, the rest dating from 1927 onwards. Many, however, were required to work with the fleet, while none was strictly adapted to anti-submarine duty. They had torpedo tubes, which were redundant, too many guns and, paradoxically, were too fast; their large engines consumed too much fuel on long Atlantic crossings. What was needed, as Churchill insisted from the outset, were smaller ships, of greater endurance, mounting fewer surface weapons but also able to carry more depth charges. The first expedient were the Hunt-class destroyers, of which eighty-six were eventually built, but they proved to need too frequent refuelling and were unstable in North Atlantic storms. Adapted American destroyers, of which fifty were swapped for the use of Caribbean bases, were more successful, though not as much liked by the Royal Navy as the ten loaned U.S. Coast Guard cutters, roomy vessels of excellent seakeeping qualities. A hasty expedient was the corvette, developed from Antarctic whalers, good seaboats, though violent rollers, but too slow, small and weakly armed to provide the backbone of the escort fleet. That was eventually supplied by the new classes of sloops and frigates, in effect small destroyers, possessing enough speed to outrun U-boats on the surface but economic in fuel consumption.[27]

While the Royal Navy was building up an adequate escort fleet, and the Royal Canadian Navy was embarking on a twenty-fold expansion to become the third largest in the world, the British were also developing and refining their anti-submarine methods. At the heart of the system lay the Operational Intelligence Centre (OIC) in the London Admiralty, divided into four sections, of which the U-boat Tracking Room was dominant; it was commanded from 1941 onwards by Rodger Winn, a recovered polio victim, who quickly established an intimate relationship with the naval team at Bletchley and proved an inspired interpreter of Enigma decrypts. The OIC was linked by secure teletype and telephone to Western Approaches Command, initially located at Plymouth, later at Liverpool, the principal receiving port for North Atlantic convoys. Western Approaches, commanded by the dramatically extrovert Admiral Sir Max Horton, was the headquarters that fought the Battle of the

Atlantic from the European side. Its control room was dominated by an enormous wall chart which showed a constantly updated display of the positions of convoys, and U-boat patrol lines and wolfpacks. "All Americans who visited British military agencies in 1941 were impressed by the degree of unification which had been achieved in the Battle of the Atlantic. From the War Cabinet to the Admiralty and Air Ministry, to Bletchley Park and the OIC and Derby House [Western Approaches], all hands worked with an extraordinary singleness of purpose."[28] There was no such efficiency on the German side. As in so many other fields of warmaking where direct comparisons can be made, British democracy proved more efficient than German dictatorship in fighting the Battle of the Atlantic. Direction of the U-boat war was characterised by suspicion and rivalry, particularly in seeking the ear of the Führer. While the British, in conditions of the strictest secrecy, included all who needed to know in the struggle, Dönitz confined his side's direction to a tiny group, located at Kerneval, in Brittany, until a commando raid alarmed him into ordering its transfer to Berlin. The result of this self-imposed exclusion from the wider German war effort was that the navy remained frozen within the technology and strategy with which it began the war—using U-boats scarcely superior to those of the Great War to conduct group attacks on the surface against oceanic convoys—while its opponents were mobilising every sort of anti-U-boat measure to defeat its Atlantic offensive.

Dönitz cannot be accused of neglecting the training of his crews. Even as the prospect of Atlantic victory slipped away from him, newly built U-boats and their fresh crews were spending as much as a year in the sheltered waters of the Baltic "working up" for battle on the convoy routes. U-boat tactics, however, remained static. Dönitz clung to the "single idea" he had conceived as a young officer in the post–Versailles Treaty navy: of using submarines as surfaced torpedo boats to hover at the limit of visibility on the flanks of an intercepted convoy and then to attack in swarms as darkness fell. He made little allowance for the ability of his opponents to vary their defensive technique. That was a grievous mistake, for the British, assisted as the Battle of the Atlantic drew out by the Canadians and Americans, varied their technique in many ways, invoked more new technologies than the Germans developed in response, and ultimately triumphed by an ingenuity of which Dönitz proved incapable.

At the outset, the British convoy escorts, weak in number, often no more than four warships for a lumbering assembly of as many as sixty merchantmen, responded to attack by running down individual U-boat contacts. They were sometimes able to drive the enemy down and sustain the attack by Asdic—the echo-sounding device developed by the Allied Submarine Detection Investigation Committee in 1918, led by the Canadian scientist R. W. Boyle and later known as sonar—but were usually forced to break contact when the convoy drew away and required their return to the protective screen. Escort successes in the early days were dispiritingly few. In 1940 they destroyed only twelve U-boats, for a loss of 133 merchantmen inbound in convoy to the British Isles.

As the convoy battles intensified, the British developed a new technique. In 1941, the newly founded escort groups—prefixed B for British, C for Canadian, later also A for American, each with a distinguishing number and consisting of 6 to 8 warships kept together under a designated commander—began to react aggressively. The British and Canadians, as they were later to instruct the Americans, knew by 1941 that the U-boats were trained to attack on the surface at night. They also knew that the worst losses, whether or not there had been an Ultra intelligence warning, usually occurred on the first night of attack. When a torpedo struck, therefore, the escorts forming the convoy screen, knowing that the attacking U-boats would seek to roam surfaced within the convoy's columns, turned inwards and fired star shell ("snowflake") to illuminate the scene. If U-boats were identified, they were engaged with gunfire or set up for ramming; if they submerged, the escorts proceeded to establish sonar contact and to drop and fire depth-charges. The purpose, besides that of hoping to hit the U-boat, was to keep it down, at slow or negative speed, thus allowing the convoy to sail out of contact, usually by making an emergency turn.

When many U-boats were in contact, however, turns could not solve the problem. The convoy simply ran into others and the battle had to be fought out. Daylight forced all U-boats to submerge, thus reducing their speed below that of the target and allowing a well-handled convoy to disappear if luck was with it. The first night of a convoy battle was notoriously the worst. As the number of escorts increased, moreover, and it became possible to form support groups, held in readiness to reinforce a hard-pressed convoy, subsequent nights also became harder for the U-boats.

As the number of escorts grew, and the skills of their crews increased, the odds began to turn their way. By the end of 1943, some escort groups had become very skilful indeed, none more so than the 2nd Escort Group, commanded by Captain F. J. Walker. He had developed a "creeping" technique, designed to overcome the loss of sonar contact which always occurred in the last 100 to 200 yards of an attack. As the distinctive "pinging" of the sonar beam on his submerged hull stopped when the attacking vessel passed ahead, a cool U-boat captain would take the opportunity for violent evasive action, often successfully. Walker perceived that by using two ships, one to keep up the sonar contact from a distance, the other to creep up silently on the target until it was overhead, the U-boat could be caught by a surprise pattern of depth-charges. The moment was signalled either by light, flag or the new TBS (talk-between-ships) radio, an American invention of which Walker made copious use.

Between 31 January and 19 February 1944, Walker's five sloops, *Woodpecker, Wild Goose, Magpie, Kite* and *Starling* (his own ship), sank six U-boats by the "creeping" method in the Western Approaches, to which Dönitz had again begun to send his boats after their withdrawal from the central North Atlantic six months earlier. The crew of the last victim, U-264, got out and were saved; the other five disappeared with all hands.

Walker's group got the chance it did because of an abundance of inward-bound convoys, which drew the U-boats into contact. Success in anti-submarine warfare almost always came in the vicinity of a convoy. Simply searching for U-boats with "hunter-killer" groups, the method favoured by Churchill in 1939–40 and by Admiral King in 1942, resulted in pointless quartering of an empty ocean. No amount of intelligence could guide anti-submarine vessels on to targets if they remained submerged, or dived when an attacker was sighted, as U-boats always did when searching for prey. The only exception to that pattern came in the middle of the war, when Dönitz began to send U-boat tankers to refuel attack boats in mid-ocean. The American escort carriers were particularly successful in destroying tankers and their clients found in groups. On 4 October 1943, aircraft from the USS *Card* found U-460 refuelling three Type VII U-boats, sank U-460 immediately and later sank U-422. Again, however, *Card* was in company with a convoy, bound for Gibraltar.

A potent source of technical intelligence was high-frequency direction-finding (Huff-Duff), a pre-war development but perfected and widely distributed only in 1943, which permitted a bearing to be taken even on the briefest U-boat transmissions and displayed visually on a screen. Since it was a passive device, giving no warning to the target of the interception, it increasingly allowed U-boats to be engaged without warning, particularly in conditions of poor visibility. It was extremely accurate over short distances. Huff-Duff, however, had teething troubles and was slow to win the confidence of escort commanders. They initially preferred high-frequency centimetric radar, which appeared at almost the same time. The various models of the 271 radar set gave clear definition up to 8,000 yards and had the additional advantage of defeating German detection devices. Because they themselves had not yet developed a centimetric system, the Germans discounted the likelihood that the British had done so, with lamentable results for U-boat survival.[29]

In the final battles around the convoys sailing the central North Atlantic route in 1943, U-boat losses reached insupportable levels. Despite forming wolfpacks as many as fifty strong, the U-boats could not break through. In the attacks on ON206 and ON520 in mid-October, only one merchantman, and that a straggler outside the escort screen, was sunk, for the loss of six U-boats in two days.

AIR ATTACK

Four of the six victims of 16–17 October 1943 were sunk by aircraft, which were increasingly proving even more deadly to U-boats than well-trained and -equipped escorts. The aircrew of the specialist anti-submarine squadrons were themselves now highly trained and had gained great experience in four years of warfare. At first their role had been little more than that of forcing down U-boats in daylight, though even the early anti-submarine aircraft were equipped with depth-charges. By 1943, however, better aircraft, with better equipment, were successfully attacking and killing U-boats at a distance from convoys. The weak spot in Dönitz's oceanic dispositions was the Bay of Biscay, for across it almost all his U-boats had to pass, from their bomb-proof bunkers in the French west coast ports, to the North Atlantic. They began by making the passage submerged during daylight hours but on the surface at night, so that they might recharge their batteries and achieve the higher speed possible on their diesels. In the first three years of the war, they made the passage safely. In the first twenty-six months only one was sunk by aircraft in the Bay, and then because Bletchley had supplied accurate intelligence and the Coastal Command aircraft engaged, a Whitley of JO2 squadron was equipped with an early model of search radar.

Coastal Command recognised that its so-called Biscay offensive was ineffectual. The aircraft were types pensioned off from Bomber Command, their search equipment was feeble, even their offensive weapons were of limited effect. During 1942 there were improvements. The most important was the development of a visual sighting device, the Leigh Light, so called after its inventor, an RAF officer. Mounted typically in a Wellington medium bomber, and steerable, it was illuminated after on-board search radar had detected the target. The pilot then flew down the beam of light, switched on only after the U-boat had lost the chance to dive deep, when depth charges were dropped either around the diving boat or into its submerging swirl.

From the spring of 1943, air attack on U-boats attempting to enter the North Atlantic either across the Bay of Biscay or, having circumnavigated the British Isles from German ports, through the Faroes–Iceland gap, was increasingly successful. On 5 May, for example, U-663 was attacked in daylight by a Sunderland flying-boat and sunk with depth charges. On 8 May a Halifax heavy bomber sank the U-boat tanker 490 with depth charges in the Bay of Biscay. On 31 May U-563 was attacked in the Bay of Biscay by a Halifax, then by another, finally by two Sunderlands which destroyed her. On the same day, U-440 was depth-charged by a Sunderland and sunk, almost on the edge of the Atlantic.[30] The destruction was not all one-way. In late 1942 Dönitz had begun to equip U-boats with extra anti-aircraft guns and issued orders to "fight it out on the surface." It was an unequal struggle, for the U-boats, particularly if caught at night in a Leigh Light beam, were far more vulnerable than their attackers. Nevertheless, considerable numbers of aircraft were hit by heavy fire when pressing home their attacks, as the pilots did with great bravery; understandably, for the sighting of a U-boat on the surface came very rarely in hundreds of hours of patrolling an empty sea. In the excitement of the moment, aircraft were flown to within a few hundred feet of the surfaced U-boat, in the effort to assure that a straddle of shallow-set depth charges would send it to the bottom. Sometimes the U-boat's firepower destroyed the attacker, killing the crew immediately or leaving the survivors to die slowly in life rafts. The same, of course, happened to the U-boat crews whose boats were sunk by air attack. Unlike those forced to abandon ship near convoys, who stood the chance of being rescued by an escort—as the famous ace Otto Kretschmer was on 9 March 1942, by HMS *Walker*, with most of his crew of U-99—sinking by an aircraft in the open sea led either to immediate death or to a longer agony in a lifejacket. The same went for the crews of unsuccessful aircraft forced to ditch, either because of offensive flak from the U-boat "fighting it out on the surface" or as a result of interception by German fighter aircraft based on the west coast of France. A particularly grisly fate awaited damaged models of the earlier twin-engined Wellington, which could not sustain altitude if one of its engines was knocked out. Clay Blair calculates that twenty-nine aircraft were shot down by their intended victims, though three were sunk while defending themselves.

On balance, however, the advantage lay with the aircraft. In the concerted Bay of Biscay offensive of April–September 1943, thirty U-boats were sunk and nineteen damaged and forced back to port. "Fighting it out on the surface" proved a suicidal tactic, even when U-boats sailed in groups, and it had to be abandoned. During 1943, moreover, even more lethal air-to-surface (and sub-surface) weapons appeared, including air-launched rocket projectiles, typically fired by the ancient Swordfish biplanes flown off British escort carriers, and 6-pounder anti-tank guns mounted in land-based Mosquitoes. Both could penetrate the U-boat pressure hull, with disastrous effect. The dominant air-launched weapon, however, proved eventually to be the so-called Mark 24 Mine, cover name for an American acoustic torpedo which homed on the propeller noises—cavitation—of a submerged U-boat. U-boats began to receive the equivalent *Zaunkönig* acoustic torpedo during 1943, which homed on the engine and propeller noises of escorts; several were sunk or seriously damaged until a decoy noise maker, towed astern, was developed. U-boats could not use decoys, since their safety lay in keeping silent. Because the Mark 24 Mine was successfully kept secret, U-boats attempting to make off submerged after being surprised by aircraft fell victim. The Mark 24s "were curious weapons to use . . . there was always an anxious time-lag before any sign of a result, sometimes as long as 13 minutes. When it came it might consist of nothing more than a brief disturbance of the sea's surface."[31]

What occurred beneath the surface does not bear contemplation. Unlike a depth charge, which could damage a U-boat badly enough to force it to surface without wrecking the boat, the Mark 24 was an impact weapon, which ruptured the pressure hull, causing it to flood too rapidly for any hope of escape by the crew. Airmen who achieved a kill were bound to experience a sense of satisfaction at the culmination of a search process which occupied hundreds of hours of fruitless flying time; but even the most hardened must have shuddered at the thought of the reality of their victims' fate. There were one or two accounts by survivors of their experience in leaving a U-boat making its way to the bottom, and they tell of a screaming mass of humanity, fighting for egress against the inrush of water, all comradeship lost, every man for himself, imbued by blind and selfish panic.

By 1944, the U-boats were fighting a lost battle. Dönitz's efforts to deploy boats capable of cruising undetected and at long range—the snorkel models, which could breathe submerged, and the Walther electro-boats, running on hydrogen peroxide, which did not need to breathe externally at all—were unsuccessful. The snorkel, hated by the crews for its tendency to burst their eardrums, so reduced the boat's speed submerged as to negate its theoretical advantages, while the Walther boats were too shoddily built to achieve the results of which they should have been capable. After four years of war, and despite episodes of offensive success which had shaken governments on both sides of the Atlantic, the U-boats were a beaten force.

The pre-war elite, which had opened the campaign against Allied shipping, were dead or captured. Their successors, who had hit British shipping so hard in 1940–41, and had rampaged along the American coast in 1942, had suffered the same fate. There were few survivors of the great convoy battles of early 1943. The new commanders and new crews of 1944 paid a terrible price for their inexperience. Dönitz never relaxed his standards of training. Even so, U-boats remorselessly fell victim to the increasingly efficient British, Canadian and American escort groups and their associated air squadrons. Of the 591 U-boats which sailed between May 1943 and May 1945, 138 were sunk either in their first voyage or under a new captain, often within a few days of putting to sea. Of the seventeen U-boats which attempted to escape from Germany to Norway in May 1945, all but one newly commissioned, all were sunk.[32]

What had begun as an unequal struggle between an inadequate fleet of British escort ships, with primitive detection devices and crude underwater weapons, had swelled during the course of the war into a major anti-submarine campaign, prosecuted on the Allied side by an ever-larger armada of British, Canadian and American destroyers, sloops, frigates, corvettes and, critically, escort aircraft carriers, supported by an eventually very large complement of land-based aircraft. In the course of six years of bitter conflict, the Allies had introduced a succession of increasingly efficient detection devices and underwater weapons, including sonar able to register depth as well as bearing, centimetric radar and high-frequency direction-finding, a wide variety of depth charges, filled eventually with the very powerful Torpex explosive, and several models of forward-firing contact weapons, particularly Hedgehog and Squid. The Americans had also developed the deadly Mark 24 Mine ("Wandering Annie"), to be dropped from aircraft.

In response, the U-boat arm, though greatly expanded in number, from 57 in 1939 to a total of 1,153 built by May 1945, scarcely developed at all. The Types VII and IX, the former undersized, the latter unhandy, supplied most of its ocean-going strength throughout the war; their experimental successors were all unsatisfactory in one way or another. Their weapons, after early faults had been corrected, scarcely improved either; only the *Zaunkönig* acoustic torpedo was a real improvement, and by the time of its appearance the wolfpacks had been driven from the convoy routes. The U-boats' passive detection devices—hydrophones, able to detect ship noises at long distance—were excellent, but no active detection device, such as centimetric radar, useful as it would have been despite its traceable emissions, was ever developed. The U-boats' own radar-detection devices, such as Metox, were crude and came to be distrusted by their users. As a result, U-boat captains, like those of Nelson's frigates, depended throughout the war on line of sight, enhanced only by forming patrol lines in exactly the same way as Nelson had done. Radio extended their range a little by providing a captain who got a sighting with means to summon others; but because of the danger of interception, and the relatively low superiority of speed of a surfaced U-boat over that of a convoy, patrol lines had to be kept quite short. As Dönitz was frequently to comment, even the best-placed patrol line, operating against convoys controlled by a headquarters that could steer them away from identified U-boat traps, was often intercepted by boats only at the extremity, leaving the rest to make up hundreds of miles of sea room before they could achieve an attacking position. Long-range aircraft, which might have found convoys at a greater distance than a U-boat captain could see from his conning tower, were in even shorter supply on the German than the Allied side; and the German Condors not only lacked the range of the B-24 Liberator, the dominant aircraft of the Atlantic battle, but were based exclusively in France, and thus limited in coverage, while their Allied equivalents flew from Northern Ireland, Cornwall, Iceland and North America and so were eventually able, with the introduction of Very Long Range models, to oversee the whole Atlantic.

THE INTELLIGENCE BALANCE

The Allies—the British and Canadians first, the Americans later— enjoyed superior tactical intelligence throughout the Atlantic battle. They had sonar (Asdic) from the start, HF/DF and centimetric radar from 1942, aerial surveillance—of expanding range—from 1939. There were serious lapses, associated with shortages of escorts in 1939–41 and with American refusal to form convoys in the first six months of 1942. Nevertheless, while tactical intelligence collected by the U-boats was limited by a patrol line's line of vision, supplemented eventually and only close to France by that of the Condor aircraft, the convoy escorts and their associated patrol aircraft always had the edge, which increased throughout the war. Eventually the Allies' tactical advantage became overwhelming.

The question of the importance of intelligence in the Atlantic battle therefore turns on the strategic issue: B-dienst versus Bletchley. The B-dienst was a remarkable organisation. Working largely with human resources, rather than the electromechanical devices later available in England and the United States, the German decoders—their target was the Naval Cypher No. 3, a book code mathematically super-enciphered— achieved impressive success for several long periods. Having broken the old (Royal) Naval Code, a system of five-digit groups, as early as 1935, it used the clues the Code provided to attack the newer Naval Cypher and by April 1940 were reading as much as 30 per cent of British transmissions. The Naval Cypher and Code were replaced, however, in August 1940 by new versions of each, against which the B-dienst had less success. It recovered its form when Naval Cypher No. 3 was introduced in June 1941 for transatlantic communication, and read it throughout 1942; in December it was reading 80 per cent of messages sent. Between 15 December, when the British added precautions, and February 1943, it was again in the dark but then found a way back, "sometimes reading directives to convoys ten and twenty hours before the movements they ordered took place." Not until June, when an entirely new cipher was distributed, were British signals at last made secure.[33]

On their side, the British, and later the Americans of OP-20-G in Washington, experienced similar periods of light and dark. The British benefited from a series of captures during 1940 and 1941—of U-33 in February 1940, of the patrol boat VP2623 in April and, in March 1941, of the trawler *Krebs* during the commando raid on the Norwegian Lofoten islands. In May and June the weather ships *München* and *Larenburg* were seized in deliberate "cutting out" operations, while on 9 May, U-110 fell into British hands. Each yielded some material—either parts of the Enigma machine, or gridded charts, or enciphering material—which, when added to what Bletchley had been able to reconstruct from interceptions, furthered decryption. Traffic for February, May, June and July 1941 was read in part or whole, as a result, and some in real time. After August 1941 until February 1942, the *Heimisch* key, called Dolphin at Bletchley, was read at a delay of not more than thirty-six hours. The decryption was considerably assisted by the deployment of the first bombes, which greatly assisted in the identification of possible Enigma key settings. Much help was also provided by the carelessness of German operators in retransmitting Enigma messages previously enciphered in the dockyard (*Werft*) or weather ciphers, which were quite easily read.[34]

After February 1942, however, Bletchley lost its way into the U-boat traffic because of the German adoption of a new Enigma key, called Shark by the British, *Triton* by the Germans, on the first of the month. Regular reading in real time did not resume again until December 1942. This period coincided with the reciprocal B-dienst success against British naval ciphers and with a peak of German advantage in the Battle of the Atlantic. In June (129), July (136) and August (117), monthly sinkings exceeded one hundred ships. Total sinkings exceeded four million tons.[35]

Paradoxically, however, during what historians of the U-boat war denote as "the great convoy battles" in the first five months of 1943, the British were frequently reading Enigma in real time and the Trade Division was redirecting some convoys away from U-boat patrol lines. Moreover, although there were massacres, U-boat losses were also rising, eventually to unbearable levels, forcing Dönitz, in May 1943, to take his boats out of the North Atlantic and so, effectively, admit defeat.

What, therefore, are we to make of F. H. Hinsley's assertion that Bletchley achieved one of its undeniable successes in the Atlantic battle? Hinsley, a Bletchley veteran, is notably modest in the claims he makes for the British Government Code and Cipher School. He specifically rejects the view that "Bletchley won the war," and rightly so. His estimate of Bletchley's success in the war against the U-boats must be set, however, against Clay Blair's sober and heavily documented assessment that over 99 per cent of all ships forming transatlantic convoys reached their destinations safely. Out of 43,526 ships sailed (many several times, of course), 272 were sunk by U-boats. Many others were sunk but usually when sailing independently or having left convoys, either to "straggle" (fall behind) or to "romp" (sail ahead).

Allied merchant seamen paid a terrible price. Over 30,000 out of 120,000 in the British Merchant Navy alone were killed in the struggle against the U-boats. The U-boat crews, however, suffered worse: 28,000, out of an enlisted force of 40,000, died in the destruction of their submarines—which amounted to 713. Aircraft sank 204, warships 240, aircraft flying from escort carriers 39; ships and aircraft operating together sank 84; mines, accidents and other incidents—such as U-boats ramming each other—made up the difference in numbers.[36]

The outcome of the Atlantic battle, seen in perspective, suggests that intelligence, as in so many other operational circumstances, was, though significant, secondary to the age-old business of fighting the issue out. In easy times, as during the U-boats' Happy Time along the eastern American seaboard in the first six months of 1942, it was not strategic intelligence but day-to-day happenstance that yielded the victims. In the more difficult times, particularly during the "great convoy battles" of early 1943, Bletchley's ability to steer convoys out of danger frustrated many of the traps Dönitz laid, but it was the stolid endurance of the merchant seamen, sticking to convoy commodores' orders, and the dogged determination of the escort crews to fight back, that did the U-boats in.

Modern interpretations of the Battle of the Atlantic represent it as a true battle, in which one side attacked, the other accepted the challenge to defend or counterattack, and counterattack prevailed. That seems correct. The Battle of the Atlantic could have been won without the assistance of the codebreakers, greatly though they helped to tip the balance in the favour of the defenders.

———◄○►———

CHAPTER EIGHT

Human Intelligence and Secret Weapons

THE SECRET AGENT dominates popular conceptions of the intelligence world, how it works and what it yields. The image of the agent is strongly imprinted on the imagination of anyone who in childhood played the war game *l'Attaque*—a sinister civilian figure parting the grasses to spy on brightly uniformed soldiers honourably and conspicuously engaged in combat—and that image has been reinforced for over a century by the work of successful fiction writers. Joseph Conrad made the agent an enemy of society, Conan Doyle, in "The Bruce-Partington Plans," a venal creature working for money; but most writers in English represented the agent as romantic and patriotic: Kipling's Kim a servant of empire in the Great Game, John Buchan's Richard Hannay bursting with Britishness in pursuit of his country's enemies, Sapper's Bulldog Drummond to whom all foreigners were objects of suspicion. Later writers, John le Carré foremost, refined the image, admitting the dubiety of the agent's role and introducing the idea of the double agent, all too understandably in view of what the public then knew of treason among Britain's university-educated class.

The idea nevertheless persisted that "intelligence" was principally the "product"—a term popularised by le Carré—of spying. The idea survived the disclosure of the Enigma secret, which revealed that the most valuable intelligence produced by the British intelligence services during the Second World War was derived from the interception and decryption of enciphered enemy signals. That was equally true of

British intelligence successes during the First World War and of the American intelligence effort in 1942–45. British and American fiction readers were by then too enamoured of the idea of the "agent in place," however, to alter their view of the essential nature of the craft. The spy, not the eavesdropper, was established in the popular imagination as the principal source of knowledge of the enemy and his evil intentions.

The popular imagination entirely overlooked the limitations within which real agents laboured. The danger of betrayal was recognised, and that of identification by enemy counter-espionage. What was discounted was the much more oppressive burden of practicality: how to discover anything worth knowing; how, even more critically, to communicate such knowledge to the home base. Memoirs by agents of the Special Operations Executive (SOE), Britain's subversive organisation inside German-occupied Europe in 1940–45, and of OSS, the contemporary American Office of Strategic Services, reveal a picture quite at variance with that of the glamour and romance depicted by writers of fiction. SOE and OSS operatives dealt in tiny scraps of information, often of apparently trivial importance—gossip picked up in cafés, numbers of freight cars seen crossing a bridge, shoulder straps of soldiers glimpsed changing station. Such scraps, when collated, had to be recorded in comprehensible form and then sent by radio transmitter, whose operators always knew that they risked being overheard by enemy interceptors with location devices and so of being intercepted and arrested in mid-transmission.

There was little that was romantic about spying in Hitler's Europe. The business was furtive, nail-biting and burdened by the suspicion of betrayal. Many agents were betrayed. The German counter-espionage service was extremely efficient at identifying networks, breaking their members and inducing those arrested to inform against their fellow conspirators. Women proved better than men at keeping out of German clutches, because of their superior ability to remain inconspicuous and to deflect difficult questions. Many women nevertheless fell victim to the Gestapo. Men in the networks were arrested in large numbers. Their fate, that of women and men alike, was despatch to Hitler's camps.

. . .

GERMAN SECRET WEAPONS

There were gaps, nevertheless, in the Gestapo system, particularly in its ability to protect the German secret weapons programme. Because the testing of Germany's pilotless weapons, later to be known to the British as the V-1 and V-2, necessitated test flights over areas populated by non-Germans in the Baltic and Poland, and because Germany's acute shortage of labour forced the organisers of the secret weapons programme to employ non-German labour in work, particularly construction work, at the secret weapon sites, information leaked out. Over time, eyewitness reports, transmitted through networks run particularly by Poles, would yield considerable amounts of information about German secret weapon development. It was not, however, information for which the British were particularly looking at the outset. Under threat of invasion after June 1940 and heavy air attack after September, they were more concerned about the dangers of the here and now rather than those that might lie in the future.

As early as February 1939, however, seven months before the outbreak of war, the Committee for the Scientific Survey of Air Defence, chaired by the distinguished scientist Sir Henry Tizard, had decided to form an intelligence section and a young physicist, Dr. R. V. Jones, had been appointed to lead it. He was first directed to collect information about bacterial and chemical weapons, then believed to be a serious menace. In October his attention was diverted elsewhere, if only briefly. A report was received of an experimental station, located on the Baltic coast between Danzig and Königsberg in East Prussia, where the Germans were testing a "rocket shell" carrying 320 pounds of Ekracite explosive over ranges up to 300 miles.[1] The source lay in gossip and might therefore have been discounted, along with much other rumour about fantasy devices. On 4 November, however, another report arrived on Dr. Jones' desk, sent by the British naval attaché in Oslo, the capital of Norway.

About 2,000 words in length, "the Oslo Report," as it became known, was not rumour but a detailed communication, obviously written by a practising scientist, describing nine weapons or weapon systems under development in Germany. Some were conventional—a new bomber, an aircraft carrier—some were not; the report also identified sites at which the new weapons were being tested. Among the unconventional weapons mentioned were a remote-controlled anti-ship glider bomb, a pilotless aircraft, "remote-controlled shells" propelled by rocket, acoustic torpedoes and an anti-aircraft proximity fuse. One of the two test sites named was at Peenemünde on the Baltic coast.[2]

The origin of the Oslo Report was mysterious and remained so for many years. Recently, however, it has been suggested—the suggestion has not been universally accepted—that the author was a German scientist, Hans Friederich Meyer, director of research at the great German electrical company Siemens.[3] Meyer, who opposed Hitler's racial policies, had a British friend, Cobden Turner, of the British General Electric Company. Much troubled by the treatment of a half-Jewish child whose mother he knew, Meyer told Turner of the case and Turner, equally troubled, managed to obtain a visa for the girl to leave Germany for England from the head of the MI6 station at the British embassy in Berlin, Frank Foley, just before war broke out. Foley was then posted to Oslo but Turner induced Meyer, as a token of gratitude for the grant of the visa, to write what we now know as the Oslo Report. Meyer typed it up on a visit to Oslo on official business on 1–2 November 1939 and it reached R.V. Jones, without attribution, two days later.

Though Jones was astounded by the Oslo Report—he described it later as "the most amazing statement I have ever seen"—it was allowed, because it was unsubstantiated and not amplified, to lie on the file. British intelligence had other matters on its mind during 1940–42; in the scientific field particularly German advances in aerial radio navigation, radar, tank technology and underwater warfare.

Jones was nevertheless right to recognise that the Oslo Report was pregnant with warning. In one form or another, however vague, it gave advance notice of at least four weapons that were to do the Allies great harm before the war was out: in ascending order of importance, the anti-ship glider bomb (the HS293), the acoustic torpedo (*Zaunkönig*), the rocket-propelled shell, eventually to be realised as the A-4 projectile— V-2 to the British and father to all ballistic missiles—and the FZG-76, or flying-bomb, father of the modern cruise missile. While the Oslo Report slept on a British shelf in 1940–42, German scientists were busy bringing to production stage the weapons of which it warned.

British—and then American—efforts spurred them on. Had RAF Bomber Command, joined in the strategic bombing offensive by the American Eighth Air Force during 1942, not succeeded in that year in breaking through the Reich's air defences and beginning the destruction of Germany's major cities, it is possible that Hitler would not have chosen to allocate the resources necessary to make "weapons of reprisal," as he came to call them, the agents of destruction of the cities of his enemies, particularly London but also Antwerp in Belgium, that they became in 1944–45. Allied bombing not only wrought terrible devastation on German homes and factories, and on Germany's cultural heritage, and terrible disruption as well as termination of ordinary Germans' everyday lives; it also directly attacked Hitler's and his Nazi party's claim to be the protectors of the German people. While bombs rained down on Berlin, Hamburg, Cologne and his state's other centres of population, unanswered after the defeat of the Luftwaffe as a means of attack against Britain in 1941, Hitler began to burn with the urge to pay back. Göring's conventional bombers had failed him. During 1943 he turned to the promise of unconventional weapons to repay the hurt.

The British scientific intelligence service's attention was first drawn back to German secret weapons at the end of 1942, when a report reached London, from a Danish chemical engineer, of a conversation overheard in a Berlin restaurant about a rocket fired from Swinemünde—not far from the Oslo Report's Peenemünde—which was said to carry five tons of explosives over 130 miles. In February 1943 a second report, from another source, gave different capabilities to the rocket but specified that it was launched from Peenemünde.[4]

Germany has little sea coast. West of the Danish peninsula, its seaboard runs inside the Friesian islands, forming a region that provided Erskine Childers with the setting for the first serious novel of espionage, *The Riddle of the Sands*, written just before the First World War and warning the British of the Kaiser's hostile intentions against their coasts. East of the Danish peninsula, inside the Baltic, lie Germany's historic mercantile ports but also the summer holiday places of its leisured class—small fishing villages amid fir-lined dunes and white beaches. It was the holiday connection that brought Germany's secret weapons programme to the Baltic. The family of Wernher von Braun, the leading young rocket designer, had spent summers there, and when a testing site was being sought, his mother suggested that remote and sparsely populated spot.[5] Work began on the site in 1936, and by 1943, when the British began to take an increasingly close interest, the establishment had grown considerably. It consisted of two locations, Peenemünde West, under the control of the Luftwaffe, where the flying bombs were under development; and Peenemünde East, the army base where the V-2 rocket, known to the Germans as the A-4, was tested. An airfield had been built, a camp for the labour force—largely non-German—laboratories, workshops, a plant for producing liquid oxygen, an essential ingredient of the V-2's fuel which was in short supply, and a housing estate for the scientific staff.

Peter Wegener, a young scientist who had been recalled from an anti-aircraft unit on the Eastern Front to work at Peenemünde, has left a revealing picture of life there. In some ways it mirrored that of Bletchley Park. The rocket scientists were young, highly educated, often from the upper middle class. Wegener was a boarding-school product from a professional family. Von Braun was from a similar background, as was, notably, Albert Speer, who, in 1943, had risen from the position of Hitler's personal architect to be Reich armaments minister and so directly responsible for V-1 and V-2 production. Von Braun and Speer were handsome and charming young men of wide interests and considerable social assurance. Wegener was formed in the same mould. After three years in the army, only latterly as an officer, he found the transfer to the university-like surroundings of Peenemünde, among people of his own sort, men and women with doctoral degrees and civilised manners, a delightful liberation. "I never heard a harsh word: everyone helped everyone else, and good humour reigned"; little attention was paid to rank; "in fact, it was a pleasure to work in this place."[6] The description might apply to Bletchley. There was this difference. At Bletchley, despite the common-room atmosphere, the in-jokes, the amateur dramatics and the undercurrent of romance, the strictest rules of security applied. No one ever spoke about his or her work outside the immediate workplace; instant removal was the penalty. At Peenemünde, by contrast, "need to know" was not a principle. "Practically all discussions in larger groups—in the mess hall and other public places—concerned technical problems. Everybody spoke freely of his work; internal security simply did not exist. *Fachsimplen,* or shop talk, squeezed out all other discussions."

It was a recipe for leaks, all the more oddly given the reputation of the British as gossips and rule-breakers and of the Germans as humourless disciplinarians. Peenemünde clearly did leak, in a way that Bletchley, whose 10,000 initiates kept the secret intact for twenty-eight years, did not. It was leaks that alerted the British to the danger brewing at

Peenemünde, first the Danish engineer's report of careless talk in a Berlin restaurant, then in February 1943 another, mentioning Peene-münde, third, in March, serious authentication. On 22 March, two cap-tured German generals, well known to the British as a result of their part in the desert battles against the Eighth Army, were brought together in a room wired for sound. They—Generals Cruewell and von Thoma—had not seen each other for several months. In the warmth of re-encounter, they began to talk—too freely. Von Thoma spoke of a visit to a test site where he had been told by the officer in charge of "huge rock-ets" which would go ten miles into the stratosphere and had unlimited range.[7] A copy of the transcript of the Cruewell–von Thoma conversa-tion was sent to R. V. Jones, head of scientific intelligence at the Air Min-istry. He passed his concern up the chain of command, asking for authority to take charge of an investigation, the results not to be released, for fear of causing unnecessary alarm, until it was complete. When his proposal reached General Ismay, Chief of Staff to Winston Churchill in his capacity as Minister of Defence, a one-man ministry Churchill had created for himself, it was turned down. Ismay argued that a matter as important as a rocket attack on Britain must be taken out of the narrow scientific field and confided to a single investigator able to call on the widest advice. He was supported by the Chiefs of Staff, who accepted the gravity of the danger. "The fact," Ismay wrote to Churchill on 15 April, "that five reports have been received since the end of 1942 indicates a foundation of fact even if details are inaccurate."[8] The Chiefs proposed that the single investigator should be Duncan Sandys, a son-in-law of Churchill who had been invalided out of the army's only rocket regiment, an anti-aircraft unit, and, as a Member of Parliament, was currently serving as Parliamentary Secretary to the Ministry of Supply, responsible for weapons research. Churchill at once agreed—Sandys was a man of known ability and energy—and he started work on 20 April.

The previous day the Central Interpretation Unit (CIU), which examined photographic intelligence at Medmenham in Buckingham-shire, not far from Bletchley, had received orders from the Air Ministry to investigate signs of a German secret weapons programme wherever they could be found. The first area of search was to be in northwest France, since prognosis suggested that the likely weapons, a long-range gun, a rocket aircraft or "a rocket launched from a tube," would have to be based within 130 miles of London. Duncan Sandys, displaying remarkable insight, wanted the search to be cast wider. He argued, on the basis of his experience with experimental anti-aircraft rockets, that any German weapons under development would have to be tested first, well away from populated areas, not in occupied territory and near the sea. Those limiting conditions suggested a site on Germany's short coast-line and as Peenemünde had already been mentioned in agents' reports, it became the obvious target for close photographic reconnaissance.[9]

Peenemünde had been overflown by RAF photographic reconnais-sance (PR) aircraft, but in the course of the general surveillance of enemy territory, not as a specific reconnaissance target. It now began to receive close attention; Peter Wegener was later to record in his memoir of Peenemünde the frequent passage overhead, in the beautiful Baltic summer months, of Mosquito aircraft. He accepted them uncompre-hendingly as part of the scene and invested them with no menace; the skies of Germany in 1943 were full of British and American aircraft, but most were bound for Berlin.[10] On 21 April, however, Peenemünde had been photographed carefully and many of its buildings identified, wrongly as would later appear. Photographed again on 14 May, and on 12 June, more mistaken identifications were made. Not until 23 June was the evidence interpreted to yield an identification of rockets. The photographic interpreters might be forgiven; what they had first described as "objects" were only 1½ millimeters long on the film. There was also much to confuse them. Unknown to anyone in Britain, two, not one, revenge weapons were under development at Peenemünde, the fly-ing bomb as well as the rocket; while in Britain the scientific defence establishment was riven by a bitter and highly personalised dispute between clever men, some of whom denied the possibility of the rocket's existence.

THE REVENGE WEAPONS

Rockets of a primitive sort had existed for centuries and, in a form capa-ble of reaching into space, had been imagined for over a hundred years. Pilotless aircraft, cruise missiles as they would be called today, had first been envisaged before the First World War, and in 1907 a French patent had been granted for a pulse-jet engine, exactly the power unit that was to drive the German flying bomb of 1944. What could be foreseen, however, was far from realisation. That was to wait until the early twentieth century. In 1926 an American, Robert Hutchings Goddard, launched the first liquid-propelled rocket, of modest capability. A contemporary, the Romanian-German Hermann Oberth, was meanwhile extending the theory of rocket propulsion, and though his work was written rather than practical, he greatly influenced a younger German, Wernher von Braun, who is universally recognised as the father of the extra-atmospheric rocket. Von Braun, by dint of single-minded enthusiasm, attracted the attention of another monomaniac, Walter Dornberger, who had succeeded in winning funds from the German army for rocket development. In the first flush of German re-armament, during the early Hitler years, von Braun and Dornberger began the German rocket programme and located it at Peenemünde. Von Braun supplied the brains, Dornberger the practicalities. An artillery officer of the Great War, he had subsequently trained as a scientist and had, in his university years, seen in rockets a means to evade the ban imposed by the Versailles treaty on Germany's possession of heavy artillery. He had persuaded his superiors of the value of his idea, to such purpose that, starting in 1936, £25 million ($40 million in current values) was to be found from the German defence budget to pay for Peenemünde.[11]

Peenemünde began as an army establishment. The A-4 (V-2) and its predecessors were seen as equivalents of heavy guns and therefore to be commanded and operated by the army's artillery branch. After the stunning victories over Poland, France, the Low Countries and, initially, Russia in 1939–41, Hitler and his generals lost interest in rockets. It was revived only by the inception of effective RAF attacks against German cities, beginning with the destruction of Lübeck in 1942. The Peenemünde budget was increased and the Luftwaffe, conscious that development of pilotless weapons most engaged the Führer's favour, sought for one of its own, to match the army's. The Argus firm was working on the design of a cheap and crude cruise missile; but the contract was eventually given to the firm of Fieseler, producer of light aircraft, notably the ubiquitous observation and liaison aircraft, the Fieseler *Storch* (Stork). The first model was flown at the end of 1942.[12]

The Fieseler flying bomb—historically the V-1, or flying bomb, but first known as the FZG-76 (FZG stood for *Flakzielgerät*, anti-aircraft target apparatus, a cover name); later, to British confusion, as the Fi. or

Phi.103—was the simplest of aircraft. A cylinder, pointed at the nose, contained a ton of high explosive, detonated by an impact fuse. Two short wings were attached at the point of launch. At the rear, mounted above the tail assembly, was the tube for a pulse-jet, fuelled by low-grade petrol fed from an on-board tank. A shutter system caused the injected fuel to burn in regular bursts, giving the missile a speed over 400 mph, its characteristic and soon to be dreaded drone and a range of 150 to 200 miles; a simple cut-out device shut off the fuel at a selected point, leaving it to dive vertically to earth. It was launched either from a ramp (catapult, as the device was first described by the British), pushed by a reaction of hydrogen peroxide and permanganate; or, more rarely, air-launched from beneath a mother aircraft. It was reliable and cheap, costing about £150 in 1944 values.

Had it been given priority, and been mass-produced in large numbers during 1943, there is little doubt that the flying bomb would have caused terrible damage to London and other southern British cities; it might even have so disrupted shipping in British southern ports as to have set back or even prevented the launching of the cross-Channel invasion in June 1944. Initial plans, drawn up by LXV Army Corps, the formation created to operate it, were for the production of 1,400 bombs in January 1944; 3,200 by April; 4,000 by May; and a maximum of 8,000 by September. If achieved, and successfully delivered, London might have suffered the equivalent of a thousand-bomber raid every four days, far worse than anything inflicted on any German city even at the height of the strategic bombing offensive.

The production schedule was not to be met; but it might have been had not the A-4 (or V-2) programme diverted so much effort from the secret weapons effort in its totality. The A-4 was expensive (about £12,000 in 1944 values), where the V-1 was cheap; it was also complex, where the V-1 was simple. Even after a prototype had been successfully flown on 3 October 1942, only the fourth to have been tested, 65,000 separate modifications had to be carried out before reliable performance was achieved. There were setbacks of all sorts, including failures of the guidance system and disintegration of the rocket body itself; the most persistent fault, however, which took months to remedy, was explosion of the fuel. Some explosions took place soon after launch or actually on the launching stand, leaving evidence that could be fairly quickly assessed; most, however, occurred in flight, tens of miles down the test range, scattering debris far and wide or into the sea and causing von Braun and his team great difficulty in identifying the problem.

The real problem was that von Braun was attempting to design the prototype of all subsequent extra-atmospheric missiles, literally the direct ancestor of the moon rocket, which the A-4 was, while the Fieseler company, though the flying bomb was indeed also the ancestor of the modern cruise missile, was merely building a cheap, simple pilotless aircraft. The flying bomb was a potentially decisive weapon, which the A-4 certainly was not; even when perfected, it was too complex, too expensive and too difficult to mass-produce, and delivered too small a warhead to achieve a decisive result. Unfortunately for Germany, the A-4 engaged the Führer's attention. As a result, instead of giving priority to the flying bomb, or ordering the V-1 and V-2 programmes to be conducted separately, he allowed them to proceed in tandem, and so to compete against each other. Competition was harmfully heightened by the separate interests of the army and the Luftwaffe, committed respectively to the V-2 and the V-1. Albert Speer, who correctly perceived that the V-2 drew wastefully on scarce resources needed for, among other things, Germany's conventional aircraft-building programme, would have cut back or even closed down the V-2 programme; but to attempt to do so was to argue against the Führer and eventually against Himmler and his SS, who sought to gain control of the V-2 programme in pursuit of their goal of dominating the German war effort.

For over two years after the first successful test firing, work proceeded in secret, von Braun striving to perfect his brainchild, the British gradually becoming aware, through the receipt of scraps of information that eventually composed an intelligence picture, of the nature of the weapon that threatened them. It is not surprising that they were baffled for so long. The V-2 was a truly revolutionary weapon. It needed no elaborate launching system, as it was believed any long-range rocket would. It achieved stability without rotation, which was thought impossible in a long-range rocket. It had an on-board and autonomous guidance system, again thought an impossibility. Above all, it was liquid- not solid-fuelled, when conventional opinion held that only solid fuel could supply the necessary power; and it was single-stage, when convention again held that to achieve the transition from slow launch speed to high speed in ballistic trajectory two stages, or separate sections of the rocket, were essential.

The V-2, in its final form, was cylindrical, tapered at the tail, which mounted four fins, and sharply pointed at the nose. It stood 50 feet high, in its firing position, was six feet in diameter, and weighed 28,557 pounds, of which 1,630 pounds comprised the warhead of Amatol explosive and 9,565 pounds the fuel, 75 per cent ethyl alcohol in liquid oxygen. At launch, a turbine pump, driven by the decomposition of hydrogen peroxide, fed the fuel into the combustion chamber where it was ignited. Oxygen and alcohol were introduced separately, the alcohol also serving to cool the combustion chamber. There were two firing bursts. The first ignited to start the motor. When it was running smoothly, a second burst lifted the rocket off its platform. Guidance was supplied initially by four small graphite fins working inside the jet stream; later guidance was given by the four external fins. Both were controlled by gyroscopes which operated servo-motors to move their surfaces. Another device tilted the rocket into level flight when it had reached the correct altitude and yet another—originally a radio signal from the ground, later an on-board mechanism which measured velocity—cut off the motor at the point from which it had been predetermined its descent should begin. The missile eventually achieved four times the speed of sound and arrived on target—which could be only roughly predicted—without warning.[13]

Militarily, the deadliest feature of the V-2 was its launching platform, the Meillerwagen, so called after its manufacturer. The Meillerwagen was what today would be called a transporter-erector, a towed cradle which could elevate the V-2 into the vertical, placing a small conical platform under the nozzles of the jet-exhausts to receive the thrust. It was simple, efficient, inexpensive and, critically, inconspicuous, and was to bamboozle the British scientists engaged on the intelligence search for months. They believed at the outset that an extra-atmospheric rocket must either be rotated and fired from a massive tube, or multi-stage, requiring also a large, static base-platform; in either case they expected solid fuel to be the propellant, at least of the first stage if it proved multi-stage. Few admitted that it might use liquid fuel, which was thought suitable only for small, short-range rockets, and none at the outset conceived of anything like the Meillerwagen. Again, they might be forgiven. The Meillerwagen was, like the V-2 itself, a revolutionary concept; its descendants were to invest the intermediate-range missiles of the Cold War years, American and Soviet alike, with their strategic menace.

THE ARGUMENT OVER THE V-WEAPONS

By early 1943 enough evidence had accumulated in Britain of German secret weapon development to alert several of the responsible authorities to the danger of pilotless weapons—or other forms of long-range attack. The authorities included R. V. Jones, who advised the Secret Intelligence Service (MI6), as well as the Air Ministry, and who enjoyed the Prime Minister's favour as the "man who broke the beams," the Luftwaffe's radio guidance system during the Blitz of 1940; Professor C. D. Ellis, the army's scientific adviser; Dr. A. D. Crew, Controller of Projectile Development at the Ministry of Supply, the British equivalent of Albert Speer's Armaments Ministry in Germany, General Ismay, Churchill's personal military staff officer; the Joint Intelligence Sub-Committee, which co-ordinated the work of MI6, MI5 (the internal security service) and Special Operations Executive (SOE), Churchill's subversive organisation overseas, and the government's Scientific Advisory Committee. Too many fingers in the pie, no doubt; but the crucial initiates were Lord Cherwell (Professor F. A. Lindemann), Paymaster-General but holding that post as Churchill's personal and long-time scientific adviser, and Duncan Sandys, since 20 April chairman of the committee, soon to be known as the Bodyline, then Crossbow Committee, charged with overall responsibility for investigation of the secret weapon threat.

Churchill believed in "creative tension" as a principle of administrative efficiency, the fostering of rivalries between government servants to generate energy in the examination of problems and the keenest critical response. It was a sound principle, as long as normal personalities were involved. There was nothing normal about the personalities of Sandys and Cherwell, or in turn about their relationship with their overlord. Sandys was an ambitious young politician of grating disposition, who had a possessively filial attitude to the Prime Minister. Cherwell, a rich bachelor scientist of exceptional intelligence, also had a possessive attitude towards Churchill. Never quite at home in England, though he was a Fellow of the Royal Society, an Oxford professor and a resident of

Devon, he seemed unable to shrug off his sense of foreignness, the product of German birth and education, burning British patriot though he was. He had attached himself to the Prime Minister during Churchill's wilderness years, adulated him personally and jealously guarded his own status as the medium through whom Churchill received scientific advice.[14] The appointment of Sandys to head the secret weapons committee touched him to the quick. Observers noted that—discreetly heterosexual though he undoubtedly was—he trembled with an almost feminine indignation at the slight. The regrettable outcome was that, because Sandys early espoused the idea that Nazi Germany was indeed developing a long-range rocket, Cherwell grasped at every strand of his extensive scientific knowledge to decry the thought: liquid fuel was unmanageable, only a solid-fuelled rocket would work, it would have to be a multi-stage monstrosity, its launch sites would be so large as to be undisguisable, it probably existed only in the minds of unreliable foreign agents. The idea, he argued in a phrase that inextinguishably attaches to his considered advice, was "a mare's nest."[15]

The secret weapons intelligence plot, between the first deliberate overflying of Peenemünde by RAF photographic intelligence aircraft in April 1943 and the arrival of the first pilotless weapon on British soil on 13 June 1944 (a flying bomb, not a rocket), was therefore bedevilled at every turn by reasoned disagreement between the parties to the investigation. To his credit, Cherwell never dismissed the feasibility of a cruise missile (the flying bomb). Indeed, he argued that, if a pilotless weapon threat existed, it was probable that it would take a cruise missile form. Because, however, agents' reports of the V-1 came later, while evidence of the rocket threat, however vague and misleading, came earlier and more plentifully, the British were both misled and caused to disagree among themselves. The disabling weakness of the German secret weapons programme was to attempt to do too much with too little; the British were further confused by the German investment in a multi-stage long-range gun (the "high-pressure pump") and a rocket-propelled anti-aircraft missile, the *Wasserfall* (Waterfall). The weakness of the British intelligence counterattack lay in an absence of practical knowledge of rocket or cruise missile technology and so a lack of clarity in their attempt to perceive what it was they were seeking to identify. The very wealth of intelligence received during April, May, June and July 1943—from agents, prisoner-of-war interrogations and air photographs—required laborious analysis but in its diversity and imprecision provided something for anyone who had taken up an intellectual position on the nature of the threat or who denied its reality.

Among those contributing were a captured German tank technology officer who co-operated so enthusiastically with his interrogators that he was appointed a British civil servant and posted to the Ministry of Supply as "Mr. Herbert." He had information on anything he was asked about including, eventually, the German secret weapons programme. He claimed that he had been involved in the development of projectiles weighing a hundred tons, launched from either a tube or a ramp. A senior officer of the Luftwaffe experimental unit, captured in April, told of his superior, Colonel Rowehl, being summoned to see Hitler at Berchtesgaden to discuss the bombardment of Britain with rockets and jet-propelled aircraft in the coming summer. Mr. Herbert, when re-interrogated, remembered that he had witnessed the launch of a sixty-ton rocket and knew of another of twenty-five tons. He mentioned the involvement of the Askania company and of Peenemünde, and other circumstantial facts, all later proved accurate.[16] During 1–5 June, four reports were received in London that substantiated, in one way or another, information on hand. They mentioned Rechlin, the Luftwaffe experimental station; Usedom, the island on which Peenemünde was located; described it as a German army, not Luftwaffe, establishment (important, because the rocket was an army weapon); and, in the last report, described the firing of three rockets, 50 to 60 feet long, from "testing pit No. 7." Large pits, clearly visible on air photographs of Peenemünde, had puzzled the interpreters. The reference was misleading, since the rockets were fired from transporter-erectors positioned in the open, but it seemed to emphasise that the British should be interested in Peenemünde.

This pot-pourri of information merely helped to harden attitudes among the investigators in Britain, not to elucidate. The positions were as follows: Duncan Sandys was fairly certain that the Germans were developing a rocket; R. V. Jones was uncertain but had an open mind; Lord Cherwell was absolutely convinced that a rocket was not technically feasible. His argument was fiercely reasonable: take-off would demand an enormous thrust; such thrust could be supplied only by an enormous charge of solid fuel inside a very large rocket; a very large rocket would need a conspicuous launch platform, either a "gun" or a large ramp; no such structures had been identified; therefore the rocket did not exist. He dismissed the notion that the Germans might be using liquid fuel—he appears not to have studied Goddard's pre-war experiments in America and to have been unaware of recent British experiments, conducted by Isaac Lubbeck for the Shell Petroleum Company—on the grounds that it would be impossible to control the flow of gases out of the rocket, which could not therefore be guided. Cherwell was to persist in this view until the contrary evidence became incontrovertible.

Meanwhile, photograph reconnaissance of "cylindrical" or "torpedo-like" objects at Peenemünde accumulated; interpretation suggested dimensions of "38 feet by eight [in diameter]," "40 feet by 4 feet thick," "35 feet long with a blunt point" (we now know the warhead had not been fitted), "a cylinder tapered at one end and provided with 3 radial fins at the other." On 23 June a photographic mission returned with film of two "torpedo-like" objects, both 38 feet long, 6 feet in diameter and with three fins. These photographs proved critical in advancing the debate about what Peenemünde threatened.[17]

Sandys summarised the evidence in a report on 28 June. "The German long-range rocket has undoubtedly reached an advanced state of development . . . frequent firings are taking place at Peenemünde." Prisoner-of-war and agent reports implied a range of 130 miles, making it likely that it would be fired from the Pas de Calais, the part of northern France nearest London. Work of a suspicious nature—in fact the construction of a large concrete bunker—had been detected at Wissant. However, the report again grossly overestimated the rocket's weight, at between 60 and 100 tons, with a warhead of up to ten tons, and was still based on the suggestion that it was solid-fuelled.

That was unfortunate, given that so much of the assessment was correct, for the continual false judgement could only lend weight to Cherwell's dismissal of such a monster rocket's existence. On 29 June the Defence Committee (Operations) of the Cabinet met in Churchill's underground command centre in Whitehall. Sandys and Cherwell were present, so was Jones, so were the Chiefs of Staff and the Prime Minister. The meeting opened with a presentation of the most recent Peenemünde photographs, described by Sandys as providing conclusive evidence of the rocket's existence. Cherwell responded by raising again his technical doubts, his warning that the signs observed might be decoys, and concluded by suggesting that if there were a secret weapon, it was probably a pilotless aircraft. Jones, asked by Churchill to comment, then disconcerted Cherwell by coming down on the side of Sandys. Thitherto he had harboured real doubts, reinforced by the deference he owed to Cherwell as one of his former Oxford pupils. Now he declared himself convinced that the rocket existed. Cherwell raised the final objection that, if so, the "flash" of its launch must have been observed, say by Swedish fishermen in Baltic waters. Since there were no reports of "flash," there could be no rocket. His objection, however, rested on his fixed belief that launch speed could be achieved only by the detonation of a large charge of cordite. The committee chose not to consider the question of whether the Germans might have overcome the difficulty of using liquid fuel. Instead it accepted the probable evidence of the rocket and made three decisions: to continue even more vigorously the examination by every means of the area of northern France within 130 miles of London; to attack rocket-launching sites in that area as soon as they were located; and to bomb Peenemünde.

The Peenemünde raid took place on the night of 17–18 August 1943. It was mounted by 433 Stirling, Halifax and Lancaster heavy bombers of RAF Bomber Command, while eight Mosquitoes staged a diversionary attack on Berlin. Earlier in the day the U.S. Eighth Air Force had bombed Schweinfurt in southern Germany. The Germans were on the alert—a low-level British code they had cracked revealed that a night raid would take place—but they expected the target to be Bremen, another north German city, or Berlin. The sky was clear, though partly obscured by clouds over Peenemünde itself. Clouds would partially disrupt the British bombing pattern. The dropping of radar-confusing foil over Denmark by the Mosquitoes on their way to Berlin would distract the German night-fighter defence.

Soon after midnight, the Pathfinders of the RAF bomber force began dropping their indicators on Peenemünde. Some fell astray, with the result that the aiming point moved southward, away from the test area at the tip of the island. One effect of the misplaced indication was to direct heavy bombing on to the camp occupied by foreign workers, killing several hundred. Nevertheless, many hits were achieved on the laboratories, the rocket factory and the scientists' housing estate. About 120 of the scientific and technical staff were killed. In the aftermath, it was decided to move the technical facilities to Kochel, in Bavaria—Peter Wegener was one of the scientists transplanted—and the manufacture of the A-4 (V-2) to a new underground Central Works in the Harz Mountains at Nordhausen. Nordhausen was to be built and operated largely by foreign labour; but the destruction of the camp at Peenemünde not only killed apparently all the foreign workers who had supplied London with secret weapon intelligence—several were Luxemburgers—but also ended the comparatively free conditions which had allowed them to communicate with British intelligence. Nevertheless, A-4 intelligence was not completely ended. The firing range was transferred to Blizna, a remote village in southern Poland, at the confluence of the Bug and Vistula Rivers, and Polish agents' reports were to keep the Crossbow Committee, as the British committee tracking the pilotless weapons' development was known, supplied with information in the coming months.[18]

Meanwhile, perturbed by reports of "long-range guns" and suggestions that mysterious buildings in both the Pas de Calais and the Cherbourg peninsula might be connected with the enemy's secret weapons programme, the Allies were led to attack a conspicuous concrete building at Watten, in the Pas de Calais, on 27 August. It was almost completely devastated by the USAAF; later it was discovered to be a rocket store, not a launch site. Other sites, including a bunker at Siracourt and a "high-pressure pump gun" battery at Mimoyecques, were to be destroyed by precision bombing in 1944.

THE IDENTIFICATION OF THE FLYING BOMB

The transfer of A-4 (V-2) rocket testing and manufacture from the Baltic to Blizna and the Harz Mountains did not end secret weapon development at Peenemünde. The flying-bomb (FZG-76 or Fi.103) installation at Peenemünde West had not been touched by the raid of 17–18 August. Tests of the pilotless aircraft continued, yielding intelligence data—collected by both neutral Swedes and combatant Danes and Poles—which reached London in increasing volume during the last months of 1943 and the spring of 1944. That was inevitable. Unlike the rocket, whose launch could be concealed, which flew beyond the speed of sound and which disintegrated on impact, the flying bomb gave all too much evidence of its existence. Flying as it did at subsonic speed and low altitude, emitting both a highly visible jet flame from its engine and a distinctive interrupted pulse beat (hence "doodle bug," as it came to be called by Londoners when it began to bombard their city after 13 June 1944), the flying bomb attracted widespread attention in the Baltic lands it passed and in Poland, where its longer flights sometimes ended. Because its terminal speed was low, it also tended to leave plentiful evidence at its crash site, if a warhead were not fitted. Indeed, the first descriptive report of a "bomb with wings" to reach England came from the Polish intelligence network in the Paris area; large numbers of Poles had settled in France after the First World War, originally to work as coal miners.[19] Some heard the news from Poland and passed it to London, where it was received in April 1943. On 23 June, SIS (MI6) received another report, from a Luxemburger employed at Peenemünde, which "described a cigar-shaped missile fired from a cubical contrivance [and] gave its range as 150 km for certain but 250 km was possible." A week or two later, SIS received, via Switzerland, a "very dirty ragged sketch plan" of the launching of a pilotless aircraft from the same Luxemburger, who had managed to escape from the Peenemünde forced labour camp (thereby probably saving his own life).[20] SIS later informed its Swiss office that the sketch "had been invaluable for the light it threw on what was going on at Peenemünde."[21]

It is significant that these reports from within occupied Europe came from non-British sources. One of the most remarkable revelations of Hinsley's exhaustive work comes on line 8 of page 125 in Volume 2, which reads, "Even had it been practicable to maintain agents in Germany . . ." This is an extraordinary admission, for it means that although Britain had two large foreign espionage organisations at work during the war years, the long-established Secret Intelligence Service (SIS or MI6) and the war-raised Special Operations Executive (SOE, devoted to subversion as well as intelligence gathering), it had no directly employed agents inside the Reich itself. Human intelligence (humint) came either from Germans, infinitesimal in number, or from foreigners working in or able to travel to Germany, particularly Poles and Czechs. Despite being occupied in 1938 and 1939, Czechoslovakia and Poland had been able to sustain, through their governments-in-exile in Britain, intelligence services inside Nazi Europe, to gather information, to run couriers and to transmit to London. Not so Britain. Its networks inside Germany—following the disastrous Venlo incident of 9 November 1939, when the

Germans entrapped two Secret Intelligence Service officers in Belgium—were "rolled up," as those in Austria had been after the Anschluss of 1939; its network in Holland had been penetrated and compromised as early as 1935.[22] Thus, by a curious reversal of relationships, Britain became the intelligence client of two politically inert governments, the Czech and Polish, exiled to London, dependent on them for certain important sorts of signal communications inside Europe, as well as for what humint they could gather. The Secret Intelligence Service did have one presumably German contact inside the *Waffenamt*, the German army's equipment office, perhaps an unwitting informer, from whom some information was passed to London through Switzerland.[23] Intelligence was also obtained more directly from the Norwegians, who were able to sustain direct sea communications with Scotland, from the Danes and from the Dutch; the SOE networks in Holland were, however, deeply penetrated by the German Abwehr. The only regular and reliable foreign intelligence from northern Europe came via the British embassy in Stockholm, but later in the war. In the earlier stages Sweden was, as it had been throughout the First World War, distinctly pro-German.

Nevertheless, humint played a significantly greater part in the identification of the V-weapons than it did in any other sector of the British intelligence war against Germany, the result of the intrinsic visibility of the object of interest, notably the flying bomb, and of Germany's dependence on foreign workers, who were inquisitive, observant and pro-Allied, often very courageously so. What today would be called "national technical means"—then taking the form of photographic reconnaissance—also played a very important part, and, marginally but significantly, Enigma again. The Luftwaffe's wretchedly poor signals security allowed Bletchley to eavesdrop on the transmissions of a specialist regiment whose duty was to monitor secret weapon flights.

From July 1943 onwards, the intelligence intake began to refer more frequently to a pilotless aircraft, forcing London to accept that both a rocket and a flying bomb were under development simultaneously; a report from an embassy stated so explicitly on 25 July.[24] In August a collection of reports suggested that a new Luftwaffe regiment, 155 (W), under Colonel Wachtel, responsible for "radio-controlled bombs," was to be deployed to France, and that it was also to control batteries of "catapults"; 155 Regiment (W) would later be identified as the flying-bomb operating unit. On 27 August some unclear photographs of the flying bomb which had landed on the Baltic island of Bornholm, but, not identified for what it was, the photos were forwarded to London by the British military attaché in Stockholm. He added that he had heard it was to be fired by "catapult" and that its engine was manufactured by the Argus company in Berlin; both pieces of information bore upon the truth.

In Britain, meanwhile, the high-level quarrel over the "rocket" continued, Lord Cherwell seeking as before to disprove its feasibility. Concern, however, was switching to the reports of the "pilotless aircraft," and with reason, since, both by British assessment and in reality, it threatened to be the secret weapon that would impact on British territory first. Photographs taken of Peenemünde on 23 June showed four small tailless aircraft on the airfield; these were later identified as prototypes of the Me 163 rocket-propelled fighter (called P30 by the British), but unease continued. On 27 August a report received from the SIS contact in the *Waffenamt* led all involved to accept that the Germans were developing a flying bomb.[25] It remained to discuss its characteristics, the threat it offered and what countermeasures might be effective.

Evidence of the flying bomb's existence now began to thicken, first from Bletchley. Enigma decrypts of 7 and 14 September were interpreted as references to a pilotless aircraft. It was by then known that two sections of the 14th Company of the Luftwaffe Experimental Signals Regiment, known respectively as Group Wachtel—after the commander of 155 Regiment (W) and Group Insect—were observing test flights at locations near Peenemünde. Their reports, decrypted at Bletchley, clearly referred to a pilotless aircraft, not a rocket, and gave its speed as between 216 and 420 mph, range 120 miles and rate of fall 6,500 feet in 40 seconds.[26] Soon afterwards, photographs taken over Peenemünde on 13 November, and a re-examination of others taken in July and September, recorded the presence of a tiny aircraft with a wingspan of twenty feet, propelled by a jet. It was designated P20—to differentiate it from the P30 (the Me 163 rocket fighter, already identified)—and accepted to be the flying bomb so long feared to be under development by the Germans.

The nature of the flying-bomb threat had already begun to gain clarity from other directions. Photographic reconnaissance over northern France, a daily undertaking as the British and Americans accelerated their preparations for the Overlord invasion, started to reveal construction work clearly not connected with coastal defence; reports from agents to SIS and SOE also drew attention to the mysterious structures. By the end of November 1943, 82 sites had been identified, 75 in the departments of Seine-inférieure and Pas de Calais, nine in the Cherbourg peninsula. They took the form of a ramp 150 feet long, curved at one end and aligned on London. Because of their shape, the British designated them "ski sites."

Sir Stafford Cripps, Minister of Aircraft Production, who took part in the Crossbow meetings, now suggested that they should be bombed on a daily basis. Bombing would be undertaken, but not until a later date. Cripps' intervention at this stage, in late October to early November, took a less helpful turn, actually diffusing rather than sharpening the focus on essentials. An extremely clever man but a barrister, not a scientist, Cripps attempted in a paper of 2 November to reconcile the differences between the Sandys and the Cherwell schools—between, that is to say, those who believed that the Germans were developing both a rocket and a flying bomb and those who denied the feasibility of a rocket but were prepared to admit the possibility of a flying bomb. Cripps' dissection of the evidence was meticulous; but, unfortunately, in his conclusions he suggested that it allowed for four, not two, possibilities, in order of priority: (1) a larger HS 293 (the glider bomb mentioned in the Oslo Report and already used by the Germans to sink the Italian battleship *Roma*); (2) pilotless aircraft; (3) a rocket smaller than the A-4 (V-2); (4) the A-4.

The Cripps report's conclusions, eccentric as they were, had the effect of drawing attention back to the Sandys point of view, in particular to a memorandum submitted by Sandys to Cripps at the outset of his enquiry. Sandys suggested that the ski sites might not, as was thought by Cripps and others, be potential rocket launching sites but might be connected with the flying bomb. By the end of November the weight of opinion at the Air Ministry (which had assumed responsibility for flying-bomb intelligence) and at the Central Interpretation Unit (where aerial photographs were examined) had moved to that view. It was confirmed on the night of 1 December 1943, when a re-examination of photographs taken recently at Peenemünde disclosed, on a ramp known to have been built in 1942, "a tiny cruciform shape set exactly on the lower end of the inclined rails—a midget aircraft actually in the position for launching."[27]

The photograph stilled argument, except from Cherwell. Though it vindicated his long-held view that, if the Germans were developing a pilotless weapon, it would turn out to be a flying bomb—he continued to deny the existence of a rocket—he still quibbled. He denied that it would carry a warhead of more than half a ton, he thought estimates of the number likely to be launched against London were exaggerated, since he doubted that the Germans could produce more than 650 automatic pilots a month (in fact the V-1 guidance system was much more crude) and so on. His was, nevertheless, an increasingly discredited voice. Official Britain now dedicated itself to three anti-flying-bomb measures. First, to establish conclusively what threat it offered. Second, to destroy the ski sites. Third, to organise an active defence, with guns, fighters, balloon barrages and any other device that would bring a V-1 down.

Technical assessment was bedevilled for some months by the belief—held by Cherwell, among others—that the flying bomb, though not an extra-atmospheric rocket, might be rocket-propelled. Then Bletchley decrypts actually perpetuated the confusion because of intercepts from Blizna, the Polish test site, referring to three different fuels, *T-stoff, Z-stoff* and E1. Only gradually did it become clear that E1 was low-grade petroleum, the fuel on which the V-1 pulse-jet actually worked, while *T-stoff* (hydrogen peroxide) was used in the launch and *Z-stoff* (sodium permanganate) was an ingredient of the rocket fuel used by the V-2, which was also test-flown from Blizna. These reports, derived from the breaking of the Luftwaffe key designated Quince at Bletchley, were decrypted during February and March 1944.

Human intelligence provided the next important advance. On 16 April the Naval Intelligence Division at the Admiralty in London received a report from the Stockholm naval attaché of the sighting, on 15 March, by the captain of a Baltic cargo ship, of two flying bombs (described as rocket projectiles) in flight. They had been fired from a shore installation ten miles away and were observed in position 54 degrees 10 minutes north, 13 degrees 46 minutes east (a little east of Peenemünde). They were the size of a small fighter, had short camouflaged wings and were "propelled at a very high speed by a rocket tube which gave approximately 300 detonations a minute" (timed by the captain's watch). Subsequent questions sent to the naval attaché elicited on 24 April the answers that a cylinder was fixed above the fuselage as a separate unit, that there were no (guidance) wires and that the noise emitted was "a series of explosions, not a continuous rumbling noise." This marvellously John Buchan observation (Richard Hannay fans would appreciate the captain's use of his watch) proved eerily accurate: the "detonations" were produced, as would later become obvious, by the automatic opening and closing of the Venetian-blind shutter which admitted air into the pulse-jet tube.[28]

At the time, though many minds must have worked on the report, the secret-weapons intelligence team failed, rather obtusely, to divine the nature of what the Swedish ship captain had seen. It took further intelligence from Sweden to reveal more. At the end of May 1944, British scientists sent to Sweden were allowed to examine two wrecked flying bombs, one dredged up from the sea bottom by the Swedish navy, another which had crashed on Swedish territory on 13 May. They also received information from the Swedes about two other crashed pilotless aircraft. Their characteristics conformed. Each was a mid-wing monoplane, with a wingspan of sixteen feet, made of steel and designed to be mass-produced, using low-grade petroleum to power an "athodyd." It was an athodyd to which Mr. Herbert had referred in his preliminary interrogation in 1943, the term describing a pulse-jet. The guidance system consisted of a rudder and two elevators, controlled by three gyroscopes, one of which worked to a compass. "It appeared to have no radio equipment." The description was accurate. The flying bomb had the crudest of guidance systems. Once launched, it flew at a predicted course and altitude, until the fuel cut-out stopped the engine at a pre-set distance, when it plunged to earth. After 8 June, when this information was passed to the Air Ministry, all important characteristics of the flying bomb were known in London, except, since the stray flying missiles lacked warheads, the weight of explosive it carried. That would soon become apparent. On 13 June the first V-1 landed on London, destroying a railway bridge in the East End. The warhead was calculated to weigh one ton.

That the flying bomb had got through was not for want of effort by the

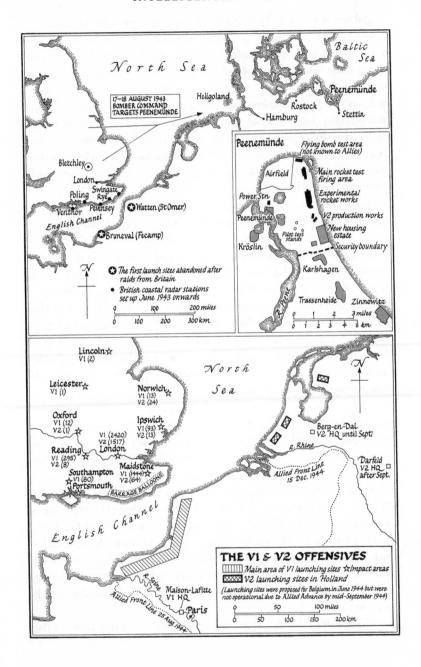

17-18 AUGUST 1943
BOMBER COMMAND
TARGETS PEENEMÜNDE

⭐ The first launch sites abandoned after raids from Britain

● British coastal radar stations set up June 1943 onwards

0 100 200 miles
0 100 200 300 km

Peenemünde

Flying bomb test area (not known to Allies)

Main rocket test firing area

Experimental rocket works

V2 production works

New housing estate

Security boundary

Pilot test stands

0 1 2 3 miles
0 1 2 3 4 5 km

Berg-en-Dal
V2 HQ until Sept.

Allied Front Line
15 Dec. 1944

Darfeld
V2 HQ after Sept.

THE V1 & V2 OFFENSIVES

▨ Main area of V1 launching sites ☆ Impact areas

▩ V2 launching sites in Holland

(Launching sites were proposed for Belgium in June 1944 but were not operational due to Allied Advance by mid-September 1944)

0 50 100 miles
0 50 100 150 200 km

Allied air forces to destroy the launching sites. The ski sites were first deliberately photographed on 3 November 1943, after a French agent reported that the contractor by whom he was employed was building structures, at eight different sites, that he could not identify. Re-examination of earlier photographs and new ones taken after 3 November revealed the existence of ninety-five such structures in northern France, and by late November their resemblance to one of the unexplained structures at Peenemünde had been noted. It was on the Peenemünde structure that the "midget aeroplane," identified on 1 December, was found by the photographic interpreters. Thereafter it was clear what the ski sites were and that they would have to be destroyed. Neither Air Marshal Harris, of Bomber Command, nor General Spaatz, of the U.S. Eighth Air Force, liked any diversion of effort from their strategic bombing offensive against German cities but higher authority insisted. On Christmas Eve 1943, 672 Flying Fortresses dropped 1,472 tons of bombs on 24 ski sites. The British and American tactical air forces continued the attacks in the spring, but in April, so concerned were the British Chiefs of Staff at the persisting threat that the strategic air forces were employed again. By 12 June 1944, the day before the flying-bomb offensive opened, 23,000 tons of bombs had been dropped, and 8,000 more on what were suspected to be storage depots; in all the tonnage exceeded by half that dropped on London during the German Blitz.[29]

The bombing proved fruitless. General Erich Heinemann, the German commander of V-1 and V-2 launch units, had always suspected that the ski sites were both too conspicuous and too vulnerable and, after the December attack, decided to abandon them and build at other places, where he laid little more than foundations, the structures to be completed from prefabricated parts at the last moment. He also gave up the large storage structures, making use of natural caves to house his stocks of secret weapons and fuel. As a deception measure, however, some repair work was done at the ski sites, while security was screwed even tighter. The result, for the Germans, was highly satisfactory. Though the sixty-six modified sites were eventually identified, by photographic reconnaissance and from agent reports, intelligence officers failed to persuade their operational superiors that attacks were deserved. It was felt that enough had been done by the bombing of the ski sites, and that the "modified sites" could be attacked if and when a pilotless weapon bombardment started. Only one attack, by fighter bombers, was launched, on 27 May.

The threat, however, had not gone away. On 10 June a Belgian source reported the passage of a hundred "rockets" by rail through Ghent towards the Franco-Belgian frontier; on the 11th new photographic intelligence revealed "much activity" at six of the modified sites, with rails being laid on ramps and buildings completed. On 12 June the British Assistant Chief of the Air Staff (Intelligence) warned the Chiefs of Staff that the Germans were making "energetic preparations to bring the pilotless aircraft sites into operation at an early date." The next day the first flying bomb fell on London. They would continue to do so until 14 January 1945.

Gun, fighter and balloon defences, deployed to a well-thought-out plan, achieved a measure of success from the outset. On 15 June, 244 flying bombs were launched but by the 21st the guns and balloons were bringing down 8 to 10 a day, fighters 30 a day, and, also because of technical faults, only about 50 a day were reaching London. By July the number had fallen to 25, by August, on one day, 14. The decrease in the number was largely due, however, to the overrunning of the launch sites as the Allied Liberation Armies advanced along the French coast. After 1 September, when Luftwaffe Regiment 155 (W) withdrew to Belgium and then Holland, most flying bombs launched against Britain were dropped from a mother aircraft. On 24 December 1944, fifty aircraft released flying bombs against Manchester, of which thirty crossed the coast and one reached the target.

It nevertheless caused more than a hundred civilian casualties. The V-1 was a highly effective weapon, hated and rightly feared by Londoners and the residents of other British cities it reached. Its approach was signalled by its distinctive pulse-jet beat, the beginning of its descent by silence, its impact by a major explosion. Because of its low terminal speed, it caused destruction over a wide area and heavy loss of life. Deaths inflicted by flying bombs are estimated at 6,184, the severely injured at 17,987. In all, 8,892 flying bombs were launched against Britain from ground ramps and 1,600 from aircraft, of which 2,419 reached the London Civil Defence Region (25 to 30 reached the ports of Southampton and Portsmouth and one Manchester) and another 1,112 landed on English soil; 3,957 were destroyed in flight, by fighters and guns about equally. Balloon barrages intercepted 231. Much of the success of the guns was due to the delivery from America of proximity fuses for their shells, devices which incorporated a miniature radar set, detonating the charge when a target was detected within effective range.[30]

THE V-2 SUCCEEDS THE V-1

By early September 1944 the British Chiefs of Staff had reason to believe that the pilotless weapon threat had passed. All launching sites within the flying bomb's known range from London had been overrun. The belief was ill-founded. The Germans would shortly turn to air launching, while Peenemünde was developing a lighter version of the V-1 that could reach England from German sites. Nevertheless, the Chiefs of Staff estimate was approximately correct. Only 235 air-launched flying bombs eluded the defenders and only 91 of the lighter model. Some caused heavy casualties; but the flying-bomb offensive was rightly judged to have passed its peak by early September.

Almost immediately, however, Peenemünde's other gift to Hitler's promise of retaliation by revenge weapons made its appearance. At 6:43 p.m. on 8 September 1944 a rocket—later known to have been launched from The Hague, in Holland—fell at Chiswick, in London, killing or seriously injuring thirteen people. A second rocket fell sixteen seconds later in Epping Forest, east of London, but caused no casualties.

The explosions might have been mistaken for those of flying bombs; but there had been no visual or radar observation of that missile, whose characteristics were by then all too well known. Moreover, very late in the day, the British intelligence establishment had at last accepted the reality of the threat. At a meeting held on 18 July, with the Prime Minister in the chair, R. V. Jones presented a paper summarising what was known of the rocket so far. The evidence consisted of reports by the Poles from Blizna of rocket firings, of similar reports from Peenemünde, of Enigma decrypts of German signals reporting flight details observed and, most tellingly, of physical evidence of a rocket misfired into Swedish territory on 13 June. Two British technical experts had been allowed to inspect the wreckage, which had subsequently been shipped to London. It was a baffling consignment, since the V-2 concerned had been used as the carrier for an experimental Waterfall anti-aircraft rocket; but the wreckage was complete enough to reveal that the rocket contained a turbo-compressor; which pointed to liquid fuel, internal guidance vanes and some radio-control equipment. Taken together, the evidence suggested that the Germans had fired between thirty and forty rockets in June and that the missile had reached a state of development "good enough, at least for a desultory bombardment of London." Churchill was furious. "We have been caught napping," he burst out, banging the table.[31]

He had some reason for displeasure. Lord Cherwell's doctrinaire dismissal of the feasibility of a rocket had caused delays; so, later on, had the tendency, typical of all intelligence bodies, for some of those involved to withhold evidence from others, on the grounds that they wished to be sure of its significance before passing it on. It was also seen, later, that not enough credibility had been attached to evidence extracted by interrogation from two prisoners of war. Eventually, too, it was realised that photographs of Peenemünde had shown rockets in their firing positions as early as 1943; they had simply not been recognised for what they were, being mistaken for "towers" connected with launching, instead of being seen as V-2s standing on end.

At the same time, it can be seen in retrospect that the British intelligence experts might be forgiven. Their fault was not obtuseness but ignorance, the result of a quite remarkable aeronautical backwardness in both Britain and the United States. Aeronautical science in both countries had achieved great success during the pre-war and war years in designing and developing highly successful fighters and bombers—of an entirely conventional type. While they, however, were building the Spitfire, Flying Fortress, Lancaster, Mosquito and P-51 Mustang, the equal or superior of their German equivalents, and the means by which Germany's cities were flattened during the strategic bombing offensive and the bomber fleets which achieved the devastation were defended, the Germans were achieving a higher and quite revolutionary level of design and development. Between 1936 and 1944 they built and flew the first practical helicopter (the Focke-Achgelis FW-61), the first turbo-jet aircraft (the Heinkel He 178), the first cruise missile (the FZG-76 or V-1) and the first extra-atmospheric rocket (the A-4 or V-2).[32] It was an astonishing achievement, largely conducted in complete secrecy. Only the small size of Germany's industrial base, compared to that of the United States, prevented it from dominating the skies during the Second World War.

Of all four achievements, helicopter, jet aircraft, cruise missile, rocket, the development of the V-2 was by far the most impressive. While Lord Cherwell, a scientist of formidable intellect, was denying that a liquid-fuelled rocket was feasible, Wernher von Braun was already perfecting his fourth model of such a missile. Having begun as a school-boy enthusiast, working entirely alone, he had engaged the support of the German army, secured funds from the German state, learnt how to generate and control huge volumes of hot gas produced by the combustion of liquids, how to insert guidance devices in the exhaust and how to moderate his rocket's rate of ascent until it could achieve a ballistic trajectory dirigible by an on-board guidance system. Between 1932 and 1942, when the first successful test firing of the A-4 was staged, it is no exaggeration to say that von Braun, still only in his thirtieth year, invented what would become both the intercontinental strategic ballistic missile and the space rocket.

Little wonder that the British scientific intelligence establishment of 1943–44, still fixed in the belief that rockets could only be propelled, and then for short distances at low speed, by solid fuel, were left to flounder among the miasma of vague agent reports, inexplicable air photographs, Enigma scraps, ill-informed prisoner-of-war interrogations which was all the intelligence machine supplied, seeking a focus. Lacking as they did the scientific and technical knowledge then enjoyed in abundance by their enemies, it is not surprising that the experts were "caught napping." They did not know, could not imagine what it was for which they were looking. They were like men from the age of the mechanical calculator striving to perceive the nature of the electronic computer.

It was very fortunate for the British that the later stages of the V-2's development proved fraught with difficulty for the Germans. Having supervised a perfect flight—with a missile not burdened by a warhead—as early as 3 October 1942, when an A-4 performed its characteristic slow-motion departure "as if being pushed by men with poles" before tilting gracefully into a ballistic trajectory and disappearing from sight at 3,000 mph, von Braun thereafter was racked with difficulty. Towards the end of 1943, when the rocket was approaching the production stage, it became apparent that between 80 and 90 per cent of firings ended in failure. Sometimes the rocket would fall back to earth at the firing point or explode when it had reached a height of only 3,000 feet, or would disintegrate on re-entering the atmosphere, or would split above the impact area, leaving the body behind while the warhead continued on course. Disintegration made assessment of the reason for failure very difficult (though providing the resistance fighters of the Polish Home Army with plentiful wreckage to forward to London; one Pole cycled 200 miles with components to reach an airstrip from which a liaison aircraft carried them back). The root of the trouble was mechanical: vibration caused breakages, particularly of electrical relays inside the rocket body. Re-entry caused shocks which broke its structure. Eventually, after 65,000 modifications, including a complete re-engineering of the nose cone containing the one-ton warhead, the rocket began to perform with reasonable consistency.

By then, however, the planned date for the inception of the "revenge" campaign had long passed. The flying-bomb operation had already been effectively defeated, by the capture of the sites from which it was launched; brilliantly effective though the V-1 was as a cheap and simple weapon, it suffered from the limitation of short range and its need for a static launching platform. The V-2 was potentially much more difficult to neutralise. Its complexity and high cost were offset by the simplicity of its launch system, which allowed it to depart from any point where a few square feet of hard surface to sustain the thrust of its exhaust gases could be built or even found. Indeed, in some respects, the Meiller-wagen, which both transported the weapon and raised it to the vertical, was as brilliant a conception as the rocket itself. Variants remain a key component of all medium-range ballistic missile systems today.

The V-2's only physical vulnerability lay in its operating crew's dependence on storage facilities and certain ancillary factories. Its production, centralised in the underground works at Nordhausen in the Harz Mountains, staffed by slave workers, was impervious to bombing; the roof was 300 feet thick. Its existence, moreover, was not discovered until late 1944, and it was recognised as an unprofitable target. When, therefore, in late July 1944 the British at last became aware of the V-2 threat, the only means available to reduce it were seen to be bombing attacks on the rocket storage centres and on the facilities producing its key constituents, particularly liquid oxygen. By then, however, most of the "large sites," as they were known to British intelligence, had already been heavily bombed and severely damaged; in any case, loss of bases in France and Belgium soon forced the V-2 units back into Holland, where, as General Dornberger had always wished, they operated from improvised sites supplied from centres deeper inside Germany on a hand-to-mouth basis. The German retreat forced the early abandonment of the two main liquid oxygen supply sources, at Liège in Belgium and Wittringen in the Saar, reducing the Germans to dependence on smaller sites difficult to identify and locate.[33]

Those surviving supplied only just enough fuel to keep up the V-2 bombardment. It sufficed, nonetheless, to sustain a diminishing delivery of rockets until 27 March 1945. What permitted the V-2 firing batteries' survival was their brilliantly simple method of operation. It took only fifteen minutes to position the Meillerwagen—it might be in a suburban street—and push the missile erect. Tanker trucks then fuelled it, while a mobile generator was linked by cable to supply power. The crew then took cover in a slit trench which had been dug while installation took place. The command team, in an armoured vehicle, finally initiated the launch procedure; fifty-four seconds after ignition the site was ready for evacuation. An hour after arrival the Meillerwagen could be on its way. Little surprise that, though the Allied Expeditionary Air Forces flew missions to catch the V-2 teams in the act, none was successful.

The only other means considered to reduce the weight of the V-2 attack was the use of deception, by a variation, much celebrated in the post-war literature of espionage, of the management of human intelligence. Throughout the war, the Germans infiltrated agents into Britain: about 70 before 1940, some being "agents in place" before war broke out; another 220 arrived during the war, of whom 120 were intended to make their way to other countries. The Germans were able to organise infiltration at this level because a steady stream of escapees from occupied Europe, 7,000 to 9,000 a year, the vast majority seeking to join their own armies-in-exile, reached Britain by one means of emigration or another.[34]

The British counter-espionage services collared almost every agent who arrived hidden among this annual stream. Only three are known to have eluded detection and only five others, once identified, refused to confess. Out of those apprehended, the British were able to form a body of double agents, some of whom had radio sets supplied by the Abwehr or other means of communication with base; one at least, code-named Tricycle by the British, was so trusted that he was allowed to journey to Lisbon during the war to consult his German controllers. Until 1944 the double agents did little more than feed German wishful thinking with reports of depressed morale in the besieged island, though some assisted preparations for D-Day by reinforcing German belief in false orders of battle. As with almost all human intelligence operations at low level, the day-to-day reporting of the double agents was of banal, mundane detail. It differed little from the material fed to controllers by the war's numerous intelligence fraudsters who, operating on their own account for one side or the other or both, had correctly detected in the appetite of intelligence organisations for information of any sort the means of making a living. *Whitaker's Almanac,* the *Encyclopaedia Britannica,* old newspapers, the BBC World Service—all grist to the world of such fantasists' "product." Eminent among the crew was the man code-named Garbo by the British, who set out to sell himself to the Germans as a pro-Nazi British resident. Operating at the outset exclusively from Lisbon, he assured his Abwehr controllers that "there are men here in Glasgow who would do anything for a litre of wine." Then, transferring his double loyalties to the British, he arrived in his notional operational area to set up a network of twenty-seven completely fictitious agents, whose expenses, paid to him in cash by the Abwehr, eventually amounted to £31,000. At the end of the war he was appointed a Member of the Order of the British Empire and retired into private life unharmed by the enforcement agencies of either of his paymasters.[35]

Garbo (in real life Juan Piyol Garcia, a Spanish citizen) supplied the Germans, when he was not allegedly organising the sabotage activities of Welsh nationalist fanatics, with enormous quantities of information, all carefully distorted by his British controllers, about the domestic affairs of the United Kingdom under German attack. It was absolutely natural, therefore, that the Germans should turn to him for first-hand information on the effects of the V-weapon campaign. The approach led to one of the most troubled passages in British intelligence management during the war. As early as January 1941, the various intelligence authorities, by then persuaded, correctly, that there were no German agents left at liberty and at work within the United Kingdom, decided to set up an organisation (soon known as the Twenty Committee, after the appropriate Roman numeral, XX, or double cross) which would deceive the German masters of controlled agents—whether they had been captured and turned or had deliberately turned themselves in—by relaying falsified information. The aim, as defined by J. A. Marriott, one of the directors of the Twenty Committee, would be to supply the Germans with "so much inaccurate information that the intelligence reports furnished by the Abwehr to the German High Command based on that information would themselves be misleading and wrong."[36]

What ensued verifies the accuracy—in spirit if not exactness—of anything the best writers of spy fiction, John le Carré foremost, portray about the workings of the organisations they describe. The inner ring of the Twenty Committee's operatives included, beside Garbo and Tricycle, who were exceptional, Brutus, a Polish air force officer escapee, and Mutt and Jeff, two Norwegian escapees who had reached Britain by boat, all three real people infiltrated under Abwehr auspices, but also the entirely fictitious Mullet and Puppet, venal British businessmen, Balloon, an army officer embittered by dismissal from the service, and Bronx and Gelatine, two ladies with friends respectively in the Foreign Office and the armed forces. The latters' friends were extraordinarily lax about the safeguarding of official papers, whose contents duly made their way back to Berlin, but all of the Twenty Committee's people did their bit. Most of what they communicated was "chicken feed"—true but useless tittle-tattle—but the identification, particularly by Garbo, of nonexistent Allied divisions before D-Day contributed significantly to the misappreciation by the German Foreign Armies West office of where the Allies intended to land.

It was with the appearance of the V-weapons, however, that the double-cross system achieved its most sophisticated effects, though at the price of heavy heart-searching at the highest level of British government. Soon after the flying bombs began to fall, it was realised that, by relaying false reports of their accuracy, it would be possible to persuade the Germans to shorten their range, thus shifting the Mean Point of Impact (MPI) south and east of Tower Bridge, believed to be the Germans' chosen MPI, towards the open countryside on the fringe of London. The Germans themselves were eager for news of where the bombs fell and had actually ordered Garbo, before the campaign began, to leave London so that he could report damage in safety. He was sent details of a reporting technique—Brutus and the fictitious Tate were similarly instructed—which would relate the explosion of the pilotless weapons to time of impact. The British realised that by giving details of flying-bomb arrivals correct as to time but wrong as to place—too far to the north or west—they could cause the Germans to shift the MPI away from London's crowded centre towards its less densely populated suburbs, so diminishing both casualties and destruction. The policy was hotly debated at Cabinet level, where allegations of "playing God" were levelled, but prevailed. It was continued during the V-2 rocket offensive and it seems to have had an effect. Ironically, during the flying-bomb offensive, the Germans were chiefly misled not by double-cross agents but by a man operating on his own account, known to the British as Ostro, located in Madrid and selling "facts" to the Germans based on newspaper reports and his own imagination. Among his achievements was a report (believed) of the destruction of Big Ben; Ostro was a self-creation of whom any spy novelist would have been proud. His information appeared to be confirmed by a Luftwaffe photographic reconnaissance flight of 6 September 1944, the first flown since January 1941, which allowed Flak Regiment 155 (W), the flying-bomb launching unit, to take credit for all the damage revealed.[37]

The double-cross system does seem to have shifted the MPI of the V-2 rockets away from the aiming point. Treasure, a mysterious double-agent under the Twenty Committee's control, sent reports during October which appear to have persuaded the Germans to shift the MPI away from east-central London down the line of the Thames estuary. The British official historian concludes that, but for that alteration, "1,300 more people would probably have been killed, 10,000 more injured and 23,000 houses damaged—to say nothing of the disruption to the economy and disruption of the country which would have resulted from the concentration of destruction which the Germans believed they were achieving between Westminster and the docks."[38]

So "spying," in the popular sense of the term, did at least play its part in British resistance to the V-weapons campaign. It was only one part, however, of a remarkably varied intelligence effort, including not only humint of several distinct sorts—the anonymous treachery of the Oslo Report, double-agency, resistance reporting and direct espionage—but also what today is called "national technical means," in the form of photographic reconnaissance, as well as sigint and a great deal of theoretical analysis.

To assess the relative importance of the different elements of the intelligence plot is not easy. Clearly the significance of the Oslo Report was very great; though less than specific in many aspects, it did indicate the trend of German military scientific research—towards, in particular, guided and pilotless weapons—and what it did not mention was as significant as what it did. It included, for example, no suggestion that Nazi

Germany was attempting to develop nuclear weapons, and given the belated, weak and diffuse espousal of a nuclear programme by the Nazi state, it was accurate in that respect. On the other hand, the Oslo Report, after arousing initial interest, was almost forgotten, for several years after 1939. Only when other intelligence, received at the end of 1942, alerted the British to rumours of pilotless weapon development was the report resurrected. Perhaps the most valuable clue it then provided was the reference to Peenemünde as a testing site.

What precipitated the search for harder evidence of pilotless weapons was, so we must believe from the official account, a piece of pure John Buchanism. Readers of *The Thirty-Nine Steps* will remember how Scudder, the hunted American, tells Richard Hannay how he detected the evil behind the Black Stone. "I got my first hint in an inn on the Achensee in Tyrol. That set me enquiring and I collected my other clues in a fur-shop in the Galician quarter of Buda, in a Stranger's Club in Vienna, and in a little bookshop off the Racknitz-Strasse in Leipsic. I completed my evidence ten days ago in Paris." According to the British official history, the first substantial report of "rockets" was received, on 18 December 1942, by SIS (MI6) from a chemical engineer "who was travelling extensively on his firm's business." Neither he nor his nationality is identified, but he had apparently overheard "a conversation in a Berlin restaurant between a Professor Fauner [a Professor Forner was known to exist] of the Berlin Technische Hochschule and an engineer, Stefan Szenassy."[39] They discussed a rocket carrying five tons of explosive with a range of 200 kilometres. The chemical engineer, under SIS prompting, produced two more reports of the rocket's characteristics, including the detail that it was tested at Swinemünde (near Peenemünde).

The "chemical engineer" then disappears from the record, reinforcing the suspicion, held by numbers of historians of Nazi Germany, that the Nazi state contained more well-placed sympathisers with the Allied cause than the danger of discovery by the Gestapo would suggest; there are numbers of "de-Nazifications" after 1945, restorations to public position, restitution of fortune that are otherwise inexplicable.[40] The chemical engineer deserved whatever reward events subsequently brought, for the convenient juxtaposition of his table with Professor Fauner's in the Berlin restaurant in December 1942 led directly to the decision to subject Peenemünde to close photographic reconnaissance, to repeated overflights and thus to the great bombing raid of 17–18 August 1943.

It has been suggested that the Peenemünde raid should have been staged earlier, should have been better organised or should have been repeated. Counsels of perfection: not until mid-1943 was the photographic evidence clear enough to identify the site as the centre of the German secret weapons programme (there were other candidate sites, at Kummersdorf and Rechlin). The raid, though missing parts of the site through failures of marking, all too familiar to Bomber Command, did terrible damage and resulted in the transfer of much of the research and production programme to other, more remote or less vulnerable sites, in central and southern Germany and deep within Poland. To repeat the operation, even had Peenemünde remained a "target-rich" objective, would have been very costly. That of 17–18 August 1943 brought the loss of 40 aircraft, out of 600, an attrition rate of 7 per cent, considerably higher than was deemed "acceptable" by Bomber Command.

Thereafter, there was little that the intelligence services could do. They had identified the threat and had directed the offence to the point of danger. Once the Germans, in reaction to the great Peenemünde raid, had removed the substance of the secret weapons programme from harm's way, the intelligence services could only, after an effort to predict the date of the opening of the campaign, attempt to put the Germans off their aim. In that, through the double-cross system, they had a certain success.

Honours in the V-weapons campaign, if that word can be used about a method of making war on civilians, go to the Germans. Both the V-1, the first cruise missile, and the V-2, the direct technical ancestor of all extra-atmospheric missiles and of the space rockets, were far in advance of any aeronautical weapon produced by their enemies in 1939–45. Wernher von Braun, who was to become an American citizen and to be celebrated as "the father of the space programme," was a scientific genius. The men who produced the V-1 were aeronautical technicians of the first class. Had Hitler had the vision to devote a proportion of Germany's scientific effort similar to that given to other weapon programmes to nuclear weapons, it is possible that, with the V-weapons, he could have won the war. The Nazi nuclear research programme was dissipated between too many competing research organisations. There was no Dornberger, no von Braun, no Peenemünde and never enough money.[41] The world, nevertheless, had a very narrow escape.

Military Intelligence Since 1945

MILITARY OPERATIONS have changed greatly since the end of the Second World War, most of all because the development of nuclear weapons has effectively prevented the major states from fighting the sort of full-scale struggles for decision which are the subject of this book. Big wars are now too dangerous for big countries to fight. That does not mean the world has become a safer place for the common man. On the contrary. It is estimated that armed conflict since 1945 has killed fifty million people, as many as died in the Second World War. Most of the victims, however, have perished in small-scale, random struggles, many scarcely to be dignified even by the name of civil war. In the last fifty years it is not the methods or weapons of 1939–45 that have harvested the major proportion of violent deaths—aerial bombardment or battles between great tank armies or the relentless grind of infantry attrition—but skirmish and all too often massacre with cheap small arms.

Even in such few major wars as have been fought, there have been few large-scale conventional battles and their number has tended to decline over time. Thus, while the Korean War of 1950–53 was almost exclusively a conflict of infantry and tank armies, and the Arab–Israeli wars of 1956–73 likewise, the biggest war of all, in Vietnam, was a protracted counter-insurgency struggle, marked by the clash of armies scarcely at all. Though the Iran–Iraq War of 1980–88 saw much heavy fighting, Iran's lack of heavy equipment and use of underage conscripts in suicide attacks made it an unequal contest bearing little resemblance to other wars of the twentieth century. In 1991 Iraq was forced to abandon its illegal occupation of Kuwait as a result of defeat in one major tank battle; but its army, more concerned to surrender than to stand its ground, cannot really be said to have given battle at all. The same can be said of its performance in the second Gulf War of 2003, in which intelligence played an important role in the targeting early on of the Iraq leadership.

That episode apart, the post-war military record yields few examples of outcomes being influenced by operational intelligence of the sort assessed in the previous chapters. Intelligence services have never been busier than they are in the nuclear world and consume more money than has ever before been spent. By far the greater proportion both of effort and funds is devoted, however, to early warning and to listening, continuous processes, intended to sustain security, not to achieve success in specific or short-term circumstances. The elaborate infrastructure of early warning—radar stations, underwater sensors, space satellite systems, radio interception towers—is enormously expensive to build, maintain and operate and so are its mobile auxiliaries, particularly airborne surveillance squadrons. The intelligence material thus collected, categorised by professionals as sigint (signals intelligence), overlapping with comint (communications intelligence) and elint (electronic intelligence), requires processing and interpretation by thousands of analysts and computer technicians. What they do and what they achieve is rarely published. The public anyhow seems indifferent to what is unquestionably the most significant sector of contemporary intelligence activity. Understandably, the complexities of intelligence technique must baffle even highly educated laymen. Only the most specialist of experts can hope to comprehend what intelligence agencies now do. It is possible, with application, for the interested general reader to follow descriptions of how the Enigma machine worked and of how the problems it presented to cryptanalysts were overcome. Modern ciphers, created through the application of enormous prime numbers to language, belong in the realm of the highest mathematics and are alleged to defy attack by even the most powerful computers yet built.

It is not surprising, therefore, that the intelligence world attracts attention only when there is a breach of security, typically in recent years by the "defection in place" of an intelligence operative who yields to greed or lust or exhibits defects of character not identified at the time of recruitment. There has been a steady trickle of such scandals, long post-dating the sensational unmasking of the "Cambridge" spies in Britain and affecting the American and Soviet services which were presumed to have been warned against such occurrences in their own ranks by the "Third" and "Fifth" Man episodes.

Public interest is also engaged by accounts of the effect of human intelligence, humint, on recent or current military operations, where such effect can be shown. Humint has unquestionably played a major part in Israel's successful efforts to hold at bay its Arab neighbours in four major wars, much minor conflict and its continuous struggle for security, for the ingathering of Jews from neighbouring lands allowed its intelligence services to recruit patriotic operatives who spoke Arabic bilingually and were able to pass as natives in their countries of former residence. It is understandable that the successes of Israeli humint remain almost completely secret. During the Vietnam War the American CIA conducted a large-scale campaign of destabilisation against the Viet Cong, largely by the targeted assassination of Viet Cong leaders in the South Vietnamese villages. Operation Phoenix remains unacknowledged; the Vietnam War was eventually lost; it would nevertheless be illuminating to know what effect Phoenix had on its conduct.

The only conventional military conflict of recent times for which a reasonably complete picture of the influence of intelligence on operations is available in all or most of its complexity—signit, elint, comint, humint and photographic or imaging intelligence—is the Falklands War of 1982, between Britain and Argentina. Rights of sovereignty over the Atlantic islands of the Falklands or Malvinas, which include such Antarctic outliers as South Georgia, Graham Land and the South Shetland, Orkney and Sandwich groups, has been disputed between Britain and Argentina since the nineteenth century. The small Falklands population is exclusively British (the other territories are effectively uninhabited) but it is a universal and deeply held belief in Argentina that the lands are theirs. Argentina has a troubled political history. Once a country of great wealth, which attracted to it over the last century large numbers of immigrants, including poor Italians seeking a better life outside Europe and an English minority who came to supply its commercial and professional class, Argentina suffered serious economic decline in the mid-twentieth century. Discontent brought to power a populist Peronist regime, so called after Colonel Juan Peron, its leader. Peronist mismanagement provoked a military coup in the 1970s. When the military junta itself became unpopular, it decided to restore its fortunes by reviving the claim to the Falklands. Recovering the Malvinas was a cause around which all Argentinians could unite.

Britain was long used to Argentina's Falklands demands. It did not take their revival in 1981–82 very seriously. Negotiations proceeded at the United Nations in New York: they were not marked by urgency, and the British found the Argentinians in reasonable mood. Unknown to Britain, however, the junta, led by General Leopoldo Galtieri, had already decided to mount an invasion at latest by October of 1982, when it was calculated that the only Royal Naval ship on station, the ice patrol vessel *Endurance*, long scheduled for retirement, would have been withdrawn. As late as March 1982, no military preparations had been made and no diplomatic crisis appeared to impend. Then what seems a chance factor altered the tempo. An Argentinian scrap reclamation party arrived at Leith in South Georgia, the Falklands dependency, declaring it was there to dismantle an old whaling station. The scrap men raised the Argentinian flag but failed to seek permission for their work from the local station of the British Antarctic Survey, the government authority. When visited, they hauled down the flag but did not regularise their presence. Constantino Davidoff, their leader, denied then and afterwards that he was sponsored by the Argentinian navy, but he is believed to have had a meeting with naval officers before landing. Once he was ashore, the British Foreign Office felt it had to act; the Ministry of Defence was more reluctant, since it regarded operations 8,000 miles from home as beyond its capabilities. Under Foreign Office pressure, a case was made to the Prime Minister, Mrs. Margaret Thatcher, who ordered *Endurance*, with a party of marines from Port Stanley, the Falklands capital, to sail for South Georgia and to await orders.

The unexpected despatch of *Endurance* perturbed the junta. If the scrap men were removed, Argentinian prestige would be damaged; but the presence of *Endurance* challenged it to military action, which it did not plan to take for several months. The Argentinians wavered, first sending a naval ship to take off most of the scrap men, then sending another with a party of Argentinian marines to "protect" those left. It was the turn of the British government to dither. It sought guidance from its own and the American intelligence services as to what Argentina intended. The signs were unclear. Budgetary economics had run down the Secret Intelligence Service (MI6) station in Buenos Aires; what signal information could be supplied by Government Communications Headquarters (GCHQ), by the American Central Intelligence Agency (CIA) and by its sister signals organisation, the National Security Agency (NSA), did not clarify the picture. The British agencies enjoyed a warm and co-operative relationship with the American agencies, based on much exchange of mutually useful material; but the CIA depended on MI6 for human intelligence, while both GCHQ and the NSA were confused by the volume of radio traffic suddenly generated in the South Atlantic by Argentinian but also Chilean vessels; the two navies were conducting a large-scale routine exercise.

Britain fell into a week-long bout of indecision; it had decided it could not tolerate any further Argentinian intervention in the affairs of its South Atlantic dependencies; but it shrank from any overt measure that would provoke Argentina to action. Eventually, the decision was taken out of its hands. On 26 March, the junta, under pressure from street demonstrations against its economic austerity programme, but even more fearful of public reaction if it appeared to back down before British diplomatic protest over the South Georgia affair, decided to advance the timetable for its invasion of the Falklands and launch the operation at once.

The Falklands were effectively undefended. Of their population of 1,800, 120 of the men belonged to the Falklands Islands Defence Force, but they were untrained and equipped only with small arms. An official British military presence was provided by Naval Party 8901, a detachment of forty Royal Marines; their number had recently been doubled by the arrival of their reliefs. Apart from *Endurance,* currently in Antarctica, there were no naval ships in the Southern Hemisphere. The Argentine armada, which began to land at dawn on 2 April, could not therefore be repelled, though it was briefly opposed. Naval Party 8901, depleted by the despatch of twelve men to reinforce South Georgia, was ordered by the governor, Sir Rex Hunt, who had been warned by London that an invasion force was at sea, to guard the airfield and the harbour. When an advance party of 150 Argentinian commandos landed, they were engaged, and in a firefight around Government House, two were killed. It was clear to Sir Rex Hunt, however, that resistance was hopeless, and after two hours, he ordered surrender. Soon afterwards the vanguard of

12,000 Argentinian troops began to land, while the Argentinian air force took control of the airfield.

The news caused an immediate and major political crisis in London. The 2nd of April was a Friday; an emergency session of Parliament, which never sits at the weekend, was called for the following day. The consensus at Westminster was that if the government could not demonstrate its willingness and ability to confront the Argentinians, it would have to resign. Fortunately for Mrs. Thatcher, a woman of iron will but untried powers of decision, she had already instituted precautionary measures. Alerted by the enormous volume of radio traffic generated by Argentinian preparations, she had ordered a submarine to sail for the South Atlantic on the previous Monday, 29 March. Much more important, indeed, as was to prove critically for the whole Falklands saga, she had on Wednesday evening ordered that a naval and military task force should be assembled to depart at once for the South Atlantic. Her desire to recapture the Falklands was never in doubt; the impetus to the decision was supplied by the arrival in her room in the House of Commons when she was consulting her ministers of the First Sea Lord, Admiral Sir Henry Leach, who gave it as his professional opinion that Britain had the power to mount such an operation and that the navy could set out by the coming weekend. He also assured the Prime Minister of victory. On return to his office he sent a signal: "The task force is to be made ready and sailed."

Its first elements departed on Monday 5 April, while its military complement was hastily assembled in Britain to follow. Three submarines, two nuclear-powered, one diesel, formed the spearhead; there were to follow, over the course of the weeks to come, 2 aircraft carriers, embarking 20 Harrier aircraft and 23 helicopters, 23 destroyers and frigates, 2 amphibious ships, 6 landing ships, 75 transports, ranging in size from large passenger liners to trawlers, and 21 tankers. The majority of the transports and tankers were "taken up from trade"—chartered or requisitioned, that is, from the merchant service.

The troops to be embarked would eventually comprise the whole of 3 Commando Brigade (40, 42, and 45 Commando, Royal Marines, 29 Commando Regiment Royal Artillery and 59 Commando Squadron Royal Engineers), attached to which were 2nd and 3rd Battalions, the Parachute Regiment, two troops of light armoured vehicles of the Blues and Royals, thirteen air defence troops, the commando logistic regiment

and the brigade's helicopter squadron. There was also a large complement of Special Forces, including three sections of the Special Boat
Squadron (SBS) and two squadrons of the Special Air Service (SAS). To
follow later was 5 Infantry Brigade (2nd Scots Guards, 1st Welsh Guards
and 1st/7th Gurkha Rifles) with some artillery and helicopters. The
Royal Air Force deployed elements of seventeen squadrons, flying fighters, bombers, helicopters, reconnaissance aircraft and air refuelling
tankers.

Refuelling, in the air and at sea, was an essential requirement, for the
task force was to operate without a land base nearer than Ascension
Island in the middle of the Atlantic. Until the airfield at Port Stanley
could be recaptured, air refuelling was less vital, for long flights over the
ocean could not be numerous. All fuel, and other supplies to the warships, however, had to be transferred ship-to-ship while under way.

The assembly of the task force was a race against time, not only
because of the need to confront the Argentinians with an armed response
as rapidly as possible but also because of the season; the onset of the
South Atlantic winter at the end of June would bring sub-Arctic weather
necessitating withdrawal from the region. Everything, from completing
dockyard maintenance to supplying the soldiers with warm clothing,
had to be done at the highest speed; at the outset it seemed that many
requirements could not be met.

It was not only the pace of material preparation that had to be forced;
so too did that of planning and intelligence gathering. The two were
intimately connected and interdependent. Britain had no base in
the region and no allies. Chile, long on bad terms with its Argentinian neighbour, was disposed to be helpful but could not risk openly
siding with Britain; most other South American countries supported
Argentina's claim to the Falklands, if only out of regional solidarity.
How was the campaign to be fought? Clearly there must be an amphibious landing but it would have to be launched from the task force's ships,
not from land. That required the navy to close up to the islands, at least
while the troops got ashore, but also to remain nearby during daylight so
that the carrier aircraft could provide support. Worryingly, the islands,
though 400 miles from the nearest stretch of Argentinian coast, were
just not far enough offshore to lie outside the range of the enemy's
land-based aircraft. The troops, once landed, would be vulnerable to air
attack. Far more worryingly, the warships and transports would also be

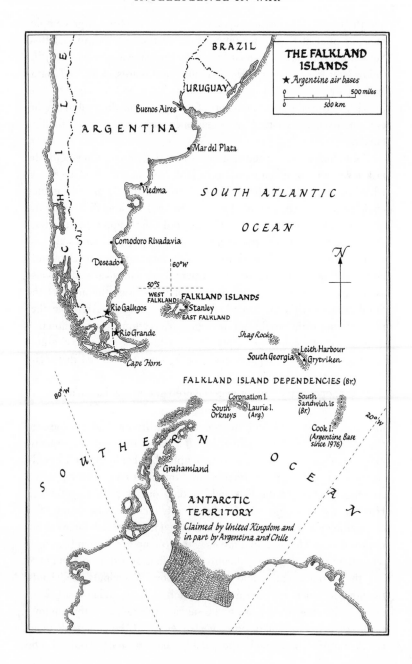

THE FALKLAND ISLANDS

★ *Argentine air bases*

500 miles
500 km

BRAZIL

URUGUAY

Buenos Aires

ARGENTINA

C H I L E

Mar del Plata

Viedma

SOUTH ATLANTIC

OCEAN

Comodoro Rivadavia

Deseado

60°W

50°S

WEST
FALKLAND

FALKLAND ISLANDS

★ Stanley

EAST FALKLAND

Rio Gallegos

Rio Grande

Cape Horn

Shag Rocks

Leith Harbour
South Georgia Grytviken

FALKLAND ISLAND DEPENDENCIES (Br.)

80°W

Coronation I.

South
Orkneys

Laurie I.
(Arg.)

South
Sandwich Is
(Br.)

Cook I.
(Argentine Base
since 1976)

20°W

S O U T H E R N   O C E A N

Grahamland

ANTARCTIC
TERRITORY

*Claimed by United Kingdom and
in part by Argentina and Chile*

N

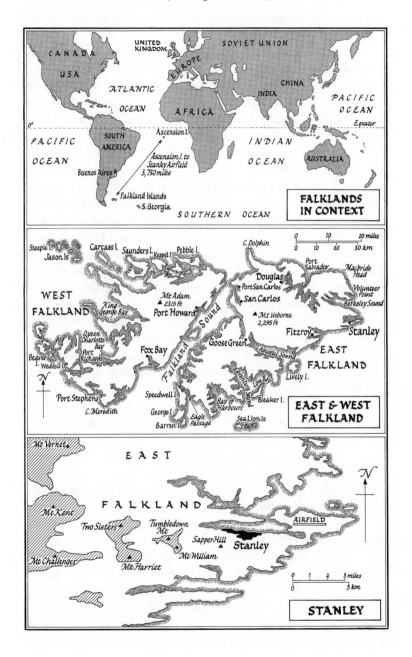

FALKLANDS IN CONTEXT

EAST & WEST FALKLAND

STANLEY

at risk, except at night, when they could stand off to the east into the broad expanse of the ocean.

How serious was the risk? That proved, both at the outset of the campaign and during its development, an embarrassingly difficult question to answer. No one in Britain really knew; no one, indeed, knew anything much that was useful about Argentina's armed forces. For reasons of economy, the Secret Intelligence Service (MI6) had closed down all but one of its stations in South America; that remaining was in Buenos Aires but its chief was too overworked to collect anything but political intelligence. The service attachés, navy, army and air, were supposed to report on their Argentinian opposite numbers; but in recent years they were more often required to act as salesmen for the British defence industries, so the excuse went afterwards; in practice, attaché appointments were final postings at the end of a middling officer's career, a farewell present for an unexceptional life. This was not particular to Argentina but the general rule; only those officers posted to the Soviet Union had the duty of acquiring intelligence and were fitted by ability and training to do so.

Yet the collection of pertinent information in any reasonably open society, which Argentina was, is not difficult and need not conflict with diplomatic propriety. Readily available service magazines contain valuable snippets of information which, if collated, quickly yield an order of battle; so do local newspapers, from stories about local men in uniform and the social affairs of locally stationed units. Service histories are also fruitful sources; units tend to occupy the same barracks for decades. Armies and navies are relatively unchanging organisations and, to anyone who takes the trouble to form a picture of their organisation, rarely conceal secrets about their location, strength or function requiring specialised intelligence scrutiny to uncover.

The archives of the Defence Intelligence Service in London ought, in short, to have contained copious and detailed reports on the Argentinian navy, army and air force in April 1982. They did not. The cupboard was almost bare. The officers of the task force have in consequence left a record of a shaming and hurried search in public libraries for such standard works as *Jane's Fighting Ships* and the Institute for Strategic Studies' *Military Balance*. Little was to be found. The *Military Balance* allots no more than two or three pages to a country the size of Argentina; *Jane's Fighting Ships* is largely a photographic album. Moreover, as the most important of Argentina's warships, the carrier *Veinticinco de Mayo*, was the ex-British HMS *Venerable,* venerable indeed since launched in 1943, and three of its largest destroyers were British-built or designed, *Jane's* could tell little the British did not know already. The marines and soldiers scanning the *Military Balance* must have been even more disheartened. It lists the barest information of numbers of units and quantities of equipment and those in separate sections; no picture of units' capabilities is discernible, therefore, while units are not named nor are their peacetime locations specified. That omission may have been seriously misleading in the frenzied days of early April 1982. The Argentinian army's three best formations were the VI, VIII and XI Mountain Brigades (Peron, incidentally, was a mountain infantry officer), which, by reason of their training and familiarity with cold climate, seemed the obvious choice for Falklands duty. Because of the junta's fear that Chile might profit from their commitment to the Falklands to strengthen its position in the disputed Cape Horn region, however, it had left the mountain brigades in their peacetime stations and decided to employ lower-grade formations drawn from the warm borders of Uruguay. GCHQ is known to have been intercepting the mountain brigades' radio traffic, confirming that they were still located in the far south even as the invasion fleet put to sea. The task force officers, apparently dependent wholly on scantily published information about the location and capability of their potential opponents, did not even know that.

The navy was quite as badly informed. Admiral Sandy Woodward, commanding the warships and transports aboard the old carrier *Hermes,* had a general picture of the risk he faced. It consisted of three elements: attack by land-based Argentinian aircraft, some of which were equipped to launch Exocet, the French-supplied sea-skimming missile (also aboard some of Woodward's ships), which was difficult to distract by electronic countermeasure and deadly if it struck home; the Argentinian surface fleet, known from radio intercepts to be at sea and organised in two groups formed respectively around the *Veinticinco de Mayo* and the ex-American heavy cruiser *Belgrano,* apparently deployed to mount a pincer movement; and Argentinian submarines. The diesel-propelled submarines were known to be difficult to detect but, it was believed, could be held at bay by the British nuclear submarines in the area; the surface fleet had been warned not to enter an "exclusion zone" proclaimed around the islands by Britain and would be attacked if it did (it did not and was attacked anyhow, by HM Submarine *Conqueror,* and *Bel-*

grano sunk); it was hoped to overcome the Exocet threat by positioning destroyers and frigates as radar pickets between the islands and Argentina to provide early warning and to distract any missiles that got through by firing "chaff," which simulated a larger target than the threatened ship.

In practice the two Argentinian diesel submarines did not manage to attack the task force; the surface fleet, partially incapacitated by equipment failure aboard the *Veinticinco de Mayo,* turned back from the exclusion zone and returned to port after the sinking of the *Belgrano.* The Exocet aircraft, by contrast, inflicted heavy damage on the task force and, with others delivering more conventional ordnance, came close to achieving a naval victory that would have secured the Falklands and humiliated Britain for decades to come.

The Argentinian air-launched Exocet, a modified version of the maritime model, known as the AM-39, was mounted on a Super Etendard aircraft, supplied by France, like the missile itself. The British believed correctly that Argentina had only five AM-39s, but wrongly that it had only one Super Etendard; the right number was five. As important as the aircraft–missile combination was the maritime reconnaissance aircraft that alerted the Super Etendards at their Rio Grande base to the presence of the task force within attack range. An antiquated American aeroplane, the SP-2H Neptune, it possessed the capability to linger beyond the horizon formed by the earth's curvature but to keep the British under radar surveillance by bobbing up over it at regular intervals. The Super Etendards, when vectored towards the target, flew at sea level, beneath British radar, until close enough for the Exocet to strike. The pilots needed to gain altitude only once or twice, and then briefly, for their own radars to acquire their targets and automatically programme the missiles to depart in the correct direction. Once launched the Exocet maintained height just above sea level by an on-board altimeter and finally homed on the target ship down the beam of its own radar.

Admiral Woodward and his staff had been wrongly informed that the Super Etendards' range was only 425 miles, too short to reach the task force east of the islands. In fact, by refuelling from one of Argentina's two KC-130 tankers, they could achieve launch positions. On 4 May, two days after the sinking of the *Belgrano,* two Super Etendards, flying from Rio Grande, approached the task force; their directing Neptune had been spotted by British radar but was thought to be searching for *Belgrano* survivors. *Glasgow* and *Coventry*, deployed as radar pickets west of the task force, caught echoes of the attacking aircraft as they rose above the horizon to correct their final approach paths. The British ships fired chaff and both Exocets, travelling only six feet above the sea, were deflected by their own course-corrections. *Sheffield*, twenty miles distant, was currently transmitting on its radio link to satellite, which prevented its hearing the warnings transmitted by its sister ships or operating its own radar. Its crew were therefore oblivious of impending danger and neither fired chaff nor manoeuvred. She was hit in the forward engine room by one of the Exocets which, though its warhead failed to explode, started a fire that eventually forced her abandonment, after heavy loss of life.

The manifestation of the Exocet threat was to exert a decisive effect both on the management of the campaign and on the intelligence effort that underlay it. Admiral Woodward at once withdrew the task force far to the east of the islands, where it was to remain until the landings began on 21 May. At the same time the Northwood joint services headquarters, from which Operation Corporate, as the campaign was code-named, was directed, began a frenzied search for means to improve intelligence collection and to strike directly at the Argentinian air menace. Of signal intelligence there was no shortage; the Argentinian army, navy and air force generated a large volume of traffic, which was intercepted not only by GCHQ, through its intercept station at Two Boats on Ascension Island, ostensibly a branch of the Cable and Wireless Company, but by the NSA, the American intelligence community having decided to lend its British partners full support at this time of need, and by a New Zealand intercept station at Waiouru.[1] The United States was also generous with satellite intelligence. The National Reconnaissance Office (NRO) had three systems in operation that could together provide electronic and imaging data, White Cloud, KH-8 and KH-11; it could also offer data from occasional overflights by the SR-71 high-altitude reconnaissance aircraft.

The limitation on the usefulness of overhead surveillance was, first, its intermittence—White Cloud made only two passes a day—but, second, that by the time it became available, the damage had been done. Overhead surveillance could have warned of the Argentinian invasion fleet setting sail, in time for the British government to have issued an ultimatum; once the fleet had arrived, it could supply little further information that was useful.

It was, among other factors, for that reason that the Northwood headquarters decided, after the shock of the first Exocet attack, to move from passive to active counter-intelligence methods. Since traditional means of warning—including satellite intelligence—had failed to avert the threat, the Ministry of Defence would be ordered to mount operations that would eliminate the risk at source. Britain's special forces would be committed to find and destroy the Exocet units in their home bases.

Special forces are a distinctively British contribution to contemporary military capability. They have their origin in Winston Churchill's directive of July 1940 to "set Europe ablaze," the immediate outcome of which was the creation of the Special Operations Executive (SOE). Churchill's belief, ill-conceived though it proved to be, was that covert attacks by irregular forces within the territory of German-occupied Europe could undermine Britain's enemy from within. He envisaged the work being done by local patriots, armed and advised by British agents. Churchill's scheme, though it did much to restore the national pride of Europe's defeated peoples, did little to weaken Nazi power. His conception of forming irregular units had an indirect result, however, that was permanently to alter the way in which states use military force. Fertilised by the idea of SOE, the British army's thinking in the middle period of the Second World War turned towards the creation of its own irregular forces, trained and equipped to operate inside enemy territory. The first such units, organised at Churchill's direct order, became the commandos, raiding forces to be landed from the sea; they had their airborne equivalent in the Parachute Regiment, which was trained and equipped to descend from aircraft behind enemy lines.

The SOE, commando and Parachute Regiment ideas coalesced to inspire free-thinking officers of the British forces in the Middle East during 1940–42 with a conception of their own: that instead of seeking to recruit civilians to fight as irregular soldiers, they should turn professionals into irregulars. The outcome was a coterie of unconventional units, the Long Range Desert Group, Popski's Private Army, the Levant Schooner Squadron, the Special Air Service. When the war came to an end, most were disbanded, to survive only as romantic memories. The

Special Air Service (SAS) found a different destiny. It had had a very successful war, attacking airfields in apparently quiet sectors of the desert and pinpoint targets in continental Europe; though stood down in 1946, it was revived—as the Malayan Scouts—to conduct covert operations against Communist terrorists in the Malayan jungle in 1948 and thereafter accumulated many other functions. By the 1980s it had become the instrument with which the army, often acting as the agent of the government, conducted covert operations against terrorists and organised criminals inside and outside the United Kingdom; it also acted as the irregular arm of the regular forces in conventional operations. Quite small—its intensely selective recruitment process limited its numbers to about 400—its effectiveness was out of all proportion to its numerical strength.

One of the functions at which it excelled was undercover observation. SAS troopers learnt how to penetrate a landscape and disappear inside it, "lying up" in "hides" for days at a time, surviving in great discomfort to bring back eye-witness accounts of enemy locations and activities. Northwood headquarters decided at the outset of Operation Corporate that, because of the paucity of intelligence derived from signal interception and overhead surveillance, it would be essential to insert SAS parties to watch and report. Those missions would shortly be enlarged to include direct attack on exposed enemy positions identified as offering critical threats to the success of the expedition.

One was decided upon at the outset. The Argentinian presence on South Georgia, though it lay 800 miles from the Falklands group, was seen as an affront; it was also soon perceived as presenting an opportunity. During the long preparatory period, as the task force moved south in stages during March and April, the government felt increasingly under pressure to allay public anxiety with news of success. The recapture of South Georgia would satisfy the requirement. A mixed party of Royal Marines and SAS was therefore embarked on HMS *Antrim* and detached to the objective. In extreme weather conditions and with inadequate equipment, the party eventually got ashore, having narrowly avoided disaster in the process, and completed their mission between 21 and 24 April. The Argentinian servicemen, who had replaced the scrap dealers, gave up easily. The marines and SAS suffered no casualties, though many had been close to death by mishap several times.

Following the South Georgia foray, the SAS, with its Royal Marines equivalent, the Special Boat Squadron (now Service), was committed directly to preliminary operations in the Falklands; at a later stage it also took a full operational part in the fighting and attempted a number of still-mysterious penetrations of the Argentinian mainland, intended to give early warning of Argentinian air strikes but also to intercept them by surprise attack.

The first major special forces mission was launched against the Falklands group in early May. Six Special Boat Squadron (SBS) teams and seven four-man SAS patrols were landed by helicopter from the fleet, the SBS tasked particularly to choose landing beaches, the SAS to gather intelligence of Argentinian deployments. One SAS patrol lay up at Bluff Cove, eventually to be chosen as a subsidiary landing place on the west coast of East Falkland, the main island, one at Darwin, near San Carlos, the initial and main landing place, three overlooking Port Stanley, the island capital on East Falkland, three on the barely inhabited West Falkland. It was there that the SAS drew first blood. On 14 May forty-five men of D Squadron, who had been guided to their destination by a patrol inserted three days earlier, landed by helicopter to strike at the airstrip on Pebble Island where the Argentinian air force had based eleven Pucara ground-attack aircraft, guarded by a hundred men. The SAS troopers were accompanied by forward observers from 29 Commando Regiment Royal Artillery to direct the fire of frigates offshore. Under the bombardment the SAS laid demolition charges which destroyed all the enemy aircraft and withdrew without loss, leaving an Argentinian officer dead and two of his men wounded.

Two independent actions by special forces followed, one on 21 May, the day of the main landing in San Carlos Water, to seize Fanning Head, which overlooked the approach, and during 25–27 May to secure lookout positions on Mount Kent, dominating Port Stanley. Both were completely successful. The Argentinians at Fanning Head were driven off by the SBS which, in the period before the main landing, also sent patrols to Campa Menta Bay, Eagle Hill, Johnson's Harbour, San Carlos and Port San Carlos.[2] On 20 May an SAS patrol had also struck a serious blow at Argentinian ability to position troops against the bridgehead, when it was secured, by finding an enemy helicopter park and destroying the four Chinooks and Pumas waiting there. The two units, 22 SAS and the SBS, continued to be involved in operations on the islands after the landings until the Argentinian surrender on 14 June.

After 4 May, however, when *Sheffield* was sunk by Exocet, the main thought of those controlling special forces was to use them in some way that would provide early warning of Exocet raids or eliminate the Super Etendards which delivered them. In either case landings on the Argentinian mainland would be required. The insertion of an SAS surveillance team was attempted by helicopter against the base at Rio Grande on the night of 17–18 May; its mission was to assess the state of the defences and then retire undetected into Chilean territory, where preparations had been made to receive it. As the helicopter landed the pilot decided that his aircraft had been detected and that he must make an escape to Chile. After a hurried flight westward, he dropped his SAS passengers to proceed on foot across the border, then landed inside Chilean territory and set fire to his machine. He and his two crew were subsequently repatriated, having unconvincingly explained their presence in Chilean airspace with the excuse that they had got lost. The SAS invaders were discovered by an undercover liaison agent, taken to Santiago and hidden there until the war was over.[3]

The second element of the scheme to eliminate the Super Etendards at Rio Grande failed because those detailed for the mission became convinced that it would end in disaster. The plan required three troops, forty-five men, to be crash-landed onto the runway in Hercules C-130 aircraft, overcome the defenders, destroy the Super Etendards, kill the pilots, whom it was hoped to trap in their quarters, and then march at high speed across country to neutral Chile. The diplomacy of the operation was dubious; so was its practicality. The soldiers' confidence was not enhanced by the discovery that the only maps of the region available dated from 1939 or had been photocopied from *The Times Atlas*. At their last briefing before departure from England, two highly experienced sergeants announced that they wished to remain behind, apparently an unprecedented event in SAS history. In the face of their doubts, the senior officer felt obliged to cancel the operation and stand the other soldiers down. Some felt the dissenters should have been dismissed; others accepted that they had reason on their side.[4]

The planners' reasons for preparing the operation, at the extreme limit of risk though it was known to be, were demonstrated on 25 May when two Super Etendards, refuelled north of the islands, approached the fleet from an unexpected direction and launched Exocets. One was distracted by chaff and fell into the sea; the second, attracted by the huge bulk of the container ship *Atlantic Conveyor*, struck home. *Conveyor* caught fire and sank, taking with it much vital heavy equipment, including three large Chinook troop-carrying helicopters, and ten Wessex, which were intended to lift the infantry forward towards Port Stanley. Their loss condemned the infantry to walk, thus seriously setting back the final stage of the ground campaign.

After the attack on *Conveyor*, however, only one Exocet remained to the Argentinians. Moreover, in fierce battles between the task force and the enemy's conventionally armed air units between 21 and 23 May, twenty-three enemy aircraft had been destroyed, taking Argentinian losses to one-third of their available strength. The Argentinian pilots had fought throughout the campaign with great courage and unexpected skill but the air battles over San Carlos Water had effectively defeated them. They were to achieve one more spectacular success, at Bluff Cove on 8 June, but by then the British ground forces were positioned on the high ground surrounding Port Stanley, whose Argentinian garrison was already showing its readiness to surrender.

There is some suggestion, unverified and unconfirmed, that the task force's ability to defend itself against air attack was reinforced during May by the insertion of another, undetected SAS surveillance mission on the Argentinian mainland and by the positioning offshore of nuclear submarines as pickets.[5] Certainly the full picture of the nature of the British early-warning system during the three weeks, 21 May–14 June, of intense fighting has not been disclosed. It cannot have succeeded by luck alone, for the air cover available was scanty, only 36 Harriers before losses, while the fleet's missile defences were patchy. The remarkable total of losses inflicted on the Argentinians, including 31 Skyhawks and 26 Mirages, speaks of a more systematic warning achievement than chance would allow.[6]

The task force suffered two grave intelligence defeats, both attributable to failures at the human level. During the subsidiary campaign to recapture South Georgia, a succession of attempts to extract an SAS party from a position made untenable by ferocious Arctic weather was saved from disaster only when a third helicopter succeeded, against every probability, in rescuing both the party and the crews of the two helicopters which had crashed in previous attempts to rescue it. The mission had been undertaken only because an army officer with exploring experience on South Georgia had assured the planners that the original mission was feasible; the episode provided an awful warning that expert information can be as flawed as any other form of intelligence. The second failure was more serious; early in the campaign a Sea Harrier from *Invincible* was shot down in an attack on the Pucara base in West Falklands (4 May); on the pilot's body, an Argentinian intelligence officer found his briefing notes, which when deciphered revealed the position from which the fleet was operating east of the Falklands. Until then it had been able to hide from the enemy in the wastes of the ocean, while keeping close enough to fight what was hoped would be a successful struggle to achieve air superiority over the islands. After 4 May, also the date when *Sheffield* was sunk by Exocet, Admiral Woodward was forced to withdraw the fleet beyond Argentinian aircraft range, and to approach the islands only when absolutely necessary.

The British had gone to war in the belief that their show of force would bring about an Argentinian withdrawal by diplomatic negotiation. After the sinking of *Sheffield* and the loss of the first Sea Harrier, they were obliged to recognise that the conflict was real; once the troops landed on 21 May, optimism grew that resistance would collapse, as the Argentinian conscripts were overcome by the superior fighting power of the British regulars. It was during the first three weeks of the campaign that the issue hung in the balance. An intelligence coup by the Argentinians, allowing them to strike one of the British carriers or a big troop-carrying ship, *Canberra* or *QE2,* with an Exocet might have shifted it their way. As it was, without access to American satellite or signal intelligence, which the British enjoyed, and with inadequate intelligence resources of their own, the Argentinians had to operate by guess and chance. Neither sufficed.

The last large war of the twentieth century, that in the Gulf against Iraq by the American-led coalition, was conducted within an intelligence environment far more favourable to the intervening force than that conditioning the Falklands War nine years earlier. The coalition was served with, besides copious and continuous sigint, frequent overflying missions yielding high-resolution photography and much electronic and sensory data, as well as satellite surveillance in all its forms. Because the Iraqis had deployed their forces beyond their own borders, in Kuwaiti territory, the coalition also had access to plentiful and exact cartography of the operational area; the combatants made no complaints at all about the quantity or quality of strategic intelligence available to them.

The acquisition of tactical intelligence in real time proved much less satisfactory. Because the Iraqi air force took refuge at an early stage in Iran, there was no need for early warning of air attack. What was required was warning of the launch of Iraqi Scud missiles, aimed at coalition forces, their Saudi bases and the territory of Israel; even more desirable was information about the Scud launchers' whereabouts. Early warning worked well, allowing the destruction of Scuds in flight on several occasions. Location of the launchers—a variant of the Meiller-wagen that had made the V-2s so difficult to attack in 1944–45—proved effectively impossible. Despite the insertion of numbers of special forces teams into Iraqi territory, no Scud launcher was found and none destroyed. Iraqi ability to hide and protect its weapons of highest value from detection by both external and internal intelligence-gathering means underlay the international crisis that began in 2002 and persists at the time of writing.

Saddam Hussein's defiance of the authority of the United Nations, by his refusal to co-operate with its weapons inspectors as required under Resolution 1441 of the Security Council, exemplifies the difficulties of obtaining intelligence about modern weapons systems even under conditions amounting to those of authorised espionage. The inspectors, though present in considerable numbers—at least a hundred—on Iraqi territory, and ostensibly enjoying unfettered freedom of movement and access, were consistently frustrated, as late as March 2003, in their efforts to uncover stocks of chemical and biological warfare materials which they had good reason to believe had not been destroyed, as was required by UN resolution, and remained hidden at a number of locations. The search for the components of nuclear warheads, which it was also strongly believed Saddam was attempting to construct, proved equally unavailing. The senior weapons inspector, Dr. Hans Blix, complained that he and his team were unable to fulfil their task—to report that Iraq had fully complied with the provisions of Resolution 1441—because they were refused full co-operation by the Iraqi authorities, particularly the freedom to interrogate in private Iraqi scientists known to be working on the weapons programme. Neither Dr. Blix nor Western anti-war protestors, who demanded more time for the inspectors to continue, seem to have made any allowance for the possibility that the objects of their search were so well concealed that whatever the apparent co-operation furnished by the Iraqis and however long investigations were protracted, his mission was bound to fail. The situation was unprecedented. A potential international law-breaker had been obliged to open his borders to officially sponsored investigators of his suspected wrong-doing and yet they remained unable to dispel the uncertainties surrounding his intentions and capabilities. In absolutely optimum conditions, in short, intelligence had failed.

Intelligence operations in the parallel "war against terror" were equally frustrated, though for different reasons. The war was misnamed, for it was so one-sided as to deprive the opponents of terrorism of any of the usual means by which one party to a conflict normally exerts pressure on the other. Al-Qaeda, the movement which had taken control of and given leadership to the diffuse forces of Islamic fundamentalist terror, has, though it means "the base" in Arabic, no identifiable base and, after the defeat of the Taliban in Afghanistan at the beginning of 2002, no territory. It is outlawed in many Muslim states, where autocratic governments fear the threat it offers, through accusations of their less than perfect adherence to the fundamentalists' conception of Islam, to established authority. The size and composition of its membership is unknown, as is the identity of its leadership, a few self-declared though elusive figureheads apart, and the structure of its command system, if one exists; it is a strength of al-Qaeda that it appears to be a coalition of like-minded but separate groups rather than a monolithic entity. Its finances, although it is known to possess large monetary resources, are mysterious, since it apparently conducts transactions by informal yet secure word-of-mouth agreements traditional within Muslim societies. It does not possess large armouries of conspicuous weapons, preferring to improvise—as by its hijacking of civilian airliners on 11 September 2001—or to make use of readily concealed means of terrorist outrage, such as plastic explosive. Like all post-1945 terrorist organisations, it appears to have learnt a great deal from the operations of the Western states' special forces during the Second World War, such as SOE and OSS, which developed and diffused most of the modern techniques of secret warfare among the resistance groups of German-occupied Europe during 1940–44; the copious literature of secret warfare against the Nazis provides the textbooks. Among the techniques described is resistance to interrogation by captured operatives, which often failed against the Gestapo, since it was prepared to use torture, but succeeds against today's Western counterterrorist organisations, culturally indisposed to employ torture and anyhow inhibited from so doing by domestic and international law. Despite the arrest and detention of hundreds of al-Qaeda operatives, reports suggest that they have successfully overcome American efforts to break down their resistance to questioning.

The only point of penetration into the world of al-Qaeda appears to have been found in its necessity to communicate. Intercommunication, as this book suggests, has almost always proved the weak link in undercover systems, whatever the methods used to make it secure. Al-Qaeda has apparently thus far trusted to the difficulty presented to Western monitoring organisations by the sheer volume of mobile and satellite telephone transmissions, seemingly hoping that its person-to-person messages will be lost among the daily billions of others. It has, fortunately, proved a false hope. Modern methods of scanning and point-targeting of transmissions allow the Western interception agencies to isolate and overhear an increasingly large number of significant messages and so to identify suspects and locate where they operate.

In the last resort, however, attacks on al-Qaeda and other fundamentalist networks will be made successful only by recourse to the oldest of all intelligence methods, direct and personal counter-espionage. Brave individuals, fluent in difficult languages and able to pass as native members of other cultures, will have to befriend and win acceptance by their own societies' enemies. It is a technique perfected by the Israelis, whose intelligence agencies enjoy the advantage of being able to recruit agents among refugees from ancient Jewish communities in Arab lands, colloquial in the speech of the countries from which they have fled but completely loyal to the state in which they have found a new home. Western states will find such recruitment more difficult. Islam imposes a powerful bond over fellow believers; even Muslim immigrants of the second or third generation, loyal to their Western countries of adoption in every other way, feel a strong aversion to what seems betrayal of co-religionists by reporting them to the authorities for religious zealotry. The problem of recruitment is acute in the United States, which lacks both Muslim communities of large size or antiquity and non-Muslim citizens with a knowledge of the appropriate languages. It may prove easier in the old imperial countries, such as Britain and France, whose intelligence agencies, particularly the British, actually have their roots in the nineteenth-century need to police their colonial dissidents and which retain a significant residue of language and other ethnographic skills.

A strange task confronts them. It diverges widely from that of Bletchley and OP-20-G, which required the highest intellectual power and rigorous dedication to the routines of radio monitoring, interception and decipherment. The masters of the new counter-intelligence will not resemble the academics and chess champions of the Enigma epic in any way at all. They will not be intellectuals, nor will they overcome their opponents by power of reason or gifts of mathematical analysis. On the contrary: it will be qualities of empathy and dissimulation that will equip them to identify, penetrate and win acceptance by the target groups. Their work will resemble that of undercover police agents who attempt to become trusted members of criminal gangs, with all the dangers and moral compromises that such a life requires. Undercover work within the terrorist groups of Northern Ireland, republican and loyalist alike, has equipped British security and specialist police bodies to understand how such undercover operations are best conducted, but the practice is always more difficult than theory and will prove particularly so with religious fanatics. Even ideological terrorists, such as the extreme nationalists of the Irish republican tradition, are sometimes susceptible to temptation or threat; republican fund-raising by blackmail and extortion has drawn the movement into crime, with corrupting effect, while its "military" ethos excludes the taking of risks that threaten the lives of "volunteers." Muslim puritans, by contrast, seem resistant to financial temptation, have demonstrated their readiness to commit suicide in furtherance of their violent aims, are committed to a code of total silence under interrogation and are bound by ties of brotherhood which have religious strength. No organisation, of course, is impervious to penetration or indestructible. All have their weak spots and weak members. It may, however, take decades for Western intelligence agencies to learn how to break in to the mysterious and alien organisations and even longer to marginalise and neutralise them.

The challenge will cast the agencies back onto methods which have come to appear outdated, even primitive, in the age of satellite surveillance and computer decryption. Kipling's Kim, who has survived into modern times only as the delightful literary creation of a master novelist, may come to provide a model of the anti-fundamentalist agent, with his ability to shed his European identity and to pass convincingly as a Muslim message-carrier, Hindu gallant and Buddhist holy man's hanger-on, far superior to any holder of a Ph.D. in higher mathematics. Buchan's Scudder, sniffing from clue to clue along a trail leading from fur shop in Buda to the back streets of Paris, shedding and adopting new disguises on the way, seems better adapted to the future world of espionage than any graduate student in regional studies. It will be ironic if the literature of imagination supplies firmer suggestions as to how the war against terrorism should be fought than academic training courses in intelligence technique provide. Ironic but not unlikely. The secret world has always occupied a halfway house between fact and fiction, and has been peopled as much by dreamers and fantasists as by pragmatists and men of reason.

The Western powers may come to count themselves fortunate that, in their time of troubles during the two world wars, the central targets of intelligence-gathering, enemy communications and secret weapons were susceptible to attack by concrete methods: overhearing, decryption and visual surveillance, together with deception in kind. They have already learnt to regret the emergence of new intelligence targets that lack any concrete form: aggressive belief systems not subject to central authority, shifting alliances of dangerous malcontents, stateless migrants disloyal to any country of settlement. It is from those backgrounds that the agents of anti-Western terrorism are recruited. Their recruiting grounds, moreover, are confusingly amorphous, disguised as they are within communities of recently arrived immigrants, many of them young men without family or documented identity, often illegal border-crossers who take on protective colouring within the large groups of "paperless" drifters merely seeking to avoid the attention of the authorities.

The United States, protected as it is by its wide oceanic frontiers and its strict and efficient border services, is certainly not impervious to terrorist penetration, as the awful events of 11 September 2001 demonstrated. The Western European states, physically contiguous to countries which hundreds of thousands of young men energetically seek to leave and constrained by their own civil rights legislation from returning illegals to their jurisdictions of origin, even if the facts can be established, are much less well defended. The security problem by which the Western European states are confronted is not only without precedent in scale or intensity but defies containment. The suspect communities grow continuously in size, the nuclei of plotters and would-be evil-doers they conceal thereby acquiring greater anonymity and freedom to prepare outrages. Financial support is not a problem, since the terrorists enjoy access to funds extracted in their countries of origin by blackmail in many forms, including straightforward protection money but also donations represented as contributions to the cause of holy war. The "war on terrorism" may be a misnomer, but it would be foolish to pretend that there is not a historic war between the "crusaders," as Muslim fundamentalists characterise the countries which descend from the kingdoms of western Christendom, and the Islamic world. It has taken many forms over more than a thousand years, and fortunes in the conflict have ebbed and flowed. A century ago it appeared to have been settled for good in favour of the West, when the region's technological superiority seemed to have reduced Islam to an irreformably backward and feeble condition. Allah, Muslims might say, is not mocked. Their certitude in the truth of their beliefs has driven those Muslims who see themselves as religious warriors to seek ways of waging holy war that outflank mere technology and promise to bring victory by the power of anti-materialist forces alone. Muslim fundamentalism is profoundly unintellectual; it is, by that token, opposed to everything the West understands by the idea of "intelligence." The challenge to the West's intelligence services is to find a way into the fundamentalist mind and to overcome it from within.

———◄O►———

The Value of Military Intelligence

W AR IS ULTIMATELY about doing, not thinking. The Macedonians
beat the Persians at Gaugamela in 331 B.C. not because they took
the enemy by surprise—Darius, the Persian emperor, spent the preliminaries of the battle attempting to bribe Alexander not to attack—but by
the ferocity of their onslaught. The Knights of St. John saved Malta from
capture by the Turks in 1565 not because they got word of their approach
but by the tenacity of their defence in a five-month siege. The British
and Indian troops repelled the Japanese attempt to invade India via
Kohima and Imphal in 1944 not because intelligence had disclosed the
enemy's plan but by stubborn, relentless, sometimes hand-to-hand combat. The Americans took Iwo Jima in 1945 not because intelligence had
revealed the lay-out of the Japanese defences—the whole tiny island
was one densely fortified position—but because the U.S. Marines, at the
cost of thousands of their own lives, inched their way forward from
bunker to bunker. In the case of none of these famous and decisive battles did thought play much of a part in bringing victory; courage and
unconsidered self-sacrifice did.

War is not an intellectual activity but a brutally physical one. War
always tends towards attrition, which is a competition in inflicting and
bearing bloodshed, and the nearer attrition approaches to the extreme,
the less thought counts. Nevertheless, few who make war at any level,
from commander to soldier in the line of battle, seek to win by attrition.
All hope for success at lesser cost. Thought offers a means of reducing

the price. It may identify weaknesses in the enemy's method of making war or in his system of defence; detailed reconnaissance of Hitler's Atlantic Wall identified the best places to land before D-Day. It may reveal defects in his armoury or suggest countermeasures to his weapons; Britain's espousal of radar before 1939 laid the basis for national survival during the Battle of Britain. It may give warning of the enemy's concealed intentions or secret devices; foreknowledge played a major part in the—partial—defeat of the flying bomb, if not the V-2 rocket, in 1944. It may unveil treachery within; during the Cold War, patient if retrospective analysis of how state secrets were betrayed to Soviet Russia, and the identity of those responsible, closed a potentially fatal gap in national security. It may disclose the nature of an enemy strategy, which threatens to strangle essential lifelines of supply, as inspired thought by a single individual in 1917 revealed how the U-boat blockade of Britain could be defeated by a simple reorganisation of shipping. It may, at its most creative, unlock a whole world of enemy secrets, as Bletchley Park's attack on the German Enigma ciphers did from 1940 onwards.

The story of the breaking of Enigma, and of Ultra, the intelligence it yielded, together with the story of Magic, the product of the American unravelling of the Japanese ciphers, is of the highest drama and the greatest importance to our understanding of the conduct of the Second World War. Without our knowledge of Ultra and Magic, it would be impossible to write the war's history; and, indeed, all history of the war written before 1974, when the Ultra secret was revealed for the first time, is flawed by reason of that gap.[1] However restrained the claims made for the influence of Ultra in bringing eventual victory—and those made by its official historian, F. H. Hinsley, are very carefully restrained—the availability of day-to-day, sometimes hour-by-hour details of the enemy's tactical control of his U-boat forces, for example, of the resupply of his ground forces in the Western Desert, sometimes of their deployment for action also, occasionally of strategic initiatives of the greatest regional significance, such as the plan to capture Crete by airborne descent in 1941, helped very greatly to win the war for the Allies and, as Hinsley demonstrates, materially shortened its course. The same is true of Magic in the Pacific. Moreover, in both theatres, the ability to overhear the enemy was an advantage the Allies enjoyed which—with certain exceptions—their opponents did not.

If there is such a thing as an ideal of military intelligence, when one side was privileged to know the other's intentions, capabilities and plan of action in place and time—how, where, what and when—while its opponent neither knew as much in return nor that his own plans were uncovered, Ultra—and Magic—occasionally met the ideal standard. The Americans before Midway were in such a position in June 1942; so were the British before the German airborne invasion of Crete in May 1941.

Yet, as we know, the British nonetheless lost the Battle of Crete. There have been several attempts to explain why, the intelligence circumstances appearing to make defeat an impossibility. It has been suggested that General Freyberg, commanding the island, believed the airborne assault to be the prelude to a later seaborne invasion, or that he was overburdened by the risk of revealing the Ultra secret, or both; in either case, misbelief or paralysing anxiety, he failed to redeploy his troops to positions which would have made the capture of Maleme airfield, the vital ground, impossible. Neither, in fact, seems to provide a complete explanation. Freyberg did fear a seaborne invasion and he was also weighed down by the need to keep the Ultra secret. He might, nevertheless, with the troops available, his highly capable and determined New Zealanders, have held the airfield had he impressed on the local commander, a brave man of proven fighting ability, the necessity of staying put and yielding not an inch. Instead, the local commander got the impression that it would be possible to retire, regroup and successfully counterattack next morning, giving his men a pause he thought they required. Next morning proved too late. In real time, the by then desperate Germans took advantage of a momentary weakening of the New Zealand defence to stage one of the most extreme do-or-die exploits in military history. They had already offered up to sacrifice the bulk of the Assault Regiment, by crash-landing its gliders into the waterless bed of the River Tavronitis, an attack the New Zealanders had largely blunted. On the morning of 21 May, they began to use the Ju52 aircraft carrying the 5th Mountain Division in almost the same way, landing them under fire on the airfield and ruthlessly ditching those hit on the runway. The death ride of the Ju52s should have resulted in disaster; but there was just not enough New Zealand fire, and that from too long a range, and just too much German recklessness. At enormous cost, in loss of both machines and lives, the Germans succeeded in building up a superiority of force at the decisive spot, seizing the airfield and using it as the point of departure for a battle-winning offensive.

The events of 20–21 May 1941 in Crete demonstrate one of the most important of all truths about the role of intelligence in warfare: that however good the intelligence available before an encounter may appear to be, the outcome, given equality of force, will still be decided by the fight; and, in a fight, determination, again given equality of force, will be the paramount factor. The New Zealanders were troops of the very first quality; Rommel, their opponent in the desert, testified that they were the best soldiers he ever met, including his own. On Crete, however, they met other soldiers who preferred collective death to defeat. The men of the 7th Airborne Division and the 5th Mountain Division were in berserker mood. It was their almost mindless courage that allowed them to prevail.

The events of 4 June 1942 at Midway provide another perspective: that, even when intelligence seems to provide the explanation of a victory, closer examination of the facts may reveal that some other factor, in that case chance, lies at the root of the matter. The Americans in 1942 were in much the same position of strategic inferiority as the British had been in 1940–41: though equipped to overhear the enemy's secret signals, they were at a severe military disadvantage by reason of recent defeats. They had lost their battle-fleet, they had lost much territory of crucial importance, and they were outnumbered in key categories of weapon systems, particularly aircraft carriers. It was greatly to their credit that, during a period of acute parsimony in defence spending, they had nevertheless succeeded in penetrating the main Japanese naval code, JN-25A, before Pearl Harbor and had had success against the more complex JN-25B by early 1942. By a combination of interception, decoding, informed speculation about Japanese intentions and, crucially, a cunning exercise in the art of the baited signal—the false revelation that Midway was suffering a water shortage—the U.S. Pacific Fleet had, as events would show, accurately persuaded itself by May 1942 that the next stage of Japanese expansion would not be westward into the Indian Ocean or southward towards Australia but eastward, from the Japanese home islands, to seize Midway, the last American-held outpost in their proximity. Covert deployment of America's only three Pacific-based aircraft carriers positioned the surviving American capital forces to take the approaching Japanese strike fleet, of four aircraft carriers, by surprise and achieve a victory.

Carrier fleets, however, consist of two elements, the ships themselves and their air groups. An air group whose carrier is sunk, while it is aloft, becomes a refugee organisation, seeking to land where it may, or to ditch if no landing place offers. A carrier without its air group is no more harmful than any cargo ship. On the morning of 4 June, the Japanese carrier striking force, surprised by five out of the six squadrons of the American carrier air groups, destroyed them all. The sixth group had got lost. Its leader, almost at the limit of fuel endurance, then spotted a Japanese destroyer, which had been detached to attack an American submarine, making speed to rejoin the main force. The white streak of its wake, on the deep blue of the ocean on a perfect Pacific day, indicated the direction he and his fellow pilots should follow. They did and, being dive-bombers arriving at 12,000 feet, while the Japanese fighters of the combat air patrol had just descended to sea level to destroy the last American torpedo-aircraft attack, found a clear run to the target. Three out of the four Japanese aircraft carriers were destroyed in five minutes.

The success of the dive-bombers made Midway a great naval victory, the greatest naval victory of all time. It was crowned later in the day by the destruction of the surviving fourth Japanese carrier. Nevertheless, it cannot be claimed that Midway was a pure intelligence victory, open and shut though the case superficially seems. The events of 4 June, up to the destruction of the fifth of the American attacking squadrons, had indeed been the outcome of decisions taken in light of an intelligence advantage; the final and decisive event, the descent of the sixth squadron on a by then defenceless Japanese carrier formation, was the result of luck. Had the U.S. submarine *Nautilus* not strayed into the path of the Japanese carriers, causing the detachment of the destroyer *Arashi* to attack it, the lost dive-bomber squadron, Bombing 6, would not have been redirected by its wake on to the target; and had *Arashi* lingered longer on its search, Bombing 6 would again not have known which way to go and would have had to turn back, mission unaccomplished, at the limit of its endurance.

There are other complexities, concerning particularly failures of reporting by Japanese reconnaissance and failure of clear thinking by the Japanese high command. Had the *Tone*'s floatplane made an earlier and more exact report when it sighted the American task force, the

Japanese carriers would have been alerted to the presence of the American carriers before their aircraft took off. Had Admiral Nagumo thought more quickly and analytically once the battle began, in particular not been distracted by the intervention of land-based aircraft from Midway, he could have initiated a much earlier attack on the American carriers, manoeuvred to a new position and avoided being caught with his flight decks cluttered with ordnance, fuel lines and fully fuelled aircraft, potential—actual, as things turned out—firebombs. Though Midway turned out to be a great American victory, in the making of which the intercept and decryption services played an essential part, it might have been exactly the opposite: a great American defeat, into which the U.S. Pacific Fleet had been drawn by the very success of its own intelligence operations.

War is the arena of chance; furthermore, nothing in war is simple. Midway bears out the truth of both of these observations. Their truth is further borne out by the course of the German cruiser campaign in the Pacific and South Atlantic in 1914. On the face of it, von Spee, with his little fleet, should have been able to make his way back to Europe from China unscathed; he might even, by careful selection of his targets, have inflicted considerable damage on his enemies' merchant shipping on the way. The vastness of the Pacific provided a perfect cloak for his movements; once in the Atlantic, an ocean half its size, a swift break for home, via the stormy waters of the northern seas, might have brought his ships back to German bases intact. The *Etappen* system, which efficiently arranged for colliers and store ships to be met at neutral harbours or remote anchorages, would have provided resupply. The ports of South America, on both coasts, teemed with German merchants and sympathisers. There promised, for von Spee and his men, the makings of a clear run home.

All the more so because of the deficiencies of contemporary wireless telegraphy. Marconi's invention, only thirteen years old, had achieved the success of an idea whose time had come. Cable telegraphy, after the first demonstration of its practicality in 1828, had taken decades to provide links between countries, even longer between continents. Not until 1850 were Britain and France connected by an undersea cable, not until 1866 Britain and North America. Thereafter the interconnections proceeded more quickly. By 1870, Britain was connected to Africa, by 1872 to India, by 1878 to Australia and New Zealand. Nevertheless, the creation of a worldwide cable network had taken fifty years. The installation of a worldwide wireless network took only ten, but it was not perfect. There were several gaps—Australia and New Zealand, for example, did not connect with either India or Africa—and signals often had to be repeated or reinforced by cable to ensure reception. The system was also liable to interference from atmospherics, lacked directionality and was easily overheard.

Radio did little to assist directly with the interception and destruction of von Spee's squadron, because he was generally scrupulous about maintaining radio silence (though an intercept at Samoa on 4 October 1914, transmitted in a broken code, revealed that he was en route from the Marquesas to Easter Island). British lapses, on the other hand, helped von Spee. It was Cradock's decision to detach *Glasgow* to Coronel to send a cable that disclosed the presence of his squadron in those waters and led to the battle. Indirectly, of course, the influence of radio on von Spee's fate was malign. Had he not decided to raid the Falklands, largely for the purpose of putting out of action its wireless station, he would not have run straight into Sturdee's hands. That was bad luck, compounded by recklessness. Had he avoided the Falklands, made his way circumspectly up the east coast of South America, picking up supplies as he went and avoiding attacks on British merchant shipping, he might have got undetected to within rapid steaming distance of home and returned to a hero's welcome. He would have had to have luck in the final stages, to avoid the cruiser patrols off the north of Scotland, but, in the winter weather of those latitudes, he might just have done it. The German battleship *Bismarck,* sailing an opposite course in May 1941, got from Germany to the North Atlantic, eluding the British Home Fleet for several days, and that in the era of radar and long-range aerial reconnaissance. Moreover, real-time intelligence was then provided by an Enigma transmission in a Luftwaffe key Bletchley could read; a senior Luftwaffe officer, with a son aboard the battleship, enquired where he might expect his son to arrive. The answer was Brest, which solved the mystery of where the ship was headed and directed the chase onto her course.

The vastness of the sea, the variety of its weather and the multiplicity of hiding places offered by its coasts and islands have always imbued precise intelligence of the enemy with the highest value in naval warfare. Fleets can disappear, single ships can navigate vast distances without ever being seen by another; indeed, in the months before Pearl

Harbor, a Japanese ship was deliberately sailed down the planned course of the attack fleet from the home islands towards Hawaii and met no other vessel during several weeks at sea. It is therefore not surprising that Nelson lost Bonaparte's fleet, several hundred strong though it was, in 1798. Anyone who has sailed the Mediterranean, a comparatively small sea, knows how long are the periods in which no other ship is seen. It is only in the vicinity of ports that companion vessels appear and they are soon sunk below the horizon by a change of course. Even coastwise sailing, the traditional means of passage in the inland sea, can be lonely. Headlands, peninsulas, islands intervene readily between one vessel and another and they are quickly lost to one another.

Nelson's loss of touch with Bonaparte's Egyptian invasion fleet, after his dismasting in the storm off Toulon, and failure to regain it, is therefore quite explicable. There were several persuasive destinations the French might have chosen, including Ireland, Spain, Naples, Constantinople and Anatolia (Turkey), as well as Egypt. The routes thither might be subsumed into two, west and east, which were mutually exclusive. Nelson took a risk, but a lesser one, in discounting the western option. Having correctly decided that Bonaparte had sailed east, a decision based not merely upon a judgement of probabilities but on the human intelligence available, including commercial agents' reports as well as gossip and rumours from within the orbit of French influence in Europe, his method of running the fugitive fleet down was quite logical. He went from place to place, searching hiding places, questioning ship captains and following up reports. His justifiable anxiety to keep Bonaparte out of Naples and Sicily led to his being foxed when the French fell on Malta; he then not only got on the trail again but got ahead of the quarry, overrunning the invasion fleet at night and arriving at Alexandria before it. His mistake, in a campaign dogged by intelligence famine, was not to wait there but, in a fever of reconsideration, to retrace his course until he picked up firm intelligence at a place which he had already visited.

The value of any study of the Nile campaign to a modern intelligence officer is to illustrate how much better provided he is than a commander of the pre-cable, pre-wireless age. Even with the telegraph and wireless, it would prove possible, as the lamentable tale of the pursuit of the *Goeben* and *Breslau* in August 1914 shows, to let an enemy get clean away. It is inconceivable, however, that Nelson could have lost Bonaparte had the Royal Navy had the signal resources available to it in the Mediterranean even sixty years later. The simplest cable system would have ensured that Nelson waited at Alexandria on the first visit and so brought the campaign to an end. Indeed, the same result could have been achieved with the resources then available, had they been in place. Had the British maintained a network of agents around the Mediterranean coastline, and kept despatch boats at friendly or neutral ports—say in Naples, Sicily, Malta, Turkish Crete and Cyprus, something not beyond the power of diplomacy to achieve, particularly with money to grease local palms—Nelson would not have had to lament his want of frigates or to use his battle fleet also as an instrument of reconnaissance. The lack of an intelligence network in the Mediterranean was, however, the outcome of a sudden surge of French power, which put weak local states in fear and perhaps was not to be reversed without a major British naval victory; which is to say that the achievement of earlier success in the campaign of the Nile depended upon fighting an earlier battle. There, precisely, was Nelson's difficulty: no intelligence, no battle. Hindsight solves Nelson's intelligence problem. In the circumstances, he did no worse than realities allowed. In a big sea, with slow ships, chasing a vanished enemy was bound to be a time-consuming business.

The vastness of the ocean also defined the circumstances in which the British fought the U-boat war. The first episode, in 1915–18, which actually brought Britain nearer starvation than the second two decades later, was eventually terminated, not by intelligence process, not by offensive methods, but by an exercise in operational analysis: a clear-thinking junior officer perceived that the formation of convoy would contain the sinking of ocean-going merchantmen within bearable limits, a notion previously rejected by the Admiralty. So it proved; sinkings declined, even though contemporary escort vessels had but the most primitive acoustic search equipment and crude anti-submarine weapons. The beginning of the second U-boat war in 1939 found the Royal Navy equipped with an active underwater search device, Asdic, later to be known as sonar, but with anti-submarine weapons scarcely improved since 1918. Moreover, the number of available escorts had declined relative to the number of essential merchant-ship sinkings. As a result, large weakly escorted convoys suffered heavy losses from the outset of hostilities. The Admiralty sought to reduce losses in a number of ways: by an accelerated programme of escort building and improvisation; by diverting aircraft from the bombing campaign to maritime escort and surveillance duties, a diversion always resisted by the RAF; by improving anti-submarine weapons and search equipment; by mining the approaches to the U-boat ports and by anti-U-boat intelligence. The intelligence campaign sought to protect convoys by diverting them away from identified U-boat positions and by directing escorts—both surface ships and anti-submarine aircraft—against individual U-boats. The principal intelligence means were radio direction-finding and the decryption of signals both from U-boats to base and vice versa. Not until May 1941, however, did Bletchley break into the U-boat traffic and its successes were offset by that of the German B-dienst, which read the British convoy code for most of 1942, and by periods of blankness, brought on by German alterations of procedure or machinery in the operation of Enigma.

Despite all intelligence difficulties and failures, rerouting of convoys was a success; only a minority of convoys were attacked and, during the long periods of bad weather that prevail in the North Atlantic winter, the U-boats often could not form wolfpacks or patrol lines, nor find convoys even when directed towards them from base.

In the last resort, however, the U-boats were defeated neither by Anglo-American intelligence success, nor by the eventual failure of the B-dienst's decryption campaign, but by battle at sea. By the spring of 1943 the combination of many new or improved measures taken by the Allies had set terms of engagement which the U-boats could not overcome. Continuous direct aerial surveillance of the convoy routes denied U-boats the freedom to cruise undetected on the surface; aggressive aerial patrolling of their exit routes from the French Atlantic ports to the high seas sank many and forced all to make their passages to war stations submerged at laboriously low speed; close protection of the convoys by escort aircraft carriers drove attacking U-boats down and resulted in frequent sinkings of those that surfaced; the multiplication of escorts, better trained and equipped to carry out group attacks, sank U-boats which found firing positions; improved radar and radio direction-finding led escorts to U-boats hovering around convoys beyond line of sight. In the end, as Dönitz was forced to admit to his own men, the balance of advantage swung so sharply against the U-boats that the German submarine fleet could be saved from destruction through attrition only by its withdrawal from the scene of action. The Battle of the Atlantic did indeed eventually become a true battle, a great naval battle extended in time and space, which was won by the Allies.

The battle against the German V-weapons was, by contrast, a real intelligence battle, in that it was intelligence that alerted the Allies to the threat and intelligence in all its forms—human, signal and imaging—that provided the beginnings of the antidote, but it ended in no such clear-cut victory. The advantage for the first four years of the Second World War ran wholly the Germans' way. Having begun to construct an extra-atmospheric rocket, capable of carrying a warhead, well before the war began, the German army had succeeded by late 1942 in solving most of the problems of launching and propelling it in flight and guiding it to its destination. Spurred into competition by the rocket programme's success, the German air force had meanwhile developed and largely perfected a cruise missile. Both weapons were greatly in advance of their time, measured against weapon development on the Allied side, where they had no counterparts.

From an intelligence point of view, the main aspect of interest aroused by the V-weapons is the difficulty the Allies encountered in taking the measure of the threat and then in deciding what countermeasures should be mobilised against it. To the lay mind, what impresses about the world of science are the openness of the scientific practitioner to new ideas and the readiness of the scientist to set aside prejudice in pursuit of fresh knowledge. Science, the layman believes, is the arena of rationality, unfettered by fixed beliefs, populated by pure intellectuals ever prepared to reject convention and depart upon a free voyage of experimental and theoretical discovery. The history of science contests that optimistic view at almost every turn. Scientists can be as prejudiced as theologians, particularly so if their pet theories are contested. No modern scientist in an influential position showed himself more prejudiced than Winston Churchill's personal scientific adviser, Professor Lindemann, whom the Prime Minister had had created Lord Cherwell. He had taken the view that long-range military rockets could work only if propelled by solid fuel, which dictated that they should be of enormous size and need highly conspicuous launch pads. He absolutely rejected the suggestion that the theoretically more compact liquid fuel could be confined and controlled as a propulsive medium. He had the mathematics to prove his point of view, and so strongly did he hold it that he used his privileged position to deride and attempt to discredit scientists junior to himself in the official hierarchy who argued the contrary.

He was, as events would painfully show, quite wrong but the evidence necessary to disprove him took precious months to accumulate. Eventually only the presentation of incontestable photographic evidence of the existence of rockets and then pilotless aircraft, later supplanted by eyewitness reports of their flight and finally by the delivery of physical fragments of the objects, drove him into admission of error. By then, fortunately, his opponents had won a hearing sufficiently strong to lead the British Chiefs of Staff to authorise a raid designed to obliterate the V-weapons centre at Peenemünde. It did not altogether achieve obliteration; it certainly did not achieve the extinction of the leading V-weapon scientists, which was one of its primary objects. Nevertheless, it set the secret weapons programme back and the delay, enhanced by the final difficulties the Germans encountered in bringing the V-1 and V-2 to an operational state, postponed their delivery against British targets beyond the opening of the D-Day invasion. This ensured that their launch sites would soon be overrun, thus negating the German expectation of postponing defeat by long-range bombardment of the invasion forces' points of departure.

The V-weapons programme has interest from another intelligence aspect—the unusual preponderance of human intelligence in influencing opinion on the other side. Human intelligence played almost no part in determining the conditions under which most of the campaigns which form case studies in this book were fought. Its importance, though paramount in the Nile campaign, when Nelson was acting as his own intelligence officer, and crucial to Stonewall Jackson in the Shenandoah Valley, was negligible during the U-boat war and quite insignificant during the campaigns of Crete and Midway. Paradoxically, in the high-technology struggle between German secret weapon scientists and their blinkered Allied opponents, human intelligence was of critical importance. The unattributable Oslo Report gave the first clue; eavesdropping, if that was what it was, by the unnamed "chemical engineer" later provided the trigger to Allied action. Thereafter, though photographic intelligence, imagery as it would now be called, supplied the earliest confirming substance, agent reports from foreign workers at Peenemünde and observations by the Polish underground provided the direct evidence that the V-weapons were actually airborne. Without those reports, and the evidence supplied by Swedish neutrals, including the sea captain with his watch, London would have lacked the picture— fairly clear as it eventually became—of what the hazily defined menace of flying bomb and supersonic rocket ultimately threatened. The intelligence attack on the V-weapons kept alive the importance, elsewhere so greatly discredited, of humint.

Humint, though the term was then unknown, also supplied the means, directly and indirectly, by which Jackson so successfully conducted his campaign against superior odds in the Shenandoah Valley in 1862. That campaign, in any large-scale military perspective, hovers uncertainly between the old and the new. In contemporaneous terms, Jackson belonged to the future, the future of the electric telegraph and the railway alongside which it usually ran. Practically, neither telegraph nor railway played anything but a tangential part in Jackson's manoeuvring. Although he eventually withdrew his Valley army to Richmond by railway, he scarcely used it to manoeuvre during his campaign of bewitchment and bewilderment; his employment of the telegraph, as a means of communication within the Valley and from it to higher command elsewhere, was intermittent. Jackson in the Valley behaved as a trusted Napoleonic subordinate might have done, superior though he was in talent to his Confederate seniors; he made his own appreciations, asked for no orders, and based his decisions on his own intelligence assessments, founded on close and local observations.

Like a pre-telegraphic and pre-railway commander, Jackson was most concerned to understand the geography of the theatre in which he was operating and to use it to his advantage. A man with an intuitive sense of ground himself—in that respect he resembled that other taciturn, relentless, hard-fighting general of the war, Ulysses S. Grant—he was greatly served by his mapmaker, Jedediah Hotchkiss, a gifted, if self-taught cartographer. In the modern world, where images of every sort abound, it is difficult to visualise the difficulties of travellers and voyagers of an earlier time, when often the only picture available of the route forward was held in the head of a fellow-traveller who had gone that way before or of a local unaccustomed to explanation. Since America east of the Appalachian chain had been settled, or at least explored and travelled, for 200 years before the Civil War's outbreak, it may seem extraordinary that much of its terrain was unmapped and indeed unknown to strangers. Such was, nevertheless, the case. Though there were turnpike roads in the Shenandoah Valley, and railroads that ran into it, the Union armies lacked maps of its topography, which was unknown in detail to their officers. Jackson, a West Virginian by upbringing, knew the outlines of the topography, but he took trouble to master the details by requesting Hotchkiss to survey the theatre and make him a military map of its most important features, particularly waterways, bridges and passes through the high ground. It was Hotchkiss' map that gave him his advantage. Jackson's succession of small, local victories, which frustrated the manoeuvres of his opponents, superior in numbers as they always were, was not the outcome of chance or recklessness but of careful calculation. He was his own intelligence officer, as Nelson had been during the Nile campaign, with the difference that, though similarly confined within a narrow zone of operations, his role as a fugitive, not a pursuer, was to mislead, confuse and avoid a decisive confrontation, rather than bring his enemy to a battle of annihilation.

All the cases studied in this book concern military intelligence in the strict sense: how the use of intelligence brought the enemy to battle on terms favourable to the intelligence victor (the Nile, the naval battle of the Falklands, Midway, the U-boat war) or spared the intelligence victor battle on unfavourable terms (the Valley); or else how the successful practice of intelligence nevertheless failed to avert an unfavourable outcome (Crete, the V-weapon campaign). Its purpose is to demonstrate that intelligence, however good, is not necessarily the means to victory; that, ultimately, it is force, not fraud or forethought, that counts. That is not the currently fashionable view. Intelligence superiority, we are constantly told, is the key to success in war, particularly the war against terrorism. It is indisputably the case that to make war without the guidance intelligence can give is to strike in the dark, to blunder about, launching blows that do not connect with the target or miss the target altogether. All that is true; without intelligence, armies and navies, as was so often the case in the age before electricity, will simply not find each other, at least not in the short term. When and if they do, the better informed force will probably fight on the more advantageous terms. Yet, having admitted the significance of the pre-vision intelligence provides, it still has to be recognised that opposed enemies, if they really seek battle, will succeed in finding each other and that, when they do, intelligence factors will rarely determine the outcome. Intelligence may be usually necessary but is not a sufficient condition of victory.

The reasons for the current overestimation of the importance of intelligence in warfare are twofold: the first is the common confusion of espionage and counter-espionage with operational intelligence proper; the second is the intermingling of operational intelligence with, and contamination by, subversion, the attempt to win military advantage by covert means.

Operational intelligence and espionage work in different time-frames. Espionage, usually but not necessarily a state activity, is a continuous process, of very great antiquity; so is its counterpart, counter-espionage. States seem to have always sought to know the secrets of each other's policy, particularly foreign but also mercantile and military policy, and to deny such secrets in return. The apparatus of espionage is common knowledge: the employment of spies, the suborning of foreign nationals in positions of confidence, the use of codes and ciphers and the maintenance of decryption and intercept services. Operational intelligence, by contrast, is specifically an activity of wartime and, at high tempo, is limited to comparatively brief periods of hostilities. The rhythm of the intelligence attack on the German V-weapons programme illustrates that: most lethargic at the outset, when the evidence was scanty and diffuse, growing intense as it became incontrovertible, then slowing again when the British, after their capture of the V-1 launch sites in northern France, wrongly persuaded themselves that the danger had been brought under control.

The intermittent pattern of operational intelligence activity is explained in part by the positions military intelligence officers occupy in the hierarchy of an army or navy. They are always subordinate to the operations staff and rarely make full careers in intelligence; indeed, most seek transfer to the operations branch, in the all too understandable hope of becoming masters rather than servants. It is difficult enough, in any case, to make a reputation as a staff officer in any branch, but while there are a number of celebrated operations officers and chiefs of staff—Berthier to Napoleon, Jodl to Hitler, Alan Brooke to Churchill— there are almost no famous intelligence officers. The best known of the Second World War, E. T. Williams, Montgomery's chief of intelligence in the Eighth Army in the desert and then in Normandy, was an Oxford don who had gone to war as a troop leader in the King's Dragoon Guards. The best known of the First World War, Sir Alfred Ewing, founder of Room 40, was a former Cambridge don who, as a civilian, became Director of Naval Education. Williams, still a young man, returned to his Oxford college after the war.[2]

Espionage and counter-espionage by contrast are, or have become in the modern world, the arena of full-time professionals. The CIA and the SIS (MI6) are organs of state and, as they evolved over time, have grown into formidable bureaucracies; the former KGB of the Soviet Union was, in at least one of its aspects, effectively a parallel government, charged to maintain the internal stability of the Soviet system as well as spy on foreign enemies and defeat foreign espionage. In all those organisations, it has been possible, indeed usual, to enter as a carefully selected recruit, to be trained, usually in a particular speciality, and to make a lifelong career. Since the career was full-time, the agencies' operatives naturally found or made activities to occupy their day-to-day working lives; and as, in practice, serious threats to state security are as intermittent as major military threats to national survival in wartime, the intelligence agencies bulked out their work by spying on each other. Indeed, if asked what spies do, the safest answer is that spies spy on spies. The parallel eavesdropping agencies—the British Government Communication Headquarters (GCHQ, descendant of Bletchley) and the American National Security Agency (NSA)—are party to serious secrets, which they pluck from the ether by interception and decrypt. At their most successful, they are able to tell their own governments the most secret business of others. They guard what they know jealously, even, paradoxically, from their companion intelligence agencies. No rivalries are more intense than those between intelligence services working, by different means, on the same side.

The disdain evinced by the "hard" agencies—NSA, GCHQ—for the "soft"—CIA, SIS—is nowhere better illustrated than by the now endlessly retold story of the Cambridge spies of the early Cold War. Donald Maclean, Guy Burgess, Kim Philby, Anthony Blunt, John Cairncross and their hangers-on were elegant young men of good family, educated at expensive schools and leading colleges, who had been seduced by the warped logic of Marxism to become Soviet agents before they joined the British Foreign Office or intelligence services. All eventually, after 1945, fell under suspicion, and three, Maclean, Burgess and later Philby, defected to the Soviet Union amid noisy media sensation. They caused great harm to their parent services and to Anglo-American trust, which took many years to restore. Indeed, for a long time the Americans took the view that the British intelligence services were fundamentally flawed, even corrupt; it was not until, much later, the Americans themselves suffered a succession of serious breaches of security inside the CIA and the military intelligence services, admittedly committed by agents who were motivated by greed rather than ideology, that relations returned to an even keel.

Yet, viewed in retrospect, the damage done by at least two of the Cambridge spies, Burgess and Philby, was superficial rather than substantial. Guy Burgess, a flamboyant homosexual and dedicated alcoholic, never rose high in the Foreign Office hierarchy. Though his background was entirely conventional—his father was a regular naval officer, and he had himself, until ill health intervened, trained as a naval cadet at Dartmouth—his personality and behaviour were not. He was an exhibitionist, a poseur, a professional rebel. Though a brilliant pupil at Eton, he wasted his time at Cambridge and had difficulty thereafter finding a job. A temporary position at the BBC led in the lax war years to a job in the Foreign Office information department; charm, reinforced by his determination to succeed in his chosen vocation as an undercover Soviet agent, then won him promotion to the post of personal assistant to the Minister of State. It did not last. His irresponsible urge to outrage the conventionally minded led to his transfer to a specialist information branch, then to the Far Eastern Department, where he continued to make a bad impression, and eventually to the British embassy in Washington. His position there was humiliatingly junior. The wonder is nevertheless that, after years of bad behaviour, the Foreign Office was still prepared to keep him on. The explanation, easily grasped by anyone who lived then, incomprehensible today, is that Burgess was protected by the indulgence felt by the well-behaved for the professional naughty boy. Their forgiveness of his excesses excused, in a sense, their own unrelenting propriety; their unwillingness to condemn absolved them of pomposity.

It is doubtful, in any case, if Burgess was ever privy to secrets that could damage his own country. The same might be said of his protégé, Kim Philby. Philby, a truly dedicated Communist convert, began life after Cambridge as a journalist but transferred at the outbreak of war, with the help of Burgess, to the subversive Special Operations Executive. Thence he migrated to the Secret Intelligence Service, which then operated under the cloak of the Foreign Office. As an intelligence officer he undoubtedly betrayed to the Russians a great deal of information about British counter-espionage and subversion and was responsible for the deaths of numbers of anti-Soviet agents, particularly Albanians and Ukrainians whom the British and Americans infiltrated behind the Iron Curtain in the early 1950s. Philby did not, however, have access to war plans or nuclear intelligence. His was a classic example of a spy spying on spies, and the atmosphere of his world is perfectly caught in the novels of John le Carré, which almost exclusively concern the operations of espionage services against each other.

Donald Maclean was a different and more serious traitor. As a promising young diplomat in the Washington embassy in 1945, he was appointed joint secretary of the Anglo-American committee on nuclear development (Combined Policy Committee) and also acquired a pass which gave him unsupervised access to the headquarters of the Atomic Energy Commission. What information he thus gained remains a matter of speculation. It was probably of less value than that supplied to Moscow by the nuclear scientists Alan Nunn May, a British citizen, and Claus Fuchs, a naturalised Briton of German origin, both committed Communists, though of much humbler social origin than the Cambridge spies. They enjoyed the advantage, however, of actually working within the nuclear laboratories at Los Alamos, where the first atomic bomb was developed, and were undoubtedly the source of the information which allowed Stalin to learn of the atomic secret before Hiroshima. Maclean, who had no scientific training, was not guilty of that betrayal. Because of the seniority of his position, however, he was undoubtedly responsible for poisoning Anglo-American trust during the early Cold War, poison that lingered for years afterwards.

The peculiar "climate" of the Cambridge spies' treason, a word chosen by the most perceptive analyst of the episode, Andrew Boyle, goes far to explain the persistent popular interest in it.[3] Not only were Burgess, Maclean and Philby privileged citizens of the society they betrayed, products of good family and its most distinguished schools and colleges; they also belonged to the social elite, knowing those who counted and at ease in the company of the fashionable and powerful. All nevertheless insisted in behaving in disreputable fashion, all three by drinking ostentatiously to excess, all three by publicly violating the sexual norms of the day: not only was Burgess a promiscuous homosexual when homosexual behaviour was still a criminal offence; Maclean, too, a married man, regularly succumbed to his homosexual impulses, while Philby, though strenuously heterosexual, treated women with cavalier selfishness. He abandoned his second wife, pregnant with their fifth child, to a lonely death by drink and drugs; he stole his third wife from a journalistic colleague after his dismissal from the secret service; he next stole Maclean's wife during their Moscow exile and finally married a Russian far younger than himself when the ex-Mrs. Maclean saw him in his true light. The Cambridge spies were not only traitors; they were also, in different but closely similar ways, monsters of egotism. No wonder that they remained for so long objects of fascination to the prurient.

Since the substance of espionage is duplicity, it should not be thought surprising that its three most notorious practitioners of modern times—they had subordinates, they also had imitators, some Soviet, some American, but none so blatantly complacent—were such unpleasant people. Treason is an intrinsically repulsive activity, so much so that it is difficult not to despise even those who, during the Nazi era and the Cold War, betrayed their countries out of devotion to universally admired ideals, such as respect for truth or democratic freedom. Because the efficient spy lies to protect himself, and evades exposure in order to advance his work, his behaviour is the opposite of what is conventionally regarded as heroic. The hero is a fighter who bares his breast to the blows of the enemy. The spy shrinks from the fight and thinks his work best done when he attracts no attention at all.

Hence a paradox. The British—and it is a peculiarly British approach to the secret world, though one also espoused by the Americans—devised during the nineteenth century a philosophy of secret warfare in which duplicity but also the heroic ethic were combined. Because Britain has always been demographically weak but strategically strong, a country of moderate population enjoying a commanding position athwart the world's most important maritime trade routes, it has naturally sought to maximise its power by mobilising what today would be called special and subversive forces in the flanks of its enemies. The practice perhaps began during the Peninsular War of 1808–14, when the British army in Portugal and Spain raised and trained locals to serve in irregular regiments under British officers; the Royal Lusitanian (Portuguese) Legion was such a body. The British also directly subsidised not only the Spanish army, such of it as survived after the political collapse of 1808, but also the bands of guerrillas which took the field in its place after the French occupation. The guerrillas never threatened to end the occupation or overturn French rule, but at the cost of dreadful suffering to the Spanish people, they succeeded in making Spain almost impossible to administer.

In India, meanwhile, the British applied a reverse technique in order to overcome disorder and restore central government. Acting nominally in support of the effete and effectively defunct Moghul emperors, they made extensive use of irregulars to put down the bands of pillagers who ransacked Moghul territory and to defeat the armies of overmighty Moghul subjects who had set up as provincial rulers in their own right. Typically, at the end of a successful campaign of pacification, they incorporated the defeated warriors into their own forces. By the mid-nineteenth century, the British were running two military establishments in India: a regular army of their own, recruited from Indians but organised on European lines, and, attached to it, a kaleidoscopic collection of irregulars, wearing local dress, observing local customs of discipline and commanded by small handfuls of British officers who had almost gone native: Shah Shujah's Contingent, the Hyderabad Contingent, the Punjab Irregular Force.

When in 1857 the Indian regulars rose in mutiny against British rule, their revolt was put down largely by mobilising the irregulars against them; and when the Indian Mutiny was over, the old regular army was almost completely replaced by the irregular forces that had rescued the Indian empire from dissolution. It retained a minimum of British officers—in 1911, the year of the Delhi Durbar, which marked ceremonially the high point of the power of the Raj, they numbered only 3,000—and they, for the most part, wore a version of native dress, spoke Indian languages and prided themselves on their immersion in the customs and culture of their soldiers.

What went for India went eventually for the rest of the British empire, which came largely to be garrisoned by their own inhabitants under the sketchiest of British control. The King's African Rifles, the Royal West African Frontier Force, the Somaliland Canal Corps, the Sudan Defence Force were native armies commanded by Britons who exerted power not by force but by imitating native habits of authority.[4] The French achieved something of the same effect in their African empire, through their organisation of the *goums* of the Moroccan mountains and the camel-riding *méharistes* of the Sahara, units even more indigenous in character than their British equivalents.[5] The French, however, never embraced the idea of imperial self-policing as comprehensively as the British did. It became a peculiarly British idea that an empire could be sustained upon the personal bond established between a local warrior and the young white officer-sportsman who had learnt his language and adopted his costume.

There was a great deal to the idea. The bonds established were very strong and were to survive the most severe tests. The British, however, took the idea too far. They convinced themselves that what worked to maintain imperial authority and even to extend imperial boundaries would work also in war against fellow Europeans. So enthralled did the late Victorians become by the ideals of empire that they persuaded themselves of the overriding appeal of those ideals to the empire's subjects. No individual was more seduced by the universality of the imperial idea than Winston Churchill. It came to him, curiously, in South Africa, during the Boer War: an attack on, and in part a rebellion by, white Afrikaners against British imperialism.

Churchill, who participated in the Boer War as both a journalist and a soldier, conceived a profound admiration for the Boer spirit. The Boers' dedication to their fight to retain the independence of their tiny republics, and their refusal to submit even when they had been objectively defeated by superior force, led him to two conclusions. The first was that, by the exercise of magnanimity, the Boers could be transformed from bitter enemies to close friends; such proved personally to be the case, for Jan Smuts, the outstanding Boer guerrilla leader, became after his people's surrender the pro-British leader of post-war South Africa and Churchill's warm political colleague. The second conclusion, which was to have less benign consequences, was that the practice of guerrilla warfare, by people of free spirit, could wear down a superior power, fetter its freedom of action, distort its strategy, and eventually force it to make great political concessions not strictly won by purely military means. This belief seems eventually to have acquired universal value in Churchill's world vision. He did not place it in context, calculating the likely reaction of a less or more ruthless enemy confronted by guerrilla action. He seems to have invested the guerrilla idea with autonomous value and come to believe that the guerrilla warrior, by the covert nature of his actions and the support he would enjoy from patriot civilians, ensured his success. Such beliefs, though founded on the Boer example, may have been reinforced by his experience of the Irish Troubles of 1918–21 and his acquaintance with another successful guerrilla leader he came to admire, Michael Collins. At any rate, by the time he became British Prime Minister in 1940, at a supreme crisis in national life, he had been involved in two large-scale guerrilla wars, one concluded successfully only with the greatest difficulty, the other undoubtedly lost, and might therefore be forgiven for holding the view that guerrilla operations were a fruitful means of undermining an offshore enemy.

"Set Europe ablaze." That was Churchill's instruction to Hugh Dalton, Minister of Economic Warfare in his 1940 government, uttered on 24 July. It was to lead to the creation of a network of subversive organisations which would penetrate the whole of Nazi-occupied Europe west of the Soviet Union, as well as the Japanese-occupied territories in the Far East. The Special Operations Executive (SOE) was the principal body; its chief task was to insert parties of agents, usually by parachute, into occupied territory, to make contact with the local resistance organisations, if they existed, to arrange for the delivery of weapons and supplies and to carry out espionage and sabotage. All were equipped with radio to maintain contact with base. In the smaller countries—Belgium, Holland, Denmark, Norway—where conditions were not suitable for guerrilla activity, the parties mainly attempted to set up reporting services (disastrously in Holland, where they were penetrated by the Germans early on and their radios used to entrap arriving agents as they landed). In France SOE organised country-wide networks of reporting agents but also trained and armed resistance bands which proliferated after the introduction of forced labour in August 1942. The French resistance, which was comparatively slow to emerge, divided from the start along ideological lines; SOE officers in the field had to play a delicate political game, since within the country itself, the Communists sought to create a secret army of their own, principally loyal to Moscow, while, outside, de Gaulle in London strove to unify the resistance and include it within his forces of Free France. In Greece and the Balkans, where there was a long tradition of local resistance to former Turkish rule, guerrilla bands formed soon after the German occupation of April to May 1941. There too, they also divided ideologically, with results disastrous for the populations. In Yugoslavia the royalist Cetniks were the group with which the SOE first made contact; their leader, Draza Mihailovic, believed, however, that his correct strategy was to build up his strength until circumstances would permit the *ustanka,* a general rising against the occupiers. His Communist opponents, the Partisans, under Josef Tito, preferred to create country-wide war, with the object of politicising the population and securing a position of power that would ensure the creation of a Communist government in the wake of the occupiers' defeat or departure. On the grounds that Tito was fighting the enemy, while the Cetniks were not, the SOE, whose Balkan directorate was heavily penetrated by British Communists, transferred its support to the Partisans in April 1943. In Greece, the SOE never gave its backing to the Communists, since Winston Churchill prudently thought it essential to keep Greece out of Stalin's orbit; nevertheless, by the ruthlessness of their internal operations, they succeeded in making themselves the dominant resistance group by 1943, and some of the arms supplied by the SOE inevitably found their way to them.

The result in Greece was civil war, which persisted long after liberation in 1944 and was not finally suppressed until 1948. Civil war was also the outcome of the Cetnik–Partisan conflict in Yugoslavia. Both conflicts led to widespread loss of civilian life, amplified by the occupiers' reprisals, which often fell on the innocent. Yugoslavia lost a higher proportion of its population than any other combatant country in the Second World War, the majority the victims of internecine violence; the Greeks also suffered heavily.

At the time, and for years afterwards, the guerrilla campaigns conducted under the auspices of the SOE within occupied Europe were celebrated as significant ingredients of the anti-Nazi war effort. The story of the SOE contributed heavily to the myth of "intelligence" as some mysterious means of war-winning, cheaper than battle and somehow more deadly, that captured the popular imagination during the early years of peace. The SOE's leading operatives—the organisers of the major networks in occupied France, the most prominent of the liaison officers dropped into the mountains of Yugoslavia and Greece—were celebrated as Second World War equivalents of Lawrence of Arabia, as glamorous as he and even more effective.

The heroism of the SOE's agents should never be diminished. Those who parachuted into France risked exposure every day they spent on operations, and the courage shown, particularly by such women as Violet Szabo and Noor Inayat Khan, swept up into the espionage world simply because they were French-speakers, humbles anyone who reads of their conduct and terrible deaths.[6] The dashing Balkan bravados, who endured bitter winters in the Yugoslav mountains and risked capture by the enemy day after day, displayed courage that was out of the ordinary also. When the balance is struck, however, the objective military value of what they achieved, measured against the consequences of their underpinnings of what were as much civil as anti-German wars, calls into question the justification for Churchill's desire to "set Europe ablaze."

Churchill's vision of a Europe-wide uprising against the German occupier—a universal *ustanka*—was fundamentally flawed, by a weakness that has distorted the theory and practice of secret war, and therefore of "intelligence," ever since. Churchill was an English gentleman, not only committed to the ideas of fair play and respect for the enemy as an honourable opponent but believing that such ideas were held by those his country fought. So they had been in the past, when European armies were commanded by other gentlemen. Not only European armies: J. F. C. Fuller, the great theorist of war and Churchill's contemporary, called his account of the Boer War of 1899–1902 *The Last of the Gentleman's Wars*. The Boers of South Africa, though determined to resist beyond the point of defeat in the open field, nevertheless conducted the guerrilla war they insisted on fighting in the aftermath by gentlemanly rules. They did not kill prisoners and they did not harm non-combatants. Though overcome after three years of resistance, they preserved their code of honour to the end.

Churchill, who as a young Member of Parliament defended the Boers in the House of Commons, though he was a veteran of the Boer War on the other side, presumed as late as 1940 that a repetition of Boer intransigence in a German-occupied Europe would evoke the same response as it had in the British-occupied Transvaal forty years earlier. He imagined that the soldiers of Nazi Germany would refrain from atrocity in the face of resistance, as his Tommy comrades-in-arms had refrained in a still-unsubdued South Africa. He had, alas, made no allowance at all for the ideological shift in continental European morality brought about by the upheaval of world war and political revolution between 1917 and 1939. He did not perceive that the overthrow of all the stabilities on which the Germans counted—monarchy and currency foremost—would usher in a regime which preached hatred against the forces of instability, primarily Communists and socialists but also deviants from traditional morality, non-German nationalists and enemies of the notion of German culture as a directing principle in continental life. He did not see that raising resistance against a regime imbued with self-righteousness, as Nazism was, would bring down vicious cruelty on those who opposed it.

Resistance, in its many forms, was an admirable movement. It kept alive in defeated and occupied countries the vision of the restoration of independence and the return to democratic life, in the longer term, when German domination would, by American and British intervention, be overthrown. In the short term, however, resistance, though preserving national honour, brought nothing but suffering to those who raised the standard and to many others who became involved unwittingly in the struggle. Resistance certainly harmed the German occupiers scarcely at all. Of the sixty German divisions garrisoning France on the eve of D-Day, none was committed to anti-resistance duty. They manned the coasts, awaiting Allied invasion, while the maintenance of internal security was left to a scattering of Gestapo units and the French police and militia. Internal security was not a German concern in the Low Countries and Scandinavia. There was no internal security problem in Czechoslovakia or even in intransigent Poland, where the Home Army observed the philosophy of Mihailovic in Yugoslavia, that of waiting upon events until circumstances favoured a national uprising; when the moment came in 1944, it was betrayed by their Russian liberators, who allowed the Germans to destroy the Polish resistance as an alternative to destroying it themselves.

In retrospect, the confusion of "resistance"—covert operations against the enemy, usually based on the concept of opposition to a totalitarian occupation or oppressive political takeover masquerading as a liberation movement—with "intelligence," properly the attack on an opponent's espionage and cipher systems, achieved nothing but harm to both. Resistance, perhaps best exemplified by the opposition of the French to the occupation of their country by the German conquerors of 1940, is entirely honourable, even if often, as French resistance largely was, ineffective. It sustains the concept of national sovereignty and keeps open the possibility of the restoration of legitimate government. Intelligence, in the sense of a national attack on an enemy's secure communication, surveillance and espionage systems, is both honourable and necessary always in wartime, now, alas, in peacetime as well.

The intermixture of resistance and intelligence in the Second World War was, however, an aberration and a particularly British one. It was eschewed by the Germans who have taken since their wars of unification in 1866–71 against the Austrians and French a highly legalistic view of the duty owed by the occupied to the occupier, a view which, by reaction, underlay their extremely harsh treatment of resistance wherever they met it: the shooting of suspected *franc-tireurs* in Belgium in 1914, several thousand of them, including women and children, and their vicious suppression of internal disorder in occupied Europe in 1939–44, ranging from transportation of those captured in France to wholesale extermination of partisans in Eastern Europe.[7] The British, by contrast, chose to foment resistance, for a variety of reasons. One was the weakness of their military position after June 1940, which encouraged them to adopt any method of warmaking that promised results. Another was their own experience, as imperialists, of rebellion in the empire, which had taught how effectively rebels could cause the dissipation of regular force. The critical reason may, however, have been that a tradition of irregular warfare ran in the British bloodstream, that of its military class at any rate. Much of the empire had been won by unconventional means, by the recruiting of tribal warriors to defeat, under the leadership of British officers, other tribal warriors, particularly in India and Africa. In the process, the British had constructed a hierarchy of most favoured nations, for military rather than trading purposes, and their names supplied the Royal Navy with those of their most powerful class of destroyers—*Sikh, Zulu, Matabele, Ashanti, Punjabi* and *Somali*. The British officers who had commanded Sikhs and Somalis admired their martial qualities, took pride in their own command of their soldiers' languages and in their understanding of their customs and believed that the combination of warrior fighting skills and European leadership made an unbeatable military mix.[8] Illogically, the irregular tradition at its most effective was personified in British eyes by the Boers, whom some of their opponents, notably Winston Churchill, chose to perceive as a white tribe.

He adopted the Boer term "commando" to denote the raiding forces he deemed should be raised to attack the flanks of Hitler's Fortress Europe in 1940; at the same time he set out, through the creation of the

SOE, to raise a Boer-style rebellion within the occupied lands. No difficulty at all was found in recruiting young officers to enter the enemy continent; their mission, to raise, arm, train and lead local resisters, lay so wholly within Britain's military tradition that volunteers abounded. Those who went to Greece, many of them distinguished classical scholars, were inspired particularly by the memory of Byron's Philhellenic mission in the Greek War of Independence against the Turks in the 1820s; something of the same mood animated those who parachuted into Yugoslavia, where the mountainous terrain, rough food, constant need to march, as well as to converse in local languages, recalled both the epic of the struggle against the Turks and the conditions of warfare on the Northwest Frontier of India. The SOE, in many of its manifestations, was a re-creation of the imperial ethic, with the difference that, since so many of its members were products of the leftward mood of the interwar Oxford and Cambridge, they could imagine themselves to be fellow "progressives" with the partisans, rather than agents of a distant imperial power.

It was all an illusion. The SOE in Western Europe did almost nothing to unlock the German grip on power within the occupied territories; fortunately, neither did it do much harm. In the Balkans, by contrast, it did very great harm indeed, supplying much of the equipment which enabled the partisans to establish Communist governments after the war, and also endorsing indirectly their right to do so. Only by a whisker was Greece spared a similar fate: had Churchill not kept his own counsel and had the murderous Greek Communists not overplayed their hand, Athens, like Belgrade, might have become a Communist capital after 1944.

The damage went wider since, by the confusion of subversion with intelligence, under the common cloak of making secret warfare, the proper intelligence community was compromised. In Britain, after the disbandment of the SOE in 1946, the Secret Intelligence Service unwisely allowed itself to be drawn into the business of subversion, with disastrous results in Albania, where the officer chosen to sponsor the anti-Communist forces was the traitor Kim Philby, and in the Baltic lands, where, as in Holland in 1941–43, the resistance came under the control of the organisation its MI6 contacts were targeting, the Russian KGB. Many anti-Communist patriots in both regions died as a result. In the United States the Central Intelligence Agency (CIA), set up in 1947 to replace the too hastily disbanded OSS of the war years, embraced both intelligence-gathering and subversive activities, separately conducted in Britain by MI6 and SOE at the outset. In a world of secrets, which does not disclose what it does or what it knows, it is not for the outsider to judge that such a joint mission was ill conceived. The character of the CIA's enemies, of whom there are many, suggests that it has right broadly on its side. In principle, however, it strikes this author that the organisation of intelligence-gathering and subversion within the same body is undesirable. Subversion is a weak way of fighting, differing from conventional warfare by the total unpredictability of its results; moreover, in a democracy, it is always liable to disavowal by legitimate authority and denunciation by authority's political opponents. Intelligence-gathering, by contrast, can yield conflict-winning outcomes and, if securely and soberly conducted, is an activity only those of ill-will can condemn.

Yet in the last resort, intelligence warfare is a weak form of attack on the enemy, also. Knowledge, the conventional wisdom has it, is power; but knowledge cannot destroy or deflect or damage or even defy an offensive initiative by an enemy unless the possession of knowledge is also allied to objective force. As David Kahn puts it simply, there is "an elemental point about intelligence . . . it is a secondary factor in war." Reflecting on the blitzkrieg defeat in 1939 of Poland, the country whose cryptanalysts broke Enigma by pure intellectual effort, an effort not matched by any other of Germany's enemies, he goes on: "all the Polish codebreaking, all the heartrending efforts and the heroic successes, had helped the Polish military not at all. Intelligence can only work through strength."[9]

Kahn's measured corrective is of the greatest importance and should be remembered by soldiers and statesmen at all times, particularly in these times of the so-called information revolution and its superhighway. Knowledge of what the enemy can do and of what he intends is never enough to ensure security, unless there are also the power and the will to resist and preferably to forestall him. How often have the rich, the well informed and the complacent known in their hearts what the future threatened. The last Abbasid Caliph no doubt suspected the fate that awaited him in Baghdad in 1258, when he cravenly surrendered himself to the Mongol Hulagu and his stranglers with their bowstrings. The soft Western democracies allowed Hitler to undermine their European security system until, almost too late, they took a stand. Contrarily, the Japanese persuaded themselves in 1941, against all the evidence and the warnings of their leading admiral, that they could attack America and survive. Foreknowledge is no protection against disaster. Even real-time intelligence is never real enough. Only force finally counts. As the civilised states begin to chart their way through the wasteland of a universal war on terrorism without foreseeable end, may their warriors shorten their swords. Intelligence can sharpen their gaze. The ability to strike sure will remain the best protection against the cloud of unknowing, prejudice and ignorance that threatens the laws of enlightenment.

———◀o▶———

REFERENCES

SELECT BIBLIOGRAPHY

INDEX

REFERENCES

CHAPTER ONE: KNOWLEDGE OF THE ENEMY

1. N. Austin and N. Rankov, *Exploration, Military and Political Intelligence in the Roman World from the Second Punic War to the Battle of Adrianople,* London, 1995, pp. 26–27, 209–10.
2. Ibid., pp. 9–10.
3. Ibid., p. 246.
4. E. Christiansen, *The Northern Crusades, the Baltic and the Catholic Frontier, 1100–1525,* London, 1980, pp. 161–63.
5. S. Runciman, *The First Crusade,* Cambridge, 1951, Book III, chapters 2 and 3, book IV, chapter 1.
6. P. Contamine, *War in the Middle Ages* (trans. M. Jones), Oxford, 1984, pp. 25–30, 219–28.
7. J. R. Alban and C. T. Allmond, "Spies and Spying in the Fourteenth Century," in J. R. Allmond, *War Literature and Politics in the Late Middle Ages,* London, 1976, pp. 73–101.
8. T. Barker, *The Military Intellectual and Battle: Raimondo Montecuccoli and the Thirty Years War,* New York, 1975, pp. 160, 242.
9. C. Duffy, *The Military Experience in the Age of Reason,* London, 1987, p. 186.
10. C. Duffy, *Frederick the Great, A Military Life,* London, 1985, pp. 59–64.
11. Austin and Rankov, p. 15.
12. For the *harkara* system and its capture by the British, see C. A. Bayly, *Empire and Information: Intelligence Gathering and Social Communication in India, 1780–1870,* Cambridge, 1996, particularly chapter 2.
13. The origin of the term "Y" is mysterious. It may derive from the symbol used to denote sound-ranging by British artillery officers during the First World War, the arms of the Y perhaps representing the sound waves received at a central interception point.

14. For the question of whether Sorge did or did not influence Soviet decision making, and was or was not believed, see F. W. Deakin and G. R. Storry, *The Case of Richard Sorge,* London, 1966, particularly chapter 13. See also Walter Laqueur, *A World of Secrets: The Uses and Limits of Intelligence,* New York, 1985, pp. 236–37, 244. Sorge, whatever his success, is nevertheless an extremely significant figure, since his character, personality and career typify those of the dedicated ideological agent at his most dangerous. Sorge was highly intelligent, very brave and completely dedicated to his beliefs, which effectively took the form of unquestioning loyalty to a country not his own.

15. Laqueur, p. 244 and footnote 20, p. 381.

16. A. Boyle, *The Climate of Treason,* London, 1979, p. 371.

CHAPTER TWO: CHASING NAPOLEON

1. Geoffrey Bennett, *Nelson the Commander,* London, 1972, p. 59.

2. Hugh Popham, A *Damned Cunning Fellow,* St. Austell, 1991, p. xiii.

3. Bennet, pp. 61–62.

4. Brian Lavery, *Nelson and the Nile,* Chatham, 1998, p. 9

5. C. de la Jonquière, *L'Expédition d'Egypte,* Paris, 1900, vol. 1, pp. 96–98.

6. H. Nicolas, *Dispatches and Letters of Nelson,* London, 1845, vol. 3, p. 17.

7. Ibid., p. 26.

8. Ibid., p. 29.

9. Ibid., p. 13.

10. Ibid., p. 30.

11. Lavery, p. 124.

12. M. Duffy, "British Naval Intelligence and Bonaparte's Egyptian Expedition of 1798," in *Mariner's Mirror,* vol. 84, no. 3, August 1998, p. 283.

13. Ibid., p. 285.

14. Lavery, p. 125.

15. G. P. B. Naish, *Navy Records Society,* 1958, vol. 100, pp. 407–9.

16. A. T. Mallon, *Life of Nelson,* Boston, 1900, vol. 1, p. 332.

17. S. E. Maffeo, *Most Secret and Confidential,* Annapolis, 2000, p. 264.

CHAPTER THREE: LOCAL KNOWLEDGE: STONEWALL JACKSON IN THE SHENANDOAH VALLEY

1. J. McPherson, *Battle Cry of Freedom: The American Civil War, Oxford History of the United States,* New York, 1988, pp. 12–13, 318–19.

2. T. Harry Williams, *Lincoln and His Generals,* London, 1952, pp. 13–14.

3. McPherson, pp. 245–46.

4. J. Waugh, *The Class of 1846*, New York, 1994, p. 264.

5. R. G. Tanner, *Stonewall in the Valley*, Mechanicsburg, Penn., 1996, pp. 3–23.

6. Williams, p. 5.

7. E. B. McElfrish, *Maps and Mapmakers of the Civil War*, New York, 1999, p. 23.

8. *The Imperial Gazetteer of India*, Oxford, 1907, vol. IV, pp. 481–507.

9. D. W. Meinig, *The Shaping of America*, New Haven, 1986, vol. 2, pp. 161–63.

10. McElfrish, p. 18.

11. C. Duffy, *Frederick the Great*, London, 1985, pp. 325–26. The practice of deeming maps to be state secrets was of considerable antiquity. King Manuel of Portugal, in the sixteenth century, threatened the death penalty to any subject caught sending abroad any chart of Cabral's vogage to India. The Spanish had already adopted the practice of weighting maps and charts so that they could be sunk if a ship was threatened with capture (by the twentieth century, the weighting of codebooks was universal in Western navies). On the value of local knowledge, acquired by everyday reconnaissance, see Machiavelli, *The Prince*, chapter XIV; the prince, he wrote, "should always be out hunting so accustoming his body to hardships and also learning some practical geography . . . This kind of ability teaches him where to locate the enemy, how to lead his army on the march and draw it up for battle." I am indebted to Dr. Paige Newmark, of Lincoln College, Oxford, for these references.

12. *Imperial Gazetteer of India*, Vol. IV, p. 499.

13. C. A. Bayly, *Empire and Information: Intelligence Gathering and Social Communication in India, 1780–1870*, Cambridge, 1996, pp. 108, 110.

14. McElfrish, p .22.

15. Tanner, p. 115.

16. McElfrish, p. 29.

17. Ibid., p. 85.

18. V. Esposito, *The West Point Atlas of American Wars*, New York, 1959, vol. 1, map 39.

19. T. Roosevelt, *Autobiography*, 1913, quoted in P. G. Tsouras, *Warrior's Words*, London, 1992.

20. Tanner, p. 117.

21. *Jackson Papers* (b), 19 March 1862, Virginia Historical Society, Richmond.

22. Tanner, p. 124.

23. M. A. Jackson, *Life and Letters of General Thomas J. Jackson*, New York, 1892, p. 248.

24. U.S. War Department, *War of the Rebellion*, IV, I , pp. 234–35.

25. Tanner, p. 194.

26. Quoted in ibid., p. 260.

27. Ibid., p. 297.

28. Ibid.

29. Ibid., p. 352.

30. R. Taylor, *Destruction and Reconstruction*, New York, 1955, p. 76.

31. Ibid., p. 438.

32. Ibid., p. 420.

References

CHAPTER FOUR: WIRELESS INTELLIGENCE

1. P. Kemp (ed.), *Oxford Companion to Ships and the Sea*, Oxford, 1976, pp. 770–71.
2. A. Hezlet, *The Electron and Sea Power*, London, 1975, p. 6.
3. Tanner, see note 5, chapter 3, pp. 417–21.
4. See J. Keegan, *The Mask of Command*, London, 1987, Chapter 3, pp. 210–12.
5. Hezlet, p. 31.
6. P. Kennedy, "Imperial Cable Communications and Strategy, 1820–1914," *EHR*, October 1971, pp. 728–52.
7. D. Kynaston, *The City of London*, London, 1995, vol. II, pp. 8, 40–41.
8. Hezlet, p. 77.
9. Ibid., p. 68.
10. Kennedy, p. 741.
11. A. Marder, *From the Dreadnought to Scapa Flow*, Oxford, 1965, vol. II, pp. 4–5
12. Ibid., II, p. 22
13. Ibid., II, p. 34
14. Quoted in J. Steinberg, *Yesterday's Deterrent*, London, 1965, p. 208.
15. P. Halpern, *A Naval History of World War I*, Annapolis, 1994, p. 65.
16. C. Burdick, *The Japanese Siege of Tsingtau*, Hamden, CT, 1976, p. 51.
17. Geoffrey Bennett, *Navel Battles of the First World War 1968*, p. 56.
18. D. Van der Vat, *The Last Corsair*, London, 1983, p. 41.
19. Bennett, p. 7.
20. Ibid., pp. 182–83.
21. Ibid., p. 78.
22. J. Corbett, *Naval Operations*, 1920, vol. I, p. 305.
23. Quoted in Bennett, p. 86.
24. Halpern, p. 36.
25. Quoted in Bennett, p. 92.
26. Halpern, p. 93.
27. Corbett, p. 344.
28. Ibid., p. 346.
29. Quoted in ibid., p. 349.
30. Quoted in ibid., p. 353.
31. Ibid., p. 357.
32. Quoted in Van der Vat, p. 61.
33. Quoted in ibid., p. 75.
34. Bennett, p. 110.
35. Ibid., p. 129.
36. K. Middlemas, *Command the Far Seas*, London, 1961, p. 194.
37. Ibid., p. 196.
38. See note 28.

CHAPTER FIVE: CRETE: FOREKNOWLEDGE NO HELP

1. See D. Showalter, *Tannenberg,* Hamden, 1991, p. 170.
2. P. Halpern, *A Naval History of World War I,* Annapolis, 1974, p. 316.
3. A. Marder, *From the Dreadnought to Scapa Flow,* vol. III, p. 42.
4. Ibid., pp. 134ff.
5. Ibid., p. 40.
6. Halpern, pp. 36–37; but see A. Lambert, *The Rules of the Game,* London, 1996, p. 49, who doubts the circumstances; the incident was certainly referred to by those in the know as "the miraculous draught of fishes," Halpern, p. 37.
7. S. Singh, *The Code Book,* London, 1999, pp. 46–51.
8. R. E. Weber, *Masked Dispatches: Cryptograms and Cryptology in American History, 1775–1900,* National Security Agency, 1993, pp. 43–44.
9. Maffeo, *Most Secret and Confidential,* Annapolis, MD, 2001, p. 83.
10. Singh, p. 120
11. S. Budiansky, *Battle of Wits,* New York, 2000, pp. 70–71.
12. R. Kippenhahn, *Code Breaking,* Woodstock, NY, 2000, pp. 28–29.
13. Singh, p. 136.
14. Ibid., pp. 134, 136.
15. Quoted in W. Kozaczuk, *Enigma,* London, 1984, p. 270.
16. Ibid., p. 277.
17. Ibid., p. 284.
18. Ibid., note 2, pp. 22–23.
19. Ibid., p. 304.
20. G. Welchman, *The Hut Six Story,* London, 1982, p. 63.
21. Ibid., p. 71.
22. R. Lewin, *Ultra Goes to War,* London, 1988, p. 47.
23. Budiansky, p. 48.
24. Welchman, pp. 76–77.
25. See Andrew Hodges, *Alan Turing: The Enigma,* London, 1992, particularly pp. 96–99 and, for Bletchley, chapter 4.
26. Welchman, p. 168.
27. F. H. Hinsley et al., *British Intelligence in the Second World War,* London, Appendix 4, vol. II, pp. 658ff.
28. Welchman, p. 98.
29. Hinsley et al., p. 657.
30. C. MacDonald, *The Lost Battle: Crete 1941,* London, 1993, pp. 11–12.
31. H. Trevor-Roper (ed.), *Hitler's War Directives,* London, 1965, pp. 68–9.
32. A. Beevor, *Crete: The Battle and the Resistance,* London, 1991, p. 76.
33. Ibid., p. 72.
34. I. Stewart, *The Struggle for Crete,* Oxford, 1966, p. 58.
35. Beevor, p. 349.
36. Ibid., p. 351–52.
37. Paul Freyberg, *Bernard Freyberg VC,* London, 1991.

38. Bennett, *Ultra and Mediterranean Strategy, 1941–45*, London, 1989, pp. 57–58.
39. Beevor, pp. 346–48.
40. Ibid., p. 105.
41. Ibid., p. 112.
42. Ibid., p. 107.
43. Quoted in ibid., p. 107.
44. MacDonald, p. 216.
45. Ibid., p. 196.
46. Stewart, pp. 317–18, 374–75.
47. MacDonald, p. 203.
48. Ibid., p. 212.
49. Bennett, p. 20.
50. Ibid., p. 19.
51. Ibid., p. 20.

CHAPTER SIX: MIDWAY: THE COMPLETE INTELLIGENCE VICTORY?

1. H. Strachan, *The First World War*, Oxford, 2001, vol. I, p. 458.
2. R. Spector, *Eagle Against the Sun*, London, 1985, p. 42.
3. Ibid., pp. 46–47.
4. H. P. Willmott, *Empires in the Balance*, London, 1982, p. 71.
5. S. Budiansky, *Battle of Wits*, New York, 2000, p. 120.
6. Ibid., p. 32ff.
7. *Pearl Harbor Revisited. United States Navy Communications Intelligence, 1924–41*, Naval Historical Center, Washington Navy Yard, 2001, p. 17.
8. R. Lewin, *The American Magic*, New York, 1982, p. 42.
9. *Pearl Harbor Revisited*, Appendix A, "Messages Intercepted Between 6 September and 4 December, 1941," pp. 53–65.
10. Spector, pp. 153–55.
11. H. Shorreck, *A Priceless Advantage*, Naval Historical Center, Washington Navy Yard, 2001, p. 9.
12. Ibid., p. 11.
13. Ibid., p. 5.
14. Ibid., p. 6.
15. Ibid., p. 8.
16. Ibid., p. 9.
17. Ibid., p. 10.
18. Spector, p. 166.
19. Shorreck, p. 10.
20. Ibid., p. 12.
21. A. Marder, *Old Friends, New Enemies: The Royal Navy and the Imperial Japanese Navy*, Oxford, vol. II, 1990, p. 93.
22. J. Winton, *Ultra in the Pacific*, London, 1993, p. 58.

23. W. Lord, *Midway: The Incredible Victory,* Ware, 2000, p. 119.
24. H. Bicheno, *Midway,* London, 2001, p. 149.

CHAPTER SEVEN: INTELLIGENCE, ONE FACTOR AMONG MANY:
THE BATTLE OF THE ATLANTIC

1. F. H. Hinsley and A. Stripp, *Codebreakers,* Oxford, 1993, p. 11.
2. Ibid., p. 12.
3. W. S. Churchill, *The Second World War,* London, 1949, p. 529.
4. P. Padfield, *Dönitz,* London, 1964, p. 101ff.
5. Ministry of Defence, *The U-Boat War in the Atlantic,* London, vol. I, 1989, p. 1.
6. Ibid., pp. 3–4.
7. J. Terraine, *Business in Great Waters,* London, 1983, p. 142.
8. Ibid., pp. 618–19.
9. *Jane's Fighting Ships,* London, 1940, p. 60ff.
10. Terraine, pp. 54, 119.
11. Padfield, p. 201.
12. Terraine, pp. 266–8.
13. Hinsley et al., *British Intelligence in the Second World War,* London, 1981 and later, Vol. 1, p. 336, Vol. 2, p. 179.
14. Ibid., Vol. 2, Appendix 4, parts 3 and 6.
15. Ibid., Vol. 2, Appendix 9, p. 681.
16. Ibid., Vol. 2, Appendix 19, pp. 751–52.
17. C. Blair, *Hitler's U-Boat War,* vol. 1, *The Hunters,* New York 1939–42, 1996, pp. 727–32, 695.
18. D. Kahn, *Seizing the Enigma,* London, 1991, pp. 211–12.
19. Ibid., chapter 16, passim.
20. Hinsley et al., vol. 3, appendix 8.
21. Kahn, chapter 20.
22. Blair, vol. 1, p. 424, vol. 2, *The Hunted. 1942–45,* p. 712.
23. Ibid., vol. 1, p. 421.
24. Ibid., vol. 2, pp. 743–44.
25. Ibid., vol. 1, p. 418.
26. Terraine, p. 629.
27. Blair, vol. 1, pp. 741–45.
28. Ibid., p. 247.
29. Ibid., vol. 2, pp. 791–92; J. Terraine, pp. 314–15.
30. Blair, vol. 2, pp. 519–20.
31. Terraine, p. 619.
32. Ministry of Defence, pp. 109-18; Blair, vol. 2, Appendix 2.
33. Kahn, pp. 211–13.
34. Hinsley et al., vol. 2, Appendix 19.
35. Blair, vol. 2, Appendix 18.
36. Ibid., vol. 2, pp. 710–11.

CHAPTER EIGHT: HUMAN INTELLIGENCE AND SECRET WEAPONS

1. D. Irving, *The Mare's Nest,* London, 1964, pp. 13–14.
2. F. Hinsley et al., *British Intelligence in the Second World War,* London, 1981 and later, vol. 1, appendix 5.
3. M. Smith, *Foley,* London, 1999.
4. Irving, p. 34.
5. P. Wegener, *The Peenemünde Wind Tunnels,* New Haven, 1996.
6. Ibid., p. 27.
7. Irving, p. 35.
8. Ibid., p. 38.
9. Ibid., p. 43.
10. Wegener, p. 10.
11. Ibid., pp. 34–40.
12. B. Collier, *The Defence of the United Kingdom,* London, 1957, pp. 353–55.
13. Irving, pp. 140–41.
14. T. Wilson, *Churchill and the Prof,* London, 1988, pp. 2–4.
15. Irving, title page.
16. Ibid., pp. 45–47, 53.
17. Hinsley et al., vol. 3, part 1, p. 369.
18. Ibid., p. 385.
19. Ibid., part 1, p. 390.
20. Ibid.
21. Ibid.
22. Hinsley et al., vol. 1, p. 57, n. 277.
23. Hinsley et al., vol. 3, part 1, p. 389.
24. Ibid., p. 379.
25. Ibid., p. 391–92.
26. Ibid., p. 402.
27. Ibid., p. 412.
28. Ibid., p. 428.
29. B. Collier, *The Battle of the V-Weapons,* London, 1964, pp. 45–46.
30. Collier, *Defence of the United Kingdom,* Appendices XLV, L.
31. Hinsley et al., vol. 3, part 1, p. 446.
32. F. H. Gibbs-Smith, *The Aeroplane,* London, 1960, chapter 14.
33. N. Longmate, *Hitler's Rockets,* London, 1985, p. 187.
34. Hinsley et al., Vol. 4, p. 184.
35. M. Howard, *British Intelligence in the Second World War,* vol. 5, 1990, pp. 18–20, 231–41. It has to be said that Garbo was a genuine anti-totalitarian and strongly pro-British.
36. Ibid., vol. 5, p. 12.
37. Ibid., pp. 177–79.
38. Ibid., p. 183.
39. Hinsley et al., vol. 3, part 1, p. 360.

40. Private information, Professor D. C. Watt.
41. D. Irving, *The Virus House*, London, 1967, passim.

EPILOGUE: MILITARY INTELLIGENCE SINCE 1945

1. N. West, *The Secret War for the Falklands*, London, 1997, pp. 20, 37–38.
2. A. Finlan, "British Special Forces and the Falklands Conflict," in *Defence and Security Analysis*, December 2002, pp. 319, 332.
3. West, p. 144.
4. Ibid., pp. 145–47.
5. Finlan, p. 826.
6. M. Hastings and S. Jenkins, *The Battle for the Falklands*, London, 1983, p. 316.

CONCLUSION: THE VALUE OF MILITARY INTELLIGENCE

1. The Ultra secret was first revealed, in a book of that title written by F. W. Winterbotham, in 1974. Winterbotham, a regular air force officer, had been head of the air section of MI6 (the Secret Intelligence Service, or SIS) and moved to Bletchley in 1939. The reason he was given permission to publish the book—which contains serious inaccuracies—is that there were official British fears of the story coming out anyhow; articles were appearing in Poland, which initiated the attack on Enigma before 1939, describing the Polish success; it was suspected that disclosures about Bletchley would shortly follow.
2. Reinhard Gehlen achieved fame as head of Foreign Armies East, branch 12 of the German General Staff, which collected intelligence about the Red Army. Since Hitler, however, disliked inconvenient facts, and Gehlen failed to insist on his accepting them, he cannot be reckoned a great intelligence officer, though he was a very efficient one. After 1945 the "Gehlen organisation" was adopted by the Americans as a source of Cold War intelligence. It later evolved into West Germany's foreign intelligence service, the *Bundersnachrichtendienst*.

   Bacler d'Albe achieved fame as intelligence officer to Napoleon, but Bonaparte, like Wellington, usually acted as his own intelligence officer. He travelled with a compact filing-cabinet of essential information, cleverly constructed to display a summary of the contents on the doors of each of its compartments. For Gehlen, see D. Kahn, *Hitler's Spies*, New York, 1978.
3. See A. Boyle, *The Climate of Treason, Five Who Spied for Russia*, London, 1979. Now somewhat outdated factually, it continues to provide the best description of the university traitors' disposition.
4. See J. Lunt, *Imperial Sunset, Frontier Soldiering in the 20th Century*, London, 1981, for such exotic forces as the Iraq Levies, the Hadrami Bedouin Legion and the Somaliland Scouts. Histories of the Indian army are many but an interesting modern one is by General S. Menezes, *Fidelity and Honour*, New Delhi 1993. General Menezes served in the Indian Army both before and after independence.

5. See A. Clayton, *France, Soldiers and Africa,* London, 1988.

6. See M. Binney, *The Women Who Lived for Danger,* London, 2002.

7. See J. Horne and A. Kramer, *German Atrocities 1914,* New Haven and London, 2001, Appendix 1.

8. See R., Kipling, *The Complete Stalky & Co,* London, 1929. " 'The surprises will begin when there is a really big row on . . . Just imagine Stalky let loose on the south side of Europe with a sufficiency of Sikhs and a reasonable prospect of loot.' " Stalky was modelled on Kipling's schoolfellow Dunsterville, who as a First World War general led a sensational intervention into the Caucasus. See Horne and Kramer, Appendix 1.

9. D. Kahn, *Seizing the Enigma,* London, 1991, p. 91.

SELECT BIBLIOGRAPHY

Sources for the case studies which form the substance of this book will be found in the chapter notes. This bibliography includes some of the more general works on intelligence which the author has found of particular value and in which he has confidence. It does not include many books often cited in bibliographies of "intelligence" which are too often sensationalist or mere compendia of intelligence gossip or speculation. It excludes most biographies and autobiographies of intelligence agents or their controllers, which are rarely reliable.

Beesly, Patrick. *Very Special Intelligence: The Story of the Admiralty's Operational Intelligence Centre 1939–45.* London, 1977. The author worked in the OIC during the Second World War and this scholarly and reliable book conveys a valuable picture of its methods and achievements. It does not cover operations in the Mediterranean or Pacific.

Bennett, Ralph. *Ultra in the West: The Normandy Campaign, 1944–5.* London, 1979; and *Ultra and Mediterranean Strategy.* London, 1989. The author, a young Cambridge history graduate, worked at Bletchley in Hut 3, which interpreted deciphered German army and air force intercepts, from February 1941 until the end of the war. He sets out to demonstrate in detail how the intercepts influenced the conduct of operations, a daunting task in which he largely succeeds. His book is one of the most original and valuable on "the Ultra secret." After the war he returned to Cambridge where he eventually became President of Magdalene College.

Boyle, Andrew. *The Climate of Treason: Five Who Spied For Russia.* London, 1979. A professional writer rather than historian, Boyle deserves attention because of his exceptional ability to portray individual character and social atmosphere. His portrait of the "Cambridge spies," Burgess, McClean and Philby particularly, are highly convincing, as is his evocation of the ethos of their public and private lives. Though now a little dated, and inaccurate in places, *The Climate of Treason* is indispensable to anyone seeking to understand the attraction of Soviet Communism to the university-educated in Britain before and after the Second World War.

Calvocoressi, Peter. *Top Secret Ultra*. London, 1980. Calvocoressi, a member of the long-established Greek community in Britain, educated at Eton and Oxford, spent 1940–45 as a Royal Air Force officer at Bletchley. His memoir is especially valuable for the picture it provides of how Bletchley worked day-to-day.

Chapman, Guy. *The Dreyfus Trials*. New York, 1972. A meticulous study, by a professional historian, of the most notorious intelligence scandal of the late nineteenth and early twentieth centuries. The long drawn-out investigation of a suspected traitor remains an object lesson in how not to conduct counter-espionage proceedings. Professor Chapman was a historian of France rather than of the intelligence world but his work is of great value to intelligence organisations everywhere.

Clark, Ronald. *The Man Who Broke Purple: The Life of the World's Greatest Cryptologist Colonel William F. Friedman*. London, 1977. Friedman has been called by David Kahn, himself the leading historian of intelligence, "the world's greatest cryptologist." Certainly his achievement in breaking the Japanese machine cipher, called PURPLE by the Americans, just before the outbreak of the Second World War, was one of the greatest cryptanalytic feats of all time. Friedman suffered a severe nervous breakdown in the aftermath but recovered sufficiently to become chief technical consultant to the National Security Agency, the principal code and cipher service of the United States.

Clayton, Aileen. *The Enemy Is Listening*. London, 1980. Clayton, a Women's Auxiliary Air Force officer, worked during the Second World War in the Middle Eastern Y Service, the organisation that intercepted and interpreted "low level" transmissions on the battlefield. Y is a neglected subject, despite its great importance, and her book is one of the few studies of it.

Cruickshank, Charles. *SOE in the Far East*. Oxford, 1983; and *SOE in Scandinavia*. Oxford, 1986. Special Operations Executive (SOE) was the subversive organisation set up by Winston Churchill in July 1940 to "set Europe ablaze." Branches were later formed in Scandinavia and the Far East, their work being described by the author in these semi-official histories.

Davidson, Basil. *Special Operations Europe: Scenes From the Anti-Nazi War*. London, 1980. Davidson served as an officer in the SOE both in its Mediterranean headquarters in Cairo and in the field in Hungary, Italy and Yugoslavia. He had strong left-wing views and was instrumental in transferring support from the royalist Cetniks in German-occupied Yugoslavia to Tito's Communist partisans. His account illuminates how easily the fostering of short-term "subversion" leads to the fomentation of civil war and atrocity, with deplorable long-term results.

Deakin, F. W. *The Embattled Mountain*. London, 1971. Deakin, later Sir William, and Warden of St. Antony's College Oxford, was a liaison officer for the SOE with Tito's partisans. His celebrated book is both a wonderful adventure story, in the

T. E. Lawrence tradition, and a chilling account of Communist ruthlessness in widening internal conflict for post-war political advantage.

Deakin, F. W. and Richard Storry. *The Case of Richard Sorge.* New York, 1966. Richard Storry, a Fellow of St. Antony's during Deakin's wardenship, was a historian of Japan and a wartime Japanese-language intelligence officer in the Far East. Their study of the most important Soviet spy to operate inside any Axis country during the Second World War brilliantly illuminates the limited usefulness even of the best placed agent.

Foot, M. R. D. *SOE in France: An Account of the Work of British Special Operations Executive in France, 1940–44.* London, 1966. The official history of the Special Operations Executive in France, by an academic historian who served as a Secret Intelligence Service officer during the Second World War. It provides an extremely detailed account of the operations of all the SOE networks in France and of their political affiliations, which were complex. Despite its scholarly objectivity, it reaches conclusions which exaggerate the military contribution made by the French resistance to Anglo-American victory in France in 1944.

Garlinski, Josef. *Intercept.* London, 1979. Garlinski is understandably concerned to set on record the pioneering achievement of his fellow Poles in breaking into Enigma traffic before the outbreak of the Second World War and of how their work contributed to the success of Bletchley.

Giskes, Herman. *London Calling North Pole.* London, 1953. Giskes was the German counter-espionage officer responsible for capturing and "turning" Dutch agents of the Special Operations Executive parachuted into the German-occupied Netherlands during 1940–43. In a highly successful counter-espionage campaign, the Germans captured almost all agents as they arrived and persuaded them to transmit back to Britain at German direction. The "England game," as the Germans called it, severely strained Dutch-British relations during the war and for some years afterwards. The episode has now been fully investigated and recounted by M. R. D. Foot in *SOE in the Netherlands,* London, 2002.

Handel, Michael, ed. *Leaders and Intelligence.* London, 1989. Handel, a professor at the U. S. Army War College, is a productive writer and editor, whose chief subject is operational intelligence. When not the principal author, he can be counted upon to assemble contributions from leading intelligence writers, such as Professor Christopher Andrew of Cambridge. All his compilations, including *War, Strategy and Intelligence,* London, 1989, and *Intelligence and Military Operations,* London, 1990, contain valuable material, bearing both upon the past and the present.

Hinsley, F. H., with E. E. Thomas, C. F. G. Ransom and R. C. Knight. *British Intelligence in the Second World War: Its Influence on Strategy and Operations,* Vol. 1, 1979; Vol. 2, 1981; Vol. 3, Part 1, 1984; Vol. 3, Part 2, 1988; Vol. 4. London, 1990. Hinsley's five volumes, the official history of British intelligence in the Second World War, are the most important single publication on the subject of their subtitle: how intelligence effects decision-making in wartime. Hinsley appears to cover almost every topic in his remit, including how Enigma was broken, how Ultra worked, how British intelligence successes and failures are to be judged in comparison with those of her enemies, and how intelligence affected the outcome of the war as a whole. His work has been criticised as "by a committee for a committee"; but that is unfair. It is an achievement of the greatest value and interest.

Hinsley, F. H. and Thripp, Alan, eds. *Codebreakers: The Inside Story of Bletchley Park,* Oxford, 1993. A fascinating collection of thirty-one essays by B. P. initiates, on such varied subjects as how the watch system worked and the building of the famous huts. An essential companion to Hinsley's official history.

Howard, Michael. *British Intelligence in the Second World War.* Vol. 5, *Strategic Deception.* London, 1990. The last volume of Hinsley's great work, by Britain's leading military historian of the twentieth century, is a fascinating account of British efforts to deceive the enemy, with mixed results but some success against Germany's secret weapons campaign.

James, William. *The Eyes of the Navy: A Biographical Study of Admiral Sir Reginald Hall.* London, 1955. Admiral Hall, known as "Blinker" in the Royal Navy because of a nervous facial tic, was the founder of the immensely successful intelligence organisation known as O.B. 40 (Old Building Room 40), by which the Admiralty achieved complete intelligence dominance over its German equivalent during the First World War. Its achievements were later compromised by boastful disclosure of its successes, particularly in cryptanalysis, in the interwar years.

Jones, R. V. *The Wizard War: British Scientific Intelligence 1939–45.* London, 1978. Jones, a young scientific civil servant, came to enjoy the favour of Winston Churchill during the Battle of Britain and afterwards because of his discovery of how the Luftwaffe used radio beams to guide its bombers to British targets. The "man who broke the beams" thereafter rose ever higher in the service, eventually outfacing Lord Cherwell in the dispute over the V-weapons threat in 1944. His account of scientific intelligence is one of the war's most valuable personal stories, though it fails to disclose why he fell into obscurity after 1945.

Kahn, David. *The Codebreakers: The Story of Secret Writing.* Rev. ed. New York, 1996. Kahn's book is a veritable encyclopaedia of cryptanalysis, superior to any other publication in the field. The original edition was published before the disclosure of the Enigma secret; the revised edition repairs the deficiency. Its great length (1,181 pages) and density will deter the casual reader but it repays the effort to persist.

Kahn, David. *Hitler's Spies: German Military Intelligence in World War II.* New York, 1978. The title is a misnomer. The book is a study of how the German military intelligence organisation worked in the field and is a rare example of an effort, by an expert, to relate intelligence inputs to operational outcomes.

Lewin, Ronald. *Ultra Goes to War.* London, 1978. Lewin's book, published four years after Winterbotham's *Ultra Secret* (1974), which first disclosed the Bletchley secret, was an attempt to correct its more serious mistakes and to set the Ultra achievement in a wider context. It remains a valuable account of the Bletchley story.

Masterman, J. C. *The Double-Cross System in the Second World War.* London, 1972. Masterman, an Oxford don who became Provost of Worcester College after the war, chaired the Double-Cross (XX) Committee during its course, a body dedicated to manipulating information so as to mislead the enemy. Its most important work was in deluding the Germans about the success of their secret weapons campaign during 1944–45.

McLachlan, Donald. *Room 39: A Study in Naval Intelligence.* London, 1968. Although published before the disclosure of the Ultra secret, and so able to refer to Bletchley only as "Station X," this has been described as "one of the best books on intelligence ever written." It is an account of the workings of the Naval Intelligence Division, by one of its officers, during the Second World War.

Powers, Thomas. *The Man Who Kept the Secrets: Richard Helms and the CIA.* New York, 1979. A biography of the Director of Central Intelligence, 1966–72, under Presidents Johnson and Nixon, by a Pulitzer Prize winner, which is also a history of the CIA from its earliest years. Cool in tone and objective in approach, it provides a wealth of information about not only intelligence procedures and operations but also about the influence of intelligence on policy and decision making.

Sweet-Escott, Bickham. *Baker Street Irregular.* London, 1965. Sweet-Escott, like Peter Calvocoressi, a graduate of Balliol College, Oxford, held a large member of staff positions in the SOE and describes its methods and many of its personalities crisply and convincingly.

Trevor-Roper, Hugh. *The Philby Affair: Espionage, Treason and Secret Services.* London, 1968. Trevor-Roper, later Regius Professor of Modern History at Oxford, Master of Peterhouse, Cambridge, and ennobled as Lord Dacre, knew Philby well and, though himself only a junior intelligence officer, provides a subtle and penetrative portrait of his ex-colleague. The book also includes an essay on Admiral Canais, head of the German Abwehr during the Second World War.

Tuchman, Barbara. *The Zimmermann Telegram.* New York, 1958. This short book made the reputation of the famous American historian. Her account of how the British Admiralty deciphered the Germans' diplomatic traffic in 1917, so revealing their efforts to persuade Mexico to attack the United States and thus bringing about America's entry into the First World War on the Allied side, is a masterpiece of intelligence history. Incomplete in part, it nevertheless stands the test of time.

Welchman, Gordon. *The Hut Six Story.* London, 1982. In 1939, Welchman was a mathematics don at Sidney Sussex College, Cambridge, one of the many recruited to join Bletchley Park at the outbreak of war. He proved highly successful at attacking Enigma and was instrumental in re-organising Bletchley to meet the challenge of all-out war. His book, besides being wholly authoritative, also provides the most comprehensible account of how Enigma worked and how Bletchley progressively broke it. Indispensable.

Winterbotham, F. W. *The Ultra Secret.* London, 1974. Winterbotham, a regular air force officer who had served with the Secret Intelligence Service, was posted to the air section of Bletchley during the war. He apparently got permission to publish this book, the first in English to disclose the Ultra Secret (though it had previously been hinted at by Trevor-Roper), because the government feared the secret was about to be broken by the Poles. Largely written from memory, *The Ultra Secret* contains many errors both of fact and interpretation.

Wohlstetter, R. *Pearl Harbor, Warning and Decision.* Stanford, 1962. Roberta Wohlstetter's examination of how Japan succeeded in mounting its surprise attack on Pearl Harbor in December 1941 is meticulous and exhaustive. Her book is widely admired by intelligence experts and, although it is not without its critics, it remains the most valuable study of the preliminaries to the outbreak of the Pacific War.

INDEX

Index

ALSO BY JOHN KEEGAN

FIELDS OF BATTLE
The Wars for North America

Spanning more than two centuries and the expanse of a continent, Keegan demonstrates how the immense spaces of North America shaped the battles and wars that were fought on its soil. He revisits fields of combat from Quebec to Little Bighorn and retraces Washington's triumph and McClellan's defeat on battlefields only a few miles apart. Once again, Keegan's scholarship gives Americans a brilliant reassessment of their military heritage.

History/0-679-74664-1

THE FIRST WORLD WAR

In this magisterial narrative, Keegan has produced the definitive account of the Great War, a cataclysm that left ten million dead. He sheds fascinating light on weaponry and technology, shows us the doomed negotiations between the monarchs and ministers of 1914, and takes us into the verminous trenches of the Western front. His panoramic account of this vast and terrible conflict is destined to take its place among the classics of world history.

History/0-375-70045-5

THE BATTLE FOR HISTORY
Re-fighting World War II

In this engaging and concise volume, Keegan evaluates books on World War II that range from general histories to biographies of the war's principal figures, from accounts of individual campaigns to studies of espionage and resistance. What emerges is an essential guide for any serious student of World War II and the riveting story of how the war has been refought by two generations of its chroniclers, as told by one of the greatest of them all.

History/War/0-679-76743-6

A HISTORY OF WARFARE

In this encyclopedically learned and immensely gripping book, one of our foremost military historians demolishes the famous dictum that war is the continuation of policy by other means. Analyzing centuries of conflict—in societies from those of the Amazon to the Balkans—Keegan unveils the deepest motives behind humanity's penchant for mass bloodshed. *A History of Warfare* is a masterpiece of military scholarship, irresistible in its style and terrifying in its implications.

History/0-679-73082-6

WAR AND OUR WORLD

Is war a natural condition of humankind? What are the origins of war? Is the modern state dependent on warfare? How does war affect the individual, combatant or noncombatant? Can there be an end to war? In a series of brilliantly concise essays, Keegan addresses these questions with a breathtaking knowledge of history and the many other disciplines that have attempted to explain the phenomenon. The themes of *War and Our World* are essential to understanding why war remains the single greatest affliction of humanity in the twenty-first century.

Military History/0-375-70520-1

VINTAGE BOOKS
Available at your local bookstore, or call toll-free to order:
1-800-793-2665 (credit cards only)